Digital Audio Watermarking Techniques and Technologies:
Applications and Benchmarks

Nedeljko Cvejic
University of Bristol, UK

Tapio Seppänen
University of Oulu, Finland

INFORMATION SCIENCE REFERENCE
Hershey · New York

Acquisitions Editor:	Kristin Klinger
Development Editor:	Kristin Roth
Senior Managing Editor:	Jennifer Neidig
Managing Editor:	Sara Reed
Copy Editor:	Ashlee Kunkel and April Schmidt
Typesetter:	Jamie Snavely
Cover Design:	Lisa Tosheff
Printed at:	Yurchak Printing Inc.

Published in the United States of America by
Information Science Reference (an imprint of IGI Global)
701 E. Chocolate Avenue, Suite 200
Hershey PA 17033
Tel: 717-533-8845
Fax: 717-533-8661
E-mail: cust@igi-pub.com
Web site: http://www.igi-pub.com/reference

and in the United Kingdom by
Information Science Reference (an imprint of IGI Global)
3 Henrietta Street
Covent Garden
London WC2E 8LU
Tel: 44 20 7240 0856
Fax: 44 20 7379 0609
Web site: http://www.eurospanonline.com

Library of Congress Cataloging-in-Publication Data

Digital audio watermarking techniques and technologies : applications and benchmarks / Nedeljko Cvejic and Tapio Seppanen, editors.

p. cm.

Summary: "Digital audio watermarking has been proposed as a new and alternative method to enforce intellectual property rights and protect digital audio from tampering. This composition of theoretical frameworks, recent research findings, and practical applications will benefit researchers and students in electrical engineering and information technology, as well as professionals working in digital audio"--Provided by publisher.

Includes bibliographical references and index.

ISBN 978-1-59904-513-9 (hc.) -- ISBN 978-1-59904-515-3 (ebook)

1. Multimedia systems--Security measures. 2. Sound--Recording and reproducing--Digital techniques. 3. Digital watermarking. I. Cvejic, Nedeljko. II. Seppänen, Tapio.

QA76.575.D533 2007

005.8'2--dc22

2007007286

British Cataloguing in Publication Data
A Cataloguing in Publication record for this book is available from the British Library.

Table of Contents

Detailed Table of Contents

This chapter gives a general overview of the audio watermarking fundamental definitions. Audio watermarking algorithms are characterized by five essential properties, namely: perceptual transparency, watermark bit rate, robustness, blind/informed watermark detection, and security. The chapter also reviews the most common signal processing manipulations that are frequently applied to the watermarked audio in order to prevent detection of the embedded watermark. Finally, several application areas for digital audio watermarking are presented and advantages of digital watermarking over standard technologies are examined.

This chapter is a comprehensive presentation of spread spectrum-based digital audio watermarking methods. The problem is viewed as the realization of a basic communications system, where the host signal presents the available channel and the watermark presents the transmitted information that needs to survive and be recovered in conditions that include noise distortion, signal transformation, standard compression, and deliberate attacks. Basic spread spectrum theory as it relates to audio watermarking is introduced followed by state-of-the-art improvements. The important synchronization problem is analyzed in detail, existing techniques are presented, and a novel, precise synchronization method is included. Finally, the role of psychoacoustics in effective watermarking is emphasized and an enhanced psychoacoustic model based on the discrete wavelet packet transform (DWPT), which ensures efficiency and transparency is included.

This chapter promotes the use of parametric synthesis models in digital audio watermarking. It argues that, because human auditory perception is not a linear process, the optimal hiding of binary data in digital audio signals should consider parametric transforms that are generally nonlinear. To support this argument, an audio watermarking algorithm based on aligning frequencies of spectral peaks to grid points is presented as a case study; its robustness is evaluated and benefits are discussed. Toward the end, research directions are suggested, including watermark-aided sound source segregation, cocktail watermarking, and counter-measure against arithmetic collusive attacks.

Digital rights management of audio signals through robust watermarking has received significant attention during the last years. Two approaches for blind robust watermarking of audio signals are presented in this chapter. Both approaches use chaotic maps for the generation of the watermark signals and a correlation-based watermark detection procedure. The first method is a zero-bit method that embeds high-frequency chaotic watermarks in the low frequencies of the discrete Fourier transform (DFT) domain, thus achieving reliable detection and robustness to a number of attacks. The second method operates on the temporal domain and robustly embeds multiple bits of information in the host signal. Experimental results are provided to demonstrate the performance of the two methods.

Based on the requirement of watermark recovery, watermarking techniques may be classified under one of three schemes: nonblind watermarking scheme, blind watermarking schemes with and without synchronization information. For the nonblind watermarking scheme, the original signal is required for extracting the watermark and hence only the owner of the original signal will be able to perform the task. For the blind watermarking schemes, the embedded watermark can be extracted even if the original signal is not readily available. Thus, the owner does not have to keep a copy of the original signal. In this chapter, three audio watermarking techniques are described to illustrate the three different schemes. The time-frequency technique belongs to the nonblind watermarking scheme; the multiple-echo hiding technique and the peak-point extraction technique fall under the blind watermarking schemes with and without synchronization information respectively.

This chapter introduces time-spread echo hiding as an advanced audio watermarking method based on echo hiding. After reviewing some relevant researches, theoretical derivation of echo hiding and time-spread echo hiding is presented. As for embedding process, differences in the structure of echo kernels between the ordinary echo hiding and the time-spread echo hiding are schematically depicted to explain advantages of the new method. Several watermark-extracting methods are introduced for the decoding process to raise the detection rate for various conditions. Performance of the method in robustness against several attacks is evaluated in terms of d' because of its statistical preciseness. Results of computer simulations tell that the time-spread echo hiding show fairly better performance than the ordinary echo hiding.

In this chapter, detailed explanations would be given on the role of echo hiding playing in audio watermarking in terms of background, functions, and applications. Additionally, a section is dedicated to discuss the various approaches proposed in the past to solve the flaws of echo hiding. Lastly, the proposed analysis-by-synthesis echo watermarking scheme based on interlaced kernels is introduced. Comparisons in audio quality and robustness performance are also looked at the proposed and conventional echo watermarking schemes.

The patchwork watermarking scheme is investigated in this chapter. The performance of this algorithm in terms of imperceptibility, robustness, and security has been shown to be satisfactory. Robustness of the patchwork algorithm to the curve-fitting attack and blind multiple-embedding attack is presented also in this chapter. Robustness against jitter attack, which is a natural enemy of this watermarking algorithm, is also studied.

In this chapter, we present an overview of our time frequency (TF) based audio watermarking methods. First, a motivation on the necessity of data authentication, and an introduction in digital rights management (DRM) to protect digital multimedia contents is presented. TF techniques provide flexible means to analyze nonstationary audio signals. We have explained the joint TF domain for watermark representation, and have employed pattern recognition schemes for watermark detection. In this chapter, we introduce two watermarking methods; embedding nonlinear and linear TF signatures as watermarking signatures. Robustness of the proposed methods against common signal manipulations is also studied in this chapter.

In this chapter, we will present a brief overview of audio watermarking with the focus on a novel audio watermarking scheme which is based on watermarks with a 'semantic' meaning and offers a method for MPEG Audio Layer 3 files which does not need the original watermark for proving copyright ownership of an audio file. This scheme operates directly in the compressed data domain, while manipulating the time and subband/channel domains and was implemented with the mind on efficiency, both in time and space. The main feature of this scheme is that it offers users the enhanced capability of proving their ownership of the audio file not by simply detecting the bit pattern that comprises the watermark itself, but by showing that the legal owner knows a hard to compute property of the watermark bit sequence ('semantic' meaning). Our discussion on the scheme is accompanied by experimental results.

The main objective of the chapter is to provide an overview of existing speech watermarking technology and to demonstrate the importance of speech processing concepts for the design and evaluation of watermarking algorithms. This chapter describes the factors to be considered while designing speech watermarking algorithms, including the choice of the domain and speech features for watermarking, watermarked signal fidelity, watermark robustness, data payload, security, and watermarking applications. The chapter presents several state-of-the-art robust and fragile speech watermarking algorithms and discusses their advantages and disadvantages.

This chapter discusses the robustness of digital audio watermarking algorithms against digital-to-analogue (D/A) and analogue-to-digital (A/D) conversions. This is an important challenge in many audio watermarking applications. We provide an overview on distortions caused by converting the signal in various scenarios. This includes environmental influences like background noise or room acoustics when taking microphone recordings, as well as sound sampling effects like quantisation noise. The aim is to show the complexity of influences that need to be taken into account. Additionally, we show test results of our own audio watermarking algorithm with respect to analogue recordings using a microphone which proves that a high robustness in this area can be achieved. To improve even the robustness of algorithms, we briefly introduce strategies against downmixing and playback speed changes of the audio signal.

Methods for evaluating the quality of watermarked objects are detailed in this chapter. It will provide an overview of subjective and objective methods usable in order to judge the influence of watermark embedding on the quality of audio tracks. The problem associated with the quality evaluation of watermarked audio data will be presented. This is followed by a presentation of subjective evaluation standards used in testing the transparency of marked audio tracks as well as the evaluation of marked items with intermediate quality. Since subjective listening tests are expensive and dependent on many not easily controllable parameters, objective quality measurement methods are discussed in the section, Objective Evaluation Standards. The section Implementation of a Quality Evaluation presents the whole process of testing the quality taking into account the methods discussed in this chapter. Special emphasis is devoted to a detailed description of the test setup, item selection, and the practical limitations. The last section summarizes the chapter.-

Digital watermarking studies have always been driven by the improvement of robustness. Most of articles of this field deal with this criterion, presenting more and more impressive experimental assessments. Some key events in this quest are the use of spread spectrum, the invention of resynchronisation schemes, the discovery of side information channel, and the formulation of the embedding and attacking strategies as a game. On the contrary, security received little attention in the watermarking community. This chapter presents a comprehensive overview of this recent concept. We list the typical applications which require a secure watermarking technique. For each context, a threat analysis is purposed. This presentation allows us to illustrate all the certainties the community has on the subject, browsing all key papers. The end of the chapter is devoted to what remains not clear, intuitions, and future trends.

Preface

Recent development of the Internet and the digital information revolution caused significant changes in the global society, ranging from the influence on the world economy to the way people nowadays communicate. Broadband communication networks and multimedia data available in a digital format (images, audio, and video) opened many challenges and opportunities for innovation. Versatile and simple-to-use software and decreasing prices of digital devices (e.g., digital photo cameras, camcorders, portable CD and mp3 players, CD and DVD recorders, laptops, etc.) have made it possible for consumers from all over the world to create, edit, and exchange multimedia data. Broadband Internet connections and almost an errorless transmission of data facilitate people to distribute large multimedia files and make identical digital copies of them.

Digital media files do not suffer from any quality loss due to multiple copying processes, such as analogue audio and VHS tapes. Furthermore, recording medium and distribution networks for analogue multimedia are more expensive. These advantages of digital media over the analogue ones transform to disadvantages with respect to the intellectual rights management because of a possibility for unlimited copying without a loss of fidelity cause a considerable financial loss for copyright holders (Cox, Miller, & Bloom, 2003; Wu & Liu, 2003; Yu, Kundur, & Lin, 2001). The ease of content modification and a perfect reproduction in digital domain have promoted the protection of intellectual ownership and the prevention of the unauthorized tampering of multimedia data to become an important technological and research issue (Kundur, 2001).

Therefore, the traditional methods for copyright protection of multimedia data are no longer sufficient. Hardware-based copy protection systems have already been easily circumvented for the analogue media. Hacking of digital media systems is even easier due to the availability of general multimedia processing platforms, for example, a personal computer. Simple protection mechanisms that were based on the information embedded into header bits of the digital file are useless because header information can easily be removed by a simple change of data format, which does not affect the fidelity of media.

Encryption of digital multimedia prevents access to the multimedia content to an individual without a proper decryption key. Therefore, content providers get paid for the delivery of perceivable multimedia, and each client that has paid the royalties must be able to decrypt a received file properly. Once the multimedia has been decrypted, it can be repeatedly copied and distributed without any obstacles. Modern software and broadband Internet provide the tools to perform it quickly and without much effort and deep technical knowledge.

Basic Definitions and Terms in Digital Watermarking

Digital watermarking has been proposed as a new, alternative method to enforce the intellectual property rights and protect digital media from tampering. It involves a process of embedding into a host signal a perceptually transparent digital signature, carrying a message about the **host signal** in order to "mark" its ownership. The digital signature is called the **digital watermark**. The digital watermark contains data that can be used in various applications, including digital rights management, broadcast monitoring, and tamper proofing. Although perceptually transparent, the existence of the watermark is indicated when watermarked media is passed through an appropriate watermark detector.

A watermark, which usually consists of a binary data sequence, is inserted into the host signal in the **watermark embedder**. Thus, a watermark embedder has two inputs; one is the watermark message (usually accompanied by a secret key) and the other is the host signal (e.g., image, video clip, audio sequence, etc.). The output of the watermark embedder is the **watermarked signal**, which cannot be perceptually discriminated from the host signal. The watermarked signal is then usually recorded or broadcasted and later presented to the **watermark detector**. The detector determines whether the watermark is present in the tested multimedia signal, and if so, what message is encoded in it. The research area of watermarking is closely related to the fields of information hiding (Anderson & Petitcolas, 1998; Johnson, Duric, & Jajodia, 2001) and steganography (Johnson & Jajodia, 1998; Katzenbeisser & Petitcolas, 1999).

Therefore, we can define **watermarking systems** as systems in which the hidden message is related to the host signal and **nonwatermarking** systems in which the message is unrelated to the host signal. On the other hand, systems for embedding messages into host signals can be divided into **steganographic systems**, in which the existence of the message is kept secret, and **nonsteganographic systems**, in which the presence of the embedded message does not have to be secret.

Audio watermarking initially started as a subdiscipline of digital signal processing, focusing mainly on convenient signal processing techniques to embed additional information to audio sequences. This included the investigation of a suitable transform domain for watermark embedding and schemes for the imperceptible modification of the host audio. Only recently watermarking has been placed to a stronger theoretical foundation, becoming a more mature discipline with a proper base in both communication modelling and information theory.

Digital Audio Watermarking and the Human Auditory System

Watermarking of audio signals is more challenging compared to the watermarking of images or video sequences due to wider dynamic range of the human auditory system (HAS) in comparison with the human visual system (HVS) (Bender, Gruhl, & Morimoto, 1996). The HAS perceives sounds over a range of power greater than 10^9:1 and a range of frequencies greater than 10^3:1. The sensitivity of the HAS to the additive white Gaussian noise (AWGN) is high as well; this noise in a sound file can be detected as low as 70 dB below ambient level. On the other hand, opposite to its large dynamic range, HAS contains a fairly small differential range, that is, loud sounds generally tend to mask out weaker sounds. Additionally, HAS is insensitive to a constant relative phase shift in a stationary audio signal and some spectral distortions interprets as natural, perceptually non-annoying ones (Bender et al., 1996).

Auditory perception is based on the critical band analysis in the inner ear where a frequency-to-location transformation takes place along the basilar membrane. The power spectra of the received sounds are not represented on a linear frequency scale but on limited frequency bands called **critical bands** (Steinebach, Petitcolas, Raynal, Dittmann, Fontaine, Seibel, et al., 2001). The auditory system is usually modelled as a bandpass filterbank, consisting of strongly overlapping bandpass filters with bandwidths around 100 Hz for bands with a central frequency below 500 Hz and up to 5000 Hz for bands placed at high frequencies. If the highest frequency is limited to 24000 Hz, 26 critical bands have to be taken into account.

Two properties of the HAS dominantly used in watermarking algorithms are **frequency (simultaneous) masking** and **temporal masking** (Steinebach et al., 2001). The concept using the perceptual holes of the HAS is taken from wideband audio coding (e.g., MPEG Compression 1, Layer 3, usually called

MP3) (Noll, 1993). In the compression algorithms, the holes are used in order to decrease the amount of the bits needed to encode audio signal, without causing a perceptual distortion to the coded audio. On the other hand, in the information hiding scenarios, masking properties are used to embed additional bits into an existing bit stream, again without generating audible noise in the audio sequence used for data hiding.

The simplest visualization of the requirements of information hiding in digital audio is so called **magic triangle** (Johnson et al., 2001). Inaudibility, robustness to attacks, and the watermark data rate are in the corners of the magic triangle. This model is convenient for a visual representation of the required trade-offs between the capacity of the watermark data and the robustness to certain watermark attacks, while keeping the perceptual quality of the watermarked audio at an acceptable level. It is not possible to attain high robustness to signal modifications and high data rate of the embedded watermark at the same time. Therefore, if a high robustness is required from the watermarking algorithm, the bit rate of the embedded watermark will be low and vice versa; high bit rate watermarks are usually very fragile in the presence of signal modifications. However, there are some applications that do not require that the embedded watermark as a high robustness against signal modifications. In these applications, the embedded data are expected to have a high data rate and to be detected and decoded using a blind detection algorithm. While the robustness against intentional attacks is usually not required, signal processing modifications, like noise addition, should not affect the covert communications (Cox et al., 2003). To qualify as steganography applications, the algorithms have to attain statistical invisibility as well.

Contents of the Book

The book is organized as follows:

Chapter I gives a general overview of the audio watermarking fundamental definitions. Audio watermarking algorithms are characterized by five essential properties, namely perceptual transparency, watermark bit rate, robustness, blind/informed watermark detection, and security. Chapter I also reviews the most common signal processing manipulations that are frequently applied to the watermarked audio in order to prevent detection of the embedded watermark. Finally, several application areas for digital audio watermarking are presented and advantages of digital watermarking over standard technologies examined.

Chapter II is a comprehensive presentation of spread spectrum-based digital audio watermarking methods. The problem is viewed as the realization of a basic communications system where the host signal presents the available channel and the watermark presents the transmitted information that needs to be recovered in conditions that include noise distortion, standard compression, and deliberate attacks. Basic spread spectrum theory as it relates to audio watermarking is introduced followed by state-of-the-art improvements. Finally, the role of psychoacoustics in effective watermarking is emphasized and an enhanced psychoacoustic model based on the discrete wavelet packet transform (DWPT) is included.

In Chapter III, the use of parametric synthesis models in digital audio watermarking is described. It argues that, because human auditory perception is not a linear process, the optimal hiding of binary data in digital audio signals should consider parametric transforms that are generally nonlinear. To support this argument, an audio watermarking algorithm based on aligning frequencies of spectral peaks to grid points is presented as a case study; its robustness is evaluated and benefits are discussed. Toward the end, the future research directions are given, including watermark-aided sound source segregation and counter-measure against arithmetic collusive attacks.

Robust audio watermarking for copyright protection and digital rights management (DRM) in general has been an active research area during the last few years. Despite the achieved progress, there are still may open issues and challenges. Two state-of-the art approaches towards blind robust audio watermarking are presented in Chapter IV. Both utilize a correlation detection scheme and watermarks generated by chaotic systems.

In Chapter V, watermarking techniques are classified as one of three schemes: nonblind watermarking scheme, blind watermarking schemes with and without synchronization information. In this chapter, three audio watermarking techniques are described to illustrate the three different schemes. The time-frequency technique belongs to the nonblind watermarking scheme; the multiple-echo hiding technique and the peak-point extraction technique fall under the blind watermarking schemes with and without synchronization information, respectively.

Chapter VI introduces time-spread echo hiding as an advanced audio watermarking method based on echo hiding. Differences in the structure of echo kernels between the ordinary echo hiding and the time-spread echo hiding are schematically depicted to explain advantages of the new method. Several watermark-extracting methods are introduced for decoding process to raise the detection rate for various conditions. Performance of the method in robustness against several attacks is evaluated in terms of **delay** because of its statistical preciseness. The time-spread echo hiding exhibited fairly better performance than the ordinary echo hiding.

In Chapter VII, detailed explanations would be given on the role of echo hiding playing in audio watermarking, in terms of background, functions, and applications. Additionally, a section is dedicated to discuss the various approaches proposed in the past to solve the flaws of echo hiding. Lastly, the proposed analysis-by-synthesis echo watermarking scheme based on interlaced kernels is introduced. Comparisons in audio quality and robustness performance are also looked at the proposed and conventional echo watermarking schemes.

The patchwork algorithm is investigated in detail in Chapter VIII. The performance of this algorithm in terms of imperceptibility and robustness has been shown to be superior. Robustness of the patchwork algorithm against the curve-fitting attack and blind multiple-embedding attack is presented as well. Toward the end, robustness against jitter attack which is a natural enemy of this watermarking algorithm is studied.

In Chapter IX, an overview of our time-frequency (TF) based audio watermarking methods is presented. TF techniques provide flexible means to analyze nonstationary audio signals. The joint TF domain for watermark representation is given and pattern recognition schemes for watermark detection were employed. In this chapter; two watermarking methods are introduced: embedding nonlinear and linear TF signatures as watermarking signatures. Robustness of the proposed methods against common signal manipulations is also studied.

In Chapter X, a brief overview of audio watermarking with the focus on a novel audio watermarking scheme is given. The scheme is based on watermarks with a "semantic" meaning and offers a method for MPEG Audio Layer 3 files which does not need the original watermark for proving copyright ownership of an audio file. The main feature of this scheme is that it offers enhanced capability of proving the ownership of the audio file not by simply detecting the bit pattern that comprises the watermark itself, but by showing that the legal owner knows a hard to compute property of the watermark bit sequence ("semantic" meaning).

The main objective Chapter XI is to provide an overview of existing speech watermarking technology and to demonstrate the importance of speech processing concepts for the design and evaluation of watermarking algorithms. This chapter describes the factors to be considered while designing speech watermarking algorithms, including the choice of the domain and speech features for watermarking, watermarked signal fidelity, watermark robustness, data payload, security, and applications. The state-of-the-art robust and fragile speech watermarking algorithms are presented and their advantages and disadvantages discussed.

Chapter XII discusses the robustness of audio watermarking algorithms against digital-to-analogue and analogue-to-digital conversions. It provides an overview on distortions caused by converting the signal in various scenarios. The aim is to show the complexity of influences that need to be taken into account. Additionally, experimental results of our author's own audio watermarking algorithm are presented, with respect to analogue recordings using a microphone which proves that a high robustness in this area can be achieved. At the end, strategies against downmixing and playback speed changes of the watermarked audio signal are presented.

Methods for evaluating the quality of watermarked objects are detailed in Chapter XIII. A presentation of subjective evaluation standards used in testing the transparency of marked audio tracks is given, as well as the evaluation of marked items with intermediate quality. Since subjective listening tests are expensive and dependent on many not easily controllable parameters, objective quality measurement methods are discussed. The process of testing the quality taking into account the methods discussed is presented as well, with particular emphasis to a detailed description of the test setup, item selection and the practical limitations.

Digital watermarking studies have always been driven by the improvement of robustness. Most of articles of this field deal with this criterion, presenting more and more impressive experimental assessments. On the contrary, security received little attention in the watermarking community. Chapter XIV presents a comprehensive overview of this recent concept. The authors list the typical applications which require a secure watermarking technique. For each context, a threat analysis is purposed. Using this presentation, the authors illustrate all the certainties the community has on the subject, the intuitions and future trends.

REFERENCES

Anderson, R., & Petitcolas, F. (1998). On the limits of steganography. *IEEE Journal on Selected Areas in Communications, 16*(4), 474–481.

Bender, W., Gruhl, D., & Morimoto, N. (1996). Techniques for data hiding. *IBM Systems Journal, 35*(3), 313–336.

Cox, I., Miller, M., & Bloom, J. (2003). *Digital watermarking*. San Francisco: Morgan Kaufmann.

Johnson, N., Duric, Z., & Jajodia, S. (2001). *Information hiding: Steganography and watermarking-attacks and countermeasures*. Boston: Kluwer Academic Publishers.

Johnson, N., & Jajodia, S. (1998). Steganalysis: The investigation of hidden information. In *Proceedings of the IEEE Information Technology Conference* (pp. 113–116). Syracuse, New York.

Katzenbeisser, S., & Petitcolas, F. (1999). *Information hiding techniques for steganography and digital watermarking*. Norwood, MA: Artech House.

Kundur, D. (2001). Watermarking with diversity: Insights and implications. *IEEE Multimedia, 8*(4), 46–52.

Noll, P. (1993). Wideband speech and audio coding. *IEEE Communications Magazine, 31*(11), 34–44.

Steinebach, M., Petitcolas, F., Raynal, F., Dittmann, J., Fontaine, C., Seibel, S. et al. (2001). Stirmark benchmark: Audio watermarking attacks. In *Proceedings of the International Conference on Information Technology: Coding and Computing* (pp. 49–54). Las Vegas, Nevada.

Wu, M., & Liu, B. (2003). *Multimedia data hiding*. New York: Springer Verlag.

Yu, H., Kundur, D., & Lin, C. (2001). Spies, thieves, and lies: The battle for multimedia in the digital era. *IEEE Multimedia, 8*(3), 8–12.

Acknowledgment

The editors would like to acknowledge the help of all involved in the collation and review process of the book, without whose support the project could not have been adequately completed. Most of the authors of chapters included in this book also served as referees for chapters written by other authors. Thanks are due to all those who provided constructive and comprehensive reviews.

Special thanks also go to the publishing team at IGI Global, whose contributions throughout the whole process from inception of the initial idea to final publication have been very valuable. In particular, the editors would like to thank our development editor, Kristin Roth, who helped immensely in the opening phase of the project and our assistant development editor, Meg Stocking who continuously sent e-mail reminders in order to keep the project on schedule.

Nedeljko Cvejic's warmest appreciation is due to his loving wife, Ana, for being a wonderful source of love, understanding, and encouragement in his life.

The compilation of this book has been a stimulating experience. We believe that this book will be a useful addition to the literature on digital watermarking, because this is the first book that focuses on audio watermarking.

At the end, we wish to thank all of the authors for their insights and excellent contributions to the book.

Nedeljko Cvejic and Tapio Seppänen
Bristol, UK and Oulu, Finland
May 2007

Chapter I
Introduction to Digital Audio Watermarking

Nedeljko Cvejic
University of Bristol, UK

Tapio Seppänen
University of Oulu, Finland

ABSTRACT

This chapter gives a general overview of the audio watermarking fundamental definitions. Audio watermarking algorithms are characterized by five essential properties, namely: perceptual transparency, watermark bit rate, robustness, blind/informed watermark detection, and security. The chapter also reviews the most common signal processing manipulations that are frequently applied to the watermarked audio in order to prevent detection of the embedded watermark. Finally, several application areas for digital audio watermarking are presented and advantages of digital watermarking over standard technologies are examined.

INTRODUCTION

The primary focus of this book is the watermarking of digital audio, that is, audio watermarking. The watermarking algorithms were primarily developed for digital images and video sequences (Bender, Gruhl, & Morimoto, 1996; Cox & Miller, 2001); interest and research in audio watermarking started slightly later (Hartung & Kutter, 1999; Swanson, Zhu, & Tewfik, 1999). In the past few years, several algorithms for the embedding and extraction of watermarks in audio sequences have been developed. All of the developed algorithms take advantage of the perceptual properties of the human auditory system (HAS) in order to add a watermark into a host signal in a perceptually transparent manner. Embedding additional information into audio sequences is a more tedious task than that of images, due to the dynamic supremacy of the HAS over human visual system (Bender et al., 1996). In addition, the amount of data that can be embedded transparently into an audio sequence is considerably lower than the amount of data that can be hidden in video sequences as an audio signal, has a dimension less than three-dimensional video files. On the other hand, many malicious attacks are against image and video watermarking algorithms (e.g., geometrical distortions, spatial scaling, etc.) cannot be implemented against audio watermarking schemes.

Requirements for Audio Watermarking Algorithms

Watermarking algorithms can be characterized by a number of defining properties (Cox, Miller, & Bloom, 2003). The relative importance of a particular property is application dependent, and in many cases the interpretation of a watermark property itself varies with the application.

Perceptual Transparency

In most of the applications, the watermark-embedding algorithm has to insert additional data without affecting the perceptual quality of the audio host signal (Bender et al., 1996; Johnston, 1998). The fidelity of the watermarking algorithm is usually defined as a perceptual similarity between the original and watermarked audio sequence. However, the quality of the watermarked audio is usually degraded, either intentionally by an adversary or unintentionally in the transmission process, before a person perceives it. In that case, it is more adequate to define the fidelity of a watermarking algorithm as a perceptual similarity between the watermarked audio and the original host audio at the point at which they are presented to a consumer.

Watermark Bit Rate

The bit rate of the embedded watermark is the number of the embedded bits within a unit of time and is usually given in bits per second (bps). Some audio watermarking applications, such as copy control, require the insertion of a serial number or author ID, with the average bit rate of up to 0.5 bps. For a broadcast monitoring watermark, the bit rate is higher, caused by the necessity of the embedding of an ID signature of a commercial within the first second at the start of the broadcast clip, with an average bit rate up to 15 bps. In some envisioned applications, for example hiding speech in audio or compressed audio stream in audio, algorithms have to be able to embed watermarks with the bit rate that is a significant fraction of the host audio bit rate, up to 150 kbps.

Robustness

The robustness of the algorithm is defined as an ability of the watermark detector to extract the embedded watermark after common signal processing procedures. Applications usually require robustness in the presence of a predefined set of signal processing modifications, so that watermark can be reliably extracted at the detection side. For example, in radio broadcast monitoring, an embedded watermark need only to survive distortions caused by the transmission process, including dynamic compression and low pass filtering, because the watermark is extracted directly from the broadcast signal. On the other hand, in some algorithms, robustness is completely undesirable and those algorithms are labelled *fragile audio watermarking* algorithms.

Blind or Informed Watermark Detection

In some applications, a detection algorithm may use the original host audio to extract a watermark from the watermarked audio sequence (informed detection). It often significantly improves the detector performance, in that the original audio can be subtracted from the watermarked copy, resulting in the watermark sequence alone. However, if *blind detection* is used, the watermark detector does not have access to the original audio, which substantially decreases the amount of data that can be hidden in the host signal. The complete process of embedding and extracting of the watermark can be modelled as a communications channel where the watermark is distorted due to the presence of strong interference and channel effects (Kirovski & Malvar, 2001). A strong interference is caused by the presence of the host audio, and channel effects correspond to signal processing operations.

Security

Watermark algorithm must be secure in the sense that an adversary must not be able to detect the presence of embedded data, let alone remove the embedded data. The security of watermark process is interpreted in the same way as the security of encryption techniques and it cannot be broken unless the authorized user has access to a secret key that controls watermark embedding. An unauthorized user should be unable to extract the data in a reasonable amount of time even if he knows that the host signal contains a watermark and is familiar with the exact watermark embedding algorithm. Security requirements vary with application, and the most stringent are in cover communications applications, and, in some cases, data are encrypted prior to embedding into host audio.

Computational Complexity

The implementation of an audio watermarking system is a tedious task, and it depends on the application involved. The principal issue from the technical point of view is the computational complexity of embedding and detection algorithms and the number of embedders and detectors used in the system. For example, in broadcast monitoring, embedding and detection must be done in real time, while in copyright protection applications time is not a crucial factor for a practical implementation. Another issue in the design of embedders and detectors, which can be implemented as hardware or software plug-ins, is the difference in the processing power of different devices (laptop, PDA, mobile phone, etc.).

ATTACKS AGAINST AUDIO WATERMARKING ALGORITHMS

Subjective quality of the watermarked signal and robustness of the embedded watermark to various modifications are general requirements for all watermarking systems. Since the requirements of robustness, inaudibility, and high capacity cannot be fulfilled simultaneously, various variations and design criteria are significant for certain applications of audio watermarking. The most important requirement addresses the inaudibility of an inserted watermark; if the quality of audio signal cannot be preserved, the method will not be accepted neither by industry nor users. When the perceptual transparency requirement has been fulfilled, the design objective is to maximize robustness inside the limits imposed by perceptual requirements, obtaining at the same time a practical watermark bit rate.

Common signal processing manipulations are frequently applied to the watermarked audio. They significantly modify frequency content and dynamics of the host signal and therefore distort the embedded watermark. Furthermore, third parties may attempt to modify the watermarked signal in order to prevent detection of the embedded data.

An example of a signal manipulation is preparation of audio material to be transmitted at a radio station. The audio sequence is first normalized and compressed to fit the loudness level of the broadcast transmission. Equalization is used as well, to optimise the quality of received audio. A denoiser (dehisser) reduces unwanted parts of the audio information and filters are used to cut off any frequency that cannot be transmitted. If a watermark is used for tracking of broadcasted commercials, it has to be robust against all the modifications described above, or the extraction will be impossible. Another case is the Internet distribution; for example, a company wants to embed watermarks as copyright protection. Thus, the watermark has to be robust against all operations usually applied to the material. In this case, the most common attack will be lossy MPEG or AAC compression, usually at high compression rates.

To evaluate the robustness of audio watermarking algorithms, attacks can be grouped by the manner in which signal manipulations distort the embedded watermark. Based on the attack models, the following group of attacks are usually identified (Steinebach, Raynal, Dittmann, Fontaine, Seibel, et al., 2001):

- **Dynamics:** These modifications change the loudness profile of the watermarked audio, with amplification and attenuation being the most basic attacks. Limiting, expansion, and compression are more complicated, as they consist of nonlinear changes depending on the input audio signal. There are even several frequency dependent compression algorithms, which only affect a part of the frequency range.
- **Filtering:** Filters cut off or amplify a selected part of the spectrum. The basic filters are high-pass and low-pass filters, but equalizers can also be seen as filters. They are used to increase or attenuate certain subbands of audio spectrum.
- **Ambience:** Audio effects simulating the presence of a room. The most common effects are reverb and delay that offer various parameters to set the quality of effect.
- **Conversion:** Watermarked audio is often subject to format changes. Mono data can be mixed up to be used in stereo environments, and stereo signal can be down mixed to mono. Sampling frequencies range from 32 kHz to 96 kHz, while sample resolution goes from 8 to 24 bits per sample.
- **Lossy compression:** Audio compression algorithms based on psychoacoustic effects (MPEG and AAC) used to reduce the size of audio files by factor 10 or more.
- **Noise:** Noise can be the result of most of the attacks described in this section, and most of the hardware components introduce noise into the signal. Adversaries usually attack the watermarked audio by adding AWGN of certain amplitude.
- **Modulation:** Modulation effects like vibrato, chorus, amplitude modulation, or flanging are usually not implemented in the broadcasting, but as most audio processing software includes such effects, they can be used as attacks to remove watermarks.
- **Time stretch and pitch shift:** These attacks either change the length of an audio sequence without changing its pitch or change the pitch of audio content without changing its length. They are used for fine tuning or fitting audio sequences into predefined time windows.
- **Sample permutations:** This group consists of the algorithms not used for audio manipulation in common environments. These attacks represent a specialized way to attack embedded watermarks in time domain.

Although a complete benchmark for the audio watermarking has not yet been implemented, there were some attempts to introduce a unified testing environment for audio watermarking algorithms (Steinebach et al., 2001). Multiple advantages of a unified third-party benchmark are obvious. First, researchers and software programmers would just provide a table of test results that would show a summary of performances of the proposed algorithm. Researchers could compare different algorithms and improve methods by adding some new features to it. Second, end users would get information whether their basic application requirements are fulfilled. Third, industry can properly estimate the risks associated with the use of a particular solution by having information about the level of reliability of the proposed solution.

AUDIO WATERMARKING APPLICATIONS

In this section, several application areas for digital watermarking will be presented and advantages of digital watermarking over standard technologies examined.

Ownership Protection

In the ownership protection applications, a watermark containing ownership information is embedded to the multimedia host signal. The watermark, known only to the copyright holder, is expected to be very robust and secure (i.e., to survive common signal processing modifications and intentional attacks), enabling the owner to demonstrate the presence of this watermark in case of dispute of ownership. Watermark detection must have a very small false alarm probability. On the other hand, ownership protection applications require a small embedding capacity of the system, because the number of bits that can be embedded and extracted with a small probability of error does not have to be large.

Proof of Ownership

It is even more demanding to use watermarks not only in the identification of the copyright ownership, but as an actual proof of ownership. The problem arises when an adversary uses editing software to replace the original copyright notice with one's own and then claims to own the copyright. In the case of early watermark systems, the problem was that the watermark detector was readily available to adversaries. As elaborated in Cox et al. (2003), anybody that can detect a watermark can probably remove it as well. Therefore, because adversaries can easily obtain a detector, they can remove owner's watermark and replace it with their own. To achieve the level of the security necessary for proof of ownership, it is indispensable to restrict the availability of the detector. When an adversary does not have the detector, the removal of a watermark can be made extremely difficult. However, even if an owner's watermark cannot be removed, an adversary might try to undermine the owner. As described in Cox et al. (2003), an adversary, using a personal own watermarking system, might be able to make it appear as if the watermark data was present in the owner's original host signal. This problem can be solved using a slight alteration of the problem statement.

Instead of a direct proof of ownership by embedding, for example, "Dave owns this image" watermark signature in the host image, algorithm will instead try to prove that the adversary's image is derived from the original watermarked image. Such an algorithm provides indirect evidence that it is more probable that the real owner owns the disputed image, because the real owner is the one who has the version from which the other two were created.

Authentication and Tampering Detection

In the content authentication applications, a set of secondary data is embedded in the host multimedia signal and is later used to determine whether the host signal was tampered. The robustness against removing the watermark or making it undetectable is not a concern as there is no such motivation from an attacker's point of view. However, forging a valid authentication watermark in an unauthorized or tampered host signal must be prevented. In practical applications, it is also desirable to locate (in time or spatial dimension) and to discriminate the unintentional modifications (e.g., distortions incurred due to moderate MPEG compression [Noll, 1993; Wu, Trappe, Wang, & Liu, 2004]) from content tampering itself. In general, the watermark embedding capacity has to be high to satisfy the need for more additional data than in ownership protection applications. The detection must be performed without the original host signal because either the original is unavailable or its integrity has yet to be established. This kind of watermark detection is usually called a *blind detection*.

Fingerprinting

Additional data embedded by a watermark in the fingerprinting applications are used to trace the originator or recipients of a particular copy of a multimedia file (Chenyu, Jie, Zhao, & Gang, 2003; Hong, Wu, Wang, & Liu, 2003; Trappe, Wu, Wang, & Liu, 2003; Wu et al., 2004). For example, watermarks carrying different serial or ID numbers are embedded in different copies of music CDs or DVDs before distributing them to a large number of recipients. The algorithms implemented in fingerprinting applications must show high robustness against intentional attacks and signal processing modifications such as lossy compression or filtering. Fingerprinting also requires good anticollusion properties of the algorithms, that is, it is not possible to embed more than one ID number to the host multimedia file; otherwise the detector is not able to distinguish which copy is present. The embedding capacity required by fingerprinting applications is in the range of the capacity needed in copyright protection applications, with a few bits per second.

Broadcast Monitoring

A variety of applications for audio watermarking are in the field of broadcasting (Termont, Stycker, Vandewege, Op de Beeck, Haitsma, Kalker, et al., 2000; Termont, De Strycker, Vandewege, Haitsma, Kalker, Maes, et al., 1999; Depovere, Kalker, Haitsma, Maes, de Strycker, Termont, et al., 1999; Kalker & Haitsma, 2000). Watermarking is an obvious alternative method of coding identification information for an active broadcast monitoring. It has the advantage of being embedded within the multimedia host signal itself rather than exploiting a particular segment of the broadcast signal. Thus, it is compatible with the already installed base of broadcast equipment, including digital and analogue communication channels. The primary drawback is that the embedding process is more complex than simply placing data into file headers. There is also a concern, especially on the part of content creators, that the watermark would introduce distortions and degrade the visual or audio quality of multimedia. A number of broadcast monitoring watermark-based applications are already available on commercial basis. These include program type identification, advertising research, and broadcast coverage research.

Copy Control and Access Control

In the copy control application, the embedded watermark represents a certain copy control or access control policy. A watermark detector is usually integrated in a recording or playback system, like in the proposed DVD copy control algorithm (Bloom, Cox, Kalker, Linnartz, Miller, & Traw, 1999) or during the development of Secure Digital Music Initiative (SDMI) (Craver & Stern, 2001). After a watermark has been detected and the content decoded, the copy control or access control policy is enforced by directing particular hardware or software operations such as enabling or disabling the record module. These applications require watermarking algorithms to be resistant against intentional attacks and signal processing modifications, to be able to perform a blind watermark detection, and to be capable of embedding a nontrivial number of bits in the host signal.

Information Carrier

The embedded watermark in this application is expected to have a high capacity and to be detected and decoded using a blind detection algorithm. While the robustness against intentional attack is not required, a certain degree of robustness against common processing like MPEG compression may be desired. A public watermark embedded into the host multimedia might be used as the link to external databases that contain certain additional information about the multimedia file itself, such as copyright information and licensing conditions. One interesting application is the transmission of metadata along with multimedia. Metadata embedded in, for example, and audio clip, may carry information about composer, soloist, and genre of music.

FUTURE RESEARCH DIRECTIONS

During the last decade, several algorithms for the embedding and extraction of watermarks in audio sequences have been developed. All of the developed audio watermarking algorithms use the perceptual properties of the human auditory system in order to add a watermark into a host signal in a perceptually transparent manner. Digital watermarking was seen as a viable alternative for copyright protection of multimedia content; one of the most important issues delaying the surge of multimedia technologies and services. In the last few years, there was some scepticism expressed and serious concerns raised about the suitability of watermarking as a solution for multimedia copyright protection and digital rights management.

After the initial hype and high expectations, digital audio watermarking has entered a phase where theoretical research based on fundamental principles of signal processing and digital communications has address many open problems. This is not a process specific to the watermarking research field; being a very novel technology digital, watermarking had to get through these phases. Important progress has been made recently that certainly made the future of watermarking look much more positive than it seemed to be a couple of years ago.

One of the main future research topics will be the comparison between traditional spread-spectrum watermarking and quantization-based techniques. Based on the theoretical analysis of watermarking as a communication channel with side information, it can be demonstrated that the quantization-based techniques have the potential to largely outperform systems based on the spread-spectrum approach. As the importance of watermark security increases in the future, watermarking researchers will have to present a watermark security evaluation, beside the perceptual distortion, the embedding rate, and the robustness of their watermarking schemes.

The further enhancement of audio watermarking methods will need to include a multidiscipline approach, which will involve researchers such as human auditory system experts, audio engineers, lawyers for copyright ownership, specialists in cryptography for assessment of watermark security, researchers in the area of audio compression and multimedia transmission, and so forth. This cooperation will facilitate the development of a framework able to develop practical and affordable watermarking technologies and capable of applying audio watermarking in the real-world scenarios, in search of "killer applications."

REFERENCES

Bender, W., Gruhl, D., & Morimoto, N. (1996). Techniques for data hiding. *IBM Systems Journal, 35*(3), 313-336.

Bloom, J., Cox, I., Kalker, T., Linnartz, J., Miller, M., & Traw, C. (1999). Copy protection for DVD video. *Proceedings of the IEEE, 87*(7), 1267-1276.

Chenyu, W., Jie, Z., Zhao, B., & Gang, R. (2003). Robust crease detection in fingerprint images. In *Proceedings of the IEEE International Conference on Computer Vision and Pattern* Recognition (pp. 505-510). Madison, Wisconsin.

Cox, I., Miller, M., & Bloom, J. (2003). *Digital watermarking.* San Francisco: Morgan Kaufmann.

Cox, I., & Miller, M. (2001). Electronic watermarking: The first 50 years. In *Proceedings of the IEEE Workshop on Multimedia Signal Processing* (pp. 225-230). Cannes, France.

Craver, S., & Stern, J. (2001). Lessons learned from SDMI. In *Proceedings of the IEEE International Workshop on Multimedia Signal Processing* (pp. 213-218). Cannes, France.

Depovere, G., Kalker, T., Haitsma, J., Maes, M., de Strycker, L., Termont, P., et al. (1999). The viva project: Digital watermarking for broadcast monitoring. In *Proceedings of the IEEE International Conference on Image Processing* (pp. 202-205). Kobe, Japan.

Hartung, F., & Kutter, M. (1999). Multimedia watermarking techniques. In *Proceedings of the IEEE, 87*(7), 1709-1107.

Hong, Z., Wu, M., Wang, Z., & Liu, K. (2003). Nonlinear collusion attacks on independent fingerprints for multimedia. In *Proceedings of the IEEE Computer Society Conference on Multimedia and Expo* (613-616). Baltimore.

Johnston, J. (1998). Estimation of perceptual entropy using noise masking criteria. In *Proceedings of the IEEE International Conference on Acoustics, Speech, and Signal Processing* (pp. 2524-2527). New York.

Kalker, T., & Haitsma, J. (2000). Efficient detection of a spatial spread-spectrum watermark in MPEG video streams. In *Proceedings of the IEEE International Conference on Image Processing* (pp. 407-410). Vancouver, British Columbia.

Kirovski, D., & Malvar, H. (2001). Spread-spectrum audio watermarking: Requirements, applications, and limitations. In *Proceedings of the IEEE International Workshop on Multimedia Signal Processing* (pp. 219-224). Cannes, France.

Noll, P. (1993). Wideband speech and audio coding. *IEEE Communications Magazine, 31*(11), 34-44.

Pan, D. (1995). A tutorial on mpeg/audio compression. *IEEE Multimedia, 2*(2), 60-74.

Steinebach, M., Petitcolas, F., Raynal, F., Dittmann, J., Fontaine, C., Seibel, S., et al. (2001). Stirmark benchmark: Audio watermarking attacks. In *Proceedings of the International Conference on Information Technology: Coding and Computing* (pp. 49-54). Las Vegas, Nevada.

Swanson, M., Zhu, B., & Tewfik, A. (1999). Current state-of-the-art, challenges and future directions for audio watermarking. In *Proceedings of the IEEE International Conference on Multimedia Computing and Systems* (pp. 19-24). Florence, Italy.

Termont, P., De Stycker, L., Vandewege, J., Op de Beeck, M., Haitsma J., Kalker, T., et al. (2000). How to achieve robustness against scaling in a real-time digital watermarking system for broadcast monitoring. In *Proceedings of the IEEE International Conference on Image Processing* (pp. 407-410). Vancouver, British Columbia.

Termont, P., De Strycker, L., Vandewege, J., Haitsma, J., Kalker, T., Maes, M., et al. (1999). Performance measurements of a real-time digital watermarking system for broadcast monitoring. In *Proceedings of the IEEE International Conference on Multimedia Computing and Systems* (pp. 220-224). Florence, Italy.

Trappe, W., Wu, M., Wang, Z., & Liu, K. (2003). Anti-collusion fingerprinting for multimedia. *IEEE Transactions on Signal Processing, 51*(4), 1069-1087.

Wu, M., Trappe, W., Wang, Z., & Liu, K. (2004). Collusion-resistant fingerprinting for multimedia. *IEEE Signal Processing Magazine, 21*(2), 15-27.

ADDITIONAL READING

Bassia, P., Pitas, I., & Nikolaidis, N. (2001). Robust audio watermarking in the time domain. *IEEE Transactions on Multimedia, 3*(2), 232-241.

Byeong-Seob, K., Nishimura, R., & Suzuki, Y. (2005). Time-spread echo method for digital audio watermarking. *IEEE Transactions on Multimedia, 7*(2), 212-221.

Cvejic, N., & Seppanen, T. (2003). Audio watermarking using attack characterisation. *Electronics Letters, 39*(13), 1020-1021.

Erkucuk, S., Krishnan, S., & Zeytinoglu, M. (2006). A robust audio watermark representation based on linear chirps. *IEEE Transactions on Multimedia, 8*(5), 925-936.

Esmaili, S., Krishnan, S., & Raahemifar, K. (2003). Audio watermarking time-frequency characteristics. *Canadian Journal of Electrical and Computer Engineering, 28*(2), 57-61.

Kim, H., & Choi, Y. (2003). A novel echo-hiding scheme with backward and forward kernels. *IEEE Transactions on Circuits and Systems for Video Technology, 13*(8), 885-889.

Larbi, S., & Jaidane-Saidane, M. (2005) Audio watermarking: A way to stationnarize audio signals. *IEEE Transactions on Signal Processing, 53*(2), 816-823.

Lemma, A., Aprea, J., Oomen, W., & Van de Kerkhof, L. (2003). A temporal domain audio watermarking technique. *IEEE Transactions on Signal Processing, 51*(4), 1088-1097.

Li, W., Xue, X., & Lu, P. (2005). Robust audio watermarking based on rhythm region detection. *Electronics Letters, 41*(4), 218-219.

Liu, Z., & Inoue, A. (2003). Audio watermarking techniques using sinusoidal patterns based on pseudorandom sequences. *IEEE Transactions on Circuits and Systems for Video Technology, 13*(8), 801-812.

Robert, A., & Picard, J. (2005). On the use of masking models for image and audio watermarking. *IEEE Transactions on Multimedia, 7*(4), 727-739.

Seok, J., & Hong, J. (2001). Audio watermarking for copyright protection of digital audio data. *Electronics Letters, 37*(1), 60-61.

Takahashi, A., Nishimura, R., & Suzuki, Y. (2005). Multiple watermarks for stereo audio signals using phase-modulation techniques. *IEEE Transactions on Signal Processing, 53*(2), 806-815.

Wang, X., & Zhao, H., (2006). A novel synchronization invariant audio watermarking scheme based on DWT and DCT. *IEEE Transactions on Signal Processing, 54*(12), 4835-4840.

Wei, L., Xiangyang, X., & Peizhong, L. (2006). Localized audio watermarking technique robust against time-scale modification. *IEEE Transactions on Multimedia, 8*(1), 60-69.

Wen-Nung, L., & Li-Chun, C. (2006). Robust and high-quality time-domain audio watermarking based on low-frequency amplitude modification. *IEEE Transactions on Multimedia, 8*(1), 46-59.

Wu, S., Huang, J., Huang, D., & Shi, Y. (2005). Efficiently self-synchronized audio watermarking for assured audio data transmission. *IEEE Transactions on Broadcasting, 51*(1), 69-76.

Yeo, I., & Kim, H. (2003). Modified patchwork algorithm: A novel audio watermarking scheme. *IEEE Transactions on Speech and Audio Processing, 11*(4), 381-386.

Zaidi, A., Boyer, R., & Duhamel, P. (2006). Audio watermarking under desynchronization and additive noise attacks. *IEEE Transactions on Signal Processing, 54*(2), 570-584.

Chapter II
Spread Spectrum for Digital Audio Watermarking

Xing He
University of Miami, Coral Gables, USA

Michael Scordilis
University of Miami, Coral Gables, USA

ABSTRACT

This chapter is a comprehensive presentation of spread spectrum-based digital audio watermarking methods. The problem is viewed as the realization of a basic communications system, where the host signal presents the available channel and the watermark presents the transmitted information that needs to survive and be recovered in conditions that include noise distortion, signal transformation, standard compression, and deliberate attacks. Basic spread spectrum theory as it relates to audio watermarking is introduced followed by state-of-the-art improvements. The important synchronization problem is analyzed in detail, existing techniques are presented, and a novel, precise synchronization method is included. Finally, the role of psychoacoustics in effective watermarking is emphasized and an enhanced psychoacoustic model based on the discrete wavelet packet transform (DWPT), which ensures efficiency and transparency is included.

INTRODUCTION

The increase in computational power and the proliferation of the Internet, witnessed over the last decade, have facilitated the production and distribution of unauthorized copies of copyrighted multimedia information. It is now easy to make a perfect digital illegal copy and distribute it overnight to millions of people via the Internet, resulting in large revenue loses to the movie and record industries every year. As a result, the problem of effective copyright protection has attracted the interest of the worldwide scientific and business communities. The most promising solution seems to be the watermarking process where the original data is marked with ownership information hidden in an imperceptible manner in the original signal.

Compared to embedding watermarks into still images, audio watermarking is significantly more challenging due to the extreme sensitivity of the human auditory system (HAS) to changes in the audio signal (Cox, Miller, & Bloom, 2002). As a result, a psychoacoustic model of the HAS is usually integrated in the audio watermarking scheme in order to ensure transparency of any changes to the original audio.

Several techniques in audio watermarking systems have been proposed in the past decade including low-bit coding (Cvejic, 2004), phase coding (He, Iliev, & Scordilis, 2004; Kuo, Johnston, Turin, & Quackenbush, 2002), echo coding (Gruhl, Lu, & Bender, 1996), patchwork coding (Bender, Gruhl, Morimoto, & Lu, 1996), and spread spectrum (Kim, 2003; Kirovski, & Attias, 2002; Kirovski & Malvar, 2001, 2003; Meel, 1999; Swanson, Zhu, Tewfik, & Boney, 1998). Among those, spread spectrum-based techniques are the most widely used because they can provide high robustness and low interference with the host signal.

In this chapter we will first illustrate the basic theory underlying the spread spectrum approach, followed by a review of the uses of spread spectrum in digital watermarking. The next section focuses on techniques that improve the detection of spread spectrum watermarking as well as the mechanisms for fighting against desynchronization attacks. It also presents a fast synchronization method for watermarking. Then we present the psychoacoustic models available for audio watermarking, and focus on one developed by He and Scordilis (2006) with an example of its incorporation in spread spectrum audio watermarking. This chapter concludes with a summary and suggests possible improvements and future trends in spread spectrum for audio watermarking. In this chapter, message or secret message refers to the original information to be inserted in the host signal, and watermark refers to the encoded or modulated message to be embedded or hidden into the host audio signal.

Figure 1. Watermarking as a communication process

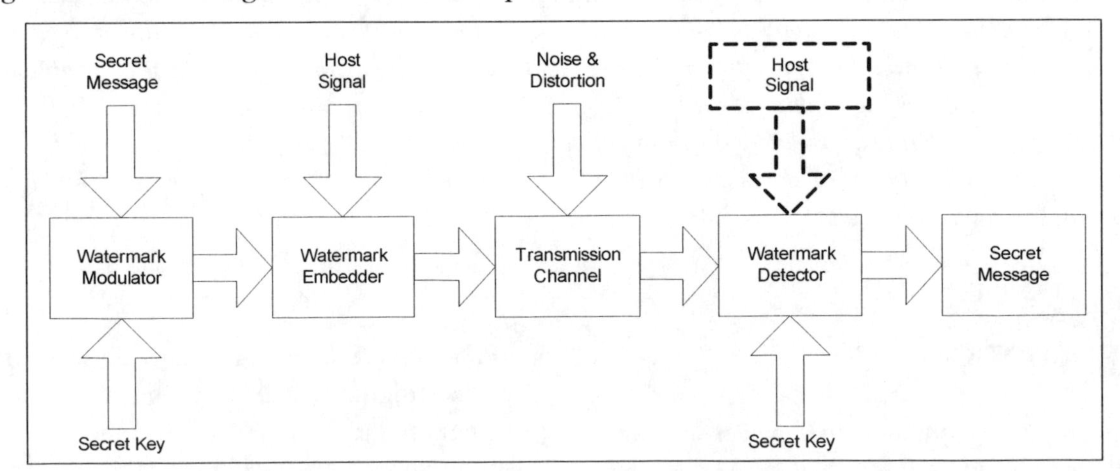

BASIC THEORY OF SPREAD SPECTRUM AND WATERMARKING

Communication Model for Watermarking

The watermarking process involves the imperceptible insertion at the encoder of information that is unrelated to the nature of the host signal and its perfect recovery at the decoder. Therefore, the host signal can be viewed as a carrier or channel used to convey the watermarked information. The watermarked signal may be subjected to a variety of processes which affect its properties, such as storing in different media, copying, noise addition, compression, sampling rate change, or even malignant attacks aiming at the destruction of the inserted watermarks. As such, watermarking can be viewed and analyzed as a communications process. Due to the similarity of watermarking and traditional communication systems, a watermarking system can be modeled from a secure data communication perspective as illustrated in Figure 1.

In Figure 1, the secret message is the watermark that the system needs to imperceptibly transmit over the transmission channel. The secret key, which is known both to the watermark embedder and the detector, is used to modulate the watermark by the watermark modulator.

The watermark embedder inserts the watermark into the host audio in an imperceptible way and sends the watermarked signal (stego-signal) to the receiver through the transmission channel, which may add noise and distortion to the transmitted signal.

The watermark detector needs to recover the secret message from the received signal with the help of the secret key. Most of the time, the received signal is a distorted version of the stego-signal. If the host audio signal is available at the detection phase, the detection is nonblind; otherwise, the detection is blind.

Let m be the secret message, k be the secret key, x be the original host signal, f be the modulation function, and w be the watermark or modulated message, then:

$$w = f(m, k) \tag{1}$$

if the watermark is independent of the host signal, or

$$w = f(m, k, x) \tag{2}$$

where the information of the host signal is incorporated with watermark modulation, which is referred as informed watermark embedding. The watermarked signal is denoted as:

$$y = x + \alpha w \tag{3}$$

where α is a strength control factor used to ensure inaudible watermark embedding.

THEORY OF SPREAD SPECTRUM TECHNOLOGY IN COMMUNICATION

There are two types of spread spectrum technologies: direct sequence spread spectrum (DSSS) and frequency hopping spread spectrum (FHSS). This section shows the basic principles of these methods as well as a comparison between them.

Direct Sequence Spread Spectrum

DSSS is the technology that employs a pseudo random noise code (PN sequence), which is independent of the signal to be transmitted, and it uses it to modulate or spread the narrow band signal into a signal with much wider bandwidth. Figure 2 and Figure 3 illustrate this process in the time and frequency domains. At the receiver side, the same PN sequence is used to despread the wideband signal back to the original narrowband signal (Meel, 1999).

Figure 2. The spread spectrum process in the time domain (Meel, 1999)

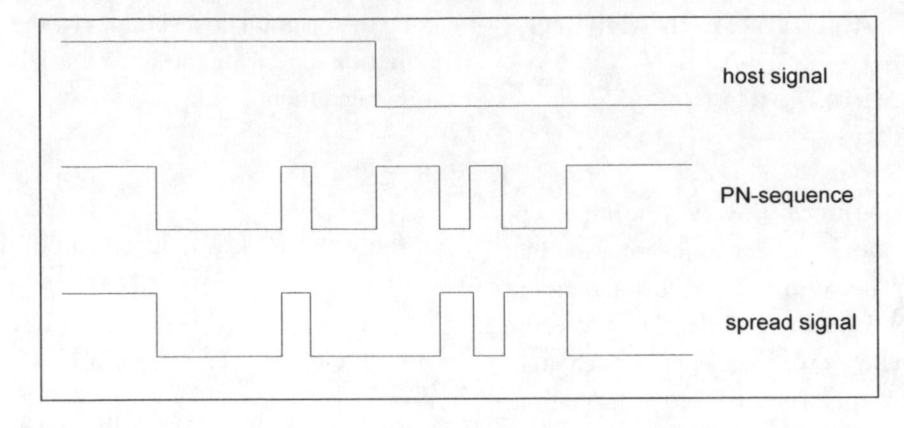

Figure 3. The spread spectrum process in the frequency domain (Chipcenter, 2006)

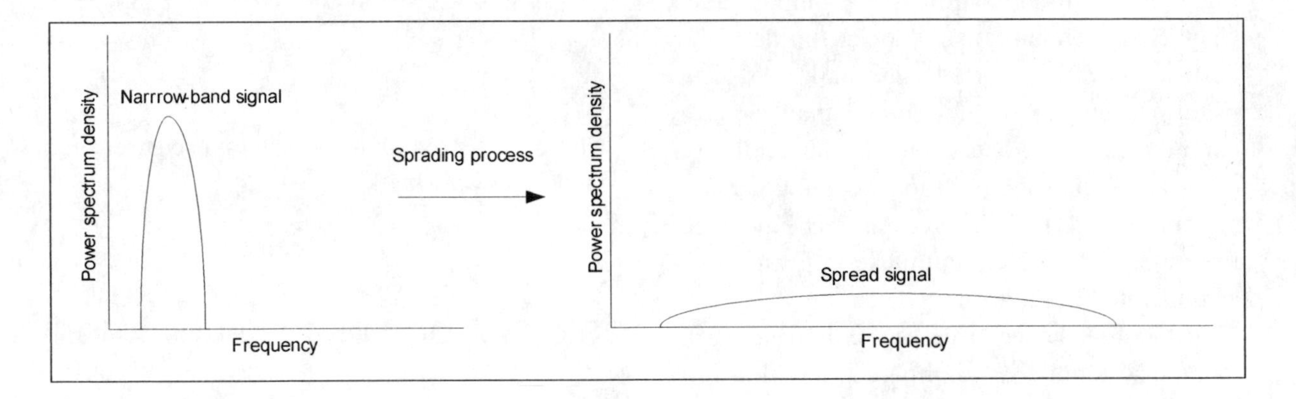

Power density is the distribution of power over frequency. Although the spread signal has the same amount of energy and carries the same information as the narrow band signal, it has much lower power density due to its much wider bandwidth. This makes it more difficult to detect the existence of the spread spectrum signal. Processing gain is the power density ratio between narrowband signal and spread spectrum signal, which in this case is usually higher than 10 (Chipcenter, 2006).

On the other hand, if the spread signal is contaminated by narrow band jam during transmission, the energy of the jam signal will spread over a much wider bandwidth when despread, resulting a relatively high signal-to-noise ratio (SNR) at the receiver which leads to easy detection of the despread signal as illustrated in Figure 4. This is the reason for the high robustness of the spread signal against a narrow band jammer.

Frequency Hopping Spread Spectrum

Another spread spectrum technology is FHSS, which does not spread the narrow band signal into a wideband signal. Instead the frequency of the carrier hops from channel to channel at different

Figure 4. De-spread signal and jammer (Chipcenter, 2006)

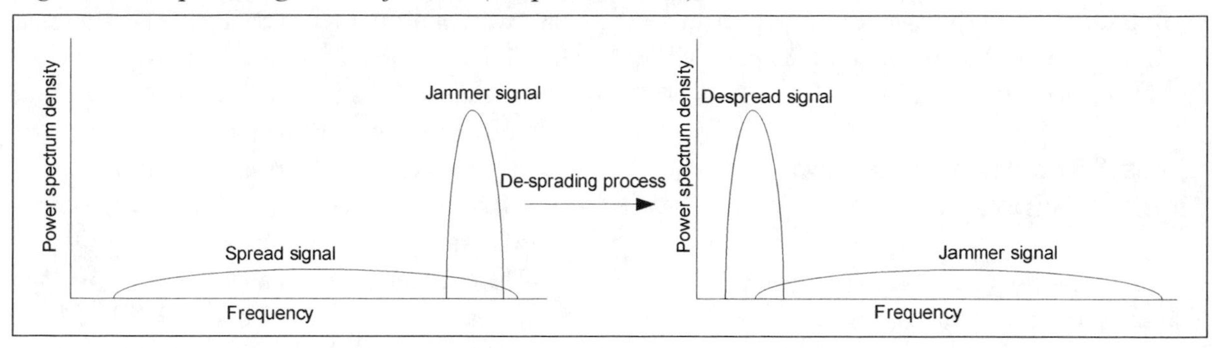

times, as shown in Figure 5, and thus it may avoid a jammer operating at some frequency. Since in FHSS the signal does not get spread, there is no processing gain. In order to achieve the same signal-to-noise ratio, the FHSS system has to output more power than DSSS system.

Since the carrier signal changes frequency from time to time, the sender and the receiver have to be tuned both in time and frequency, which makes synchronization a critical issue. Once it hops, it is very difficult, if not impossible for the receiver to resynchronize in case synchronization is lost. Before the receiver can be synchronized with the sender, it has to spend more time to search the signal and to lock on it in time and frequency. This time interval is called lock-in time. FHSS systems need much longer lock-in time than DSSS systems, which only have to synchronize with the timing of the PN sequence.

In order to make the initial synchronization successful, the hopper needs to park at a fixed frequency before hopping, such frequency is called parking frequency. If unfortunately, the jammer is also located in that frequency, the hopper will not be able to hop at all (Chipcenter, 2006; Meel, 1999).

Despite those disadvantages, the FHSS system can usually carry more data than the DSSS system because the signal is narrowband. It is also harder for opponents to attack the FHSS system than the

Figure 5. FHSS illustration (Meel, 1999)

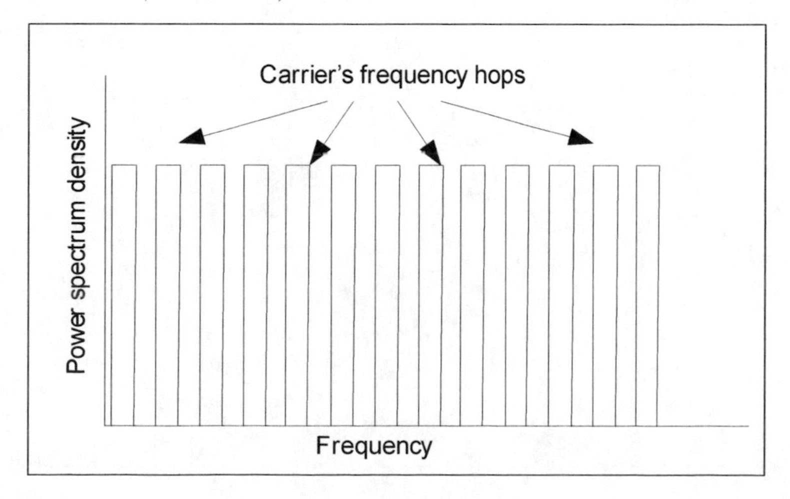

DSSS system because now the opponent has to track both time and frequency slots to achieve synchronization. A comparison between DSSS and FHSS is shown in Table 1.

Spread Spectrum for Audio Watermarking

In this section, spread spectrum technology as it applies to audio watermarking is presented. Substantial work has been carried out in this area and is presented in several key publications (Bender et al., 1996; Cox et al., 2002; Cox, Kilian, Leighton, & Shamoon, 1997; Cvejic, 2004; Gruhl et al., 1996; Meel, 1999).

Several spread spectrum-based watermarking systems have been proposed since middle the 90s when Cox first introduced spread spectrum into watermarking (Cox et al., 1997). A typical frame-based spread spectrum audio watermarking system is illustrated in Figure 6 (encoder) and Figure 7 (decoder). 'F' denotes transform functions (Fourier, discrete cosine transform (DCT), wavelet transform, etc.) and 'IF' is the appropriate inverse transform function.

The encoder operates as follows:

a. The input original audio is segmented into overlapped frames of N samples long.
b. If the watermark is to be embedded in the transform domain then each frame is converted into that domain (e.g., frequency, sub band/wavelet, or cepstral domain.)

Table 1. Comparison between DSSS and FHSS

	DSSS	FHSS
Robustness	High	High
Payload	Low	High
Needs parking frequency	No	Yes
Frequency hops	No	Yes
Synchronization	Easy	Hard
Re-synchronization	Not difficult	Very difficult
Processing gain	High	None
SNR after dispreading	High	Low
Lock-in time	Fast	Slow

Figure 6. Typical spread spectrum audio watermarking system; the encoder

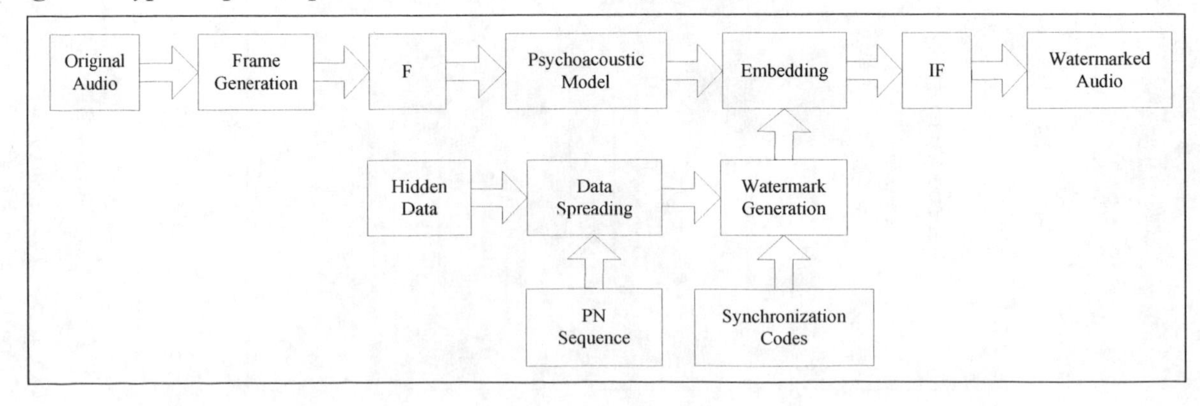

Figure 7. Typical spread spectrum audio watermarking system; the decoder

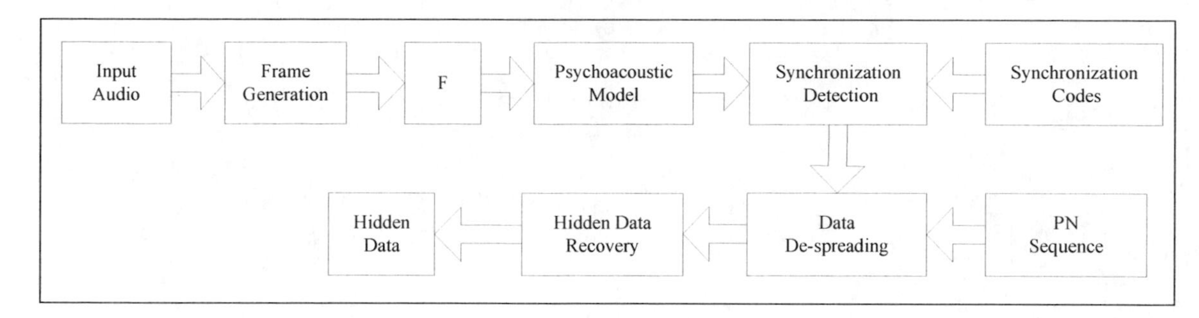

c. A psychoacoustic model is applied to determine the masking thresholds for each frame in order to render the inserted data inaudible.
d. The data are spread by a pseudo random sequence (PN sequence).
e. Synchronization codes are attached to the spread data thus producing the final watermarks to be embedded.
f. Watermark embedding into the original audio is conditional on the masking thresholds constraints.
g. If the watermark is embedded in the transform domain then inverse transform is applied on the watermarked frame to convert the watermarked audio back to the time domain.

The system decoder works in a reverse manner as in Figure 7 and its operations are:

a. The incoming audio is first segmented into overlapped frames.
b. The frame is converted into transform domain if necessary.
c. The same psychoacoustic model is applied on the data to determine the masking thresholds.
d. Synchronization is performed by searching for the synchronization codes.
e. The appropriate data are then despread in order to detect and recover any hidden data.

Analysis of Traditional SS Watermarking Systems

The core part of traditional spread spectrum watermarking is illustrated in Figure 8 where k is the secret key used by a binary pseudo random number generator (PRN) to produce a pseudo random sequence (PN sequence) u with zero mean, and whose elements are either $+\sigma_u$ or $-\sigma_u$. The sequence is then multiplied by the secret message bit m, which is either '1' or '–1,' and added to the signal. The watermarked signal s is denoted as:

$$y = x + mu \tag{4}$$

Suppose the watermark bit rate is *1/N* bits/sample, which means N transformed or time domain samples of the original signal (host signal) are needed to embed one bit of information.

The distortion D caused by embedding is denoted as

$$D = \| y - x \| = \| mu \| = \| u \| = \sigma_u^2 \tag{5}$$

where $\| \, \|$ is the norm defined as $\| x \| = \sqrt{<x, x>}$ and $<>$ is the inner product calculated as:

$$<x, u> = \sum_{i=0}^{N-1} x_i u_i \tag{6}$$

and N is the vector length of x and u.

Figure 8. Traditional spread spectrum core (Cvejic, 2004)

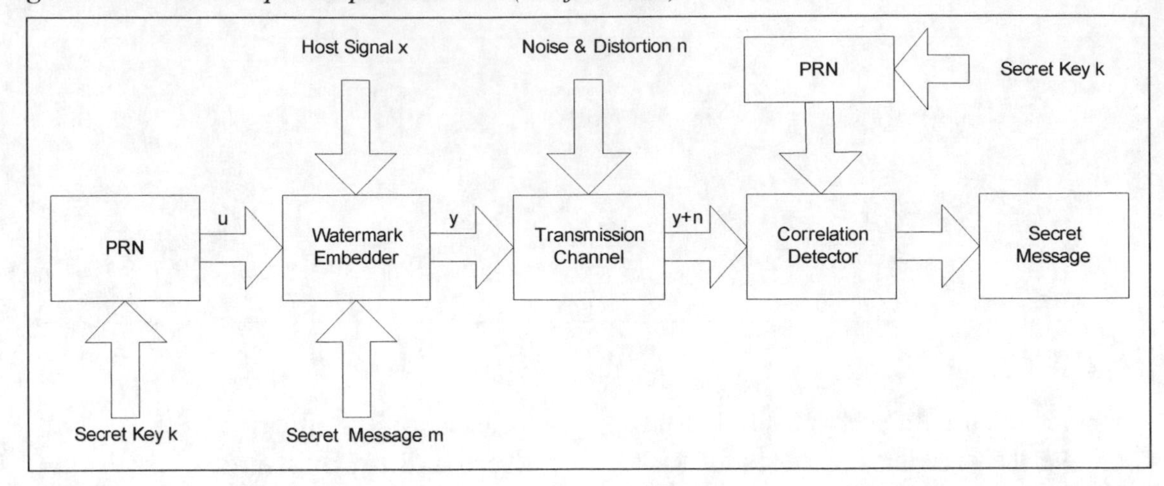

Noise n is considered to be additive in the channel and the received signal in the decoding side is modeled as:

$$\hat{y} = y + n \quad (7)$$

Normalized correlation is performed between the received signal and the PN sequence u to extract the watermark as follows:

$$r = \frac{<\hat{y},u>}{<u,u>} = \frac{<mu + x + n, u>}{\sigma_u^2} = m + x_u + n_u \quad (8)$$

where

$$x_u = \frac{<x,u>}{\sigma_u^2} \text{ and } n_u = \frac{<n,u>}{\sigma_u^2} \quad (9)$$

Usually x and n are considered Gaussian sequences as $x \sim N(0,\sigma_x^2)$ and $n \sim N(0,\sigma_n^2)$.

Uncorrelation between x and u is assumed when the lengths of both vectors are large enough and it leads to $x_u \approx 0$. The same assumption applies to n and u which leads to $n_u \approx 0$.

Therefore the recovered watermark is denoted as:

$$\hat{m} = sign(r) \quad (10)$$

Normalized correlation r is considered also a Gaussian process as:

$$r \sim N(m_r, \sigma_r^2)$$

and

$$m_r = E(r) = m\sigma_r^2 = \frac{\sigma_x^2 + \sigma_n^2}{N\sigma_u^2} \quad (11)$$

The equal error extraction probability p is given by equation (12), where $erfc()$ is the complementary error function.

Equation 12.

$$p = \Pr\{\hat{m} < 0 \mid m = 1\} = \Pr\{\hat{m} > 0 \mid m = -1\} = \frac{1}{2} erfc\left(\frac{m_r}{\sigma_r \sqrt{2}}\right) = \frac{1}{2} erfc\left(\sqrt{\frac{N\sigma_u^2}{2(\sigma_x^2 + \sigma_n^2)}}\right)$$

SURVEY OF CURRENT AUDIO WATERMARKING SCHEMES BASED ON SPREAD SPECTRUM

Basic Direct Sequence Spread Spectrum from Cox et al. (1997)

Cox et al. (1997) proposed a secure robust watermarking approach for multimedia based on spread spectrum technology. Although their system was originally proposed for watermarking images the basic idea also applies to audio watermarking.

The watermarking procedure proceeds as follows:

From host signal, a sequence of values $x = \{x_i\}$ is extracted, which is used for inserting secret message $m = \{m_i\}$. The watermarked sequence is denoted as $w' = \{w_i'\}$ and inserted back into the place of x to obtain a watermarked signal y. During transmission, possible distortion or attacks may affect y and the received signal is now denoted as $\hat{y}$, which may not be identical to y. Assume the host signal is available at the decoder side, a probably altered watermarked sequence $w^* = \{w_i^*\}$ is first extracted from the received signal $\hat{y}$. A possibly corrupted message m^* is extracted from w^* and compared to m for statistical significance (Cox et al., 1997).

During watermark embedding, a scaling parameter α which controls the watermark strength is specified and used in the watermark embedding equation. Three embedding formulae used are:

$$y = x + \alpha w \tag{13}$$

$$y = x(1 + \alpha w) \tag{14}$$

$$y = x(e^{\alpha w}) \tag{15}$$

Equation (14) is used in Cox et al. (1997) with $\alpha = 0.1$.

The similarity between the extracted watermark and the embedded watermark is measured by:

$$sim(m, m^*) = \frac{m.m^*}{\sqrt{m^*.m^*}} \tag{16}$$

Large values of $sim(m, m^*)$ are important and typically if $sim(m, m^*) > 6$, then a watermark is detected in the received signal.

Time Domain Spread Spectrum Watermarking Scheme by Cvejic, Keskinarkaus, and Seppänen (2001)

Cvejic et al. (2001) proposed an alternative audio spread spectrum watermarking scheme in the

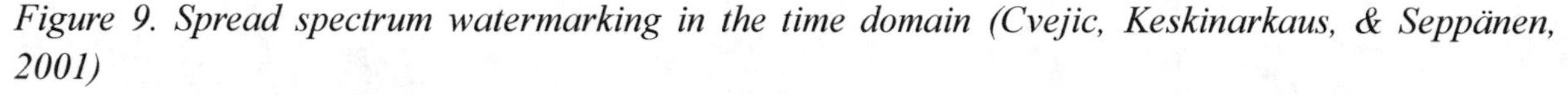

Figure 9. Spread spectrum watermarking in the time domain (Cvejic, Keskinarkaus, & Seppänen, 2001)

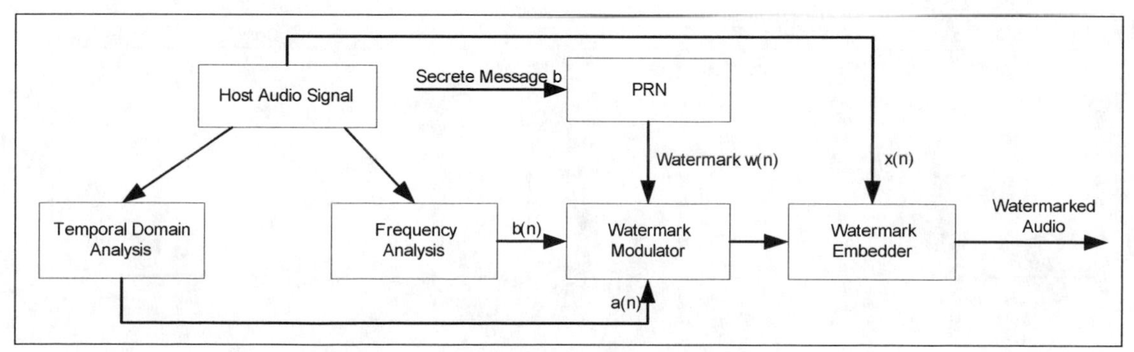

time domain. The embedding part is illustrated in Figure 9.

Time domain masking properties of the HAS are used to maximize the amplitude of the watermark to be embedded, which increases robustness during detection, while still keeping the watermark imperceptible and maintaining high perceptual audio quality.

A frequency analysis block counts the signal zero crossing rate in the basic block interval and derives the high frequency spectral information. The power of the embedded watermark is proportional to the amount and strength of high frequency components in the spectrum. The purpose of this frequency domain analysis is to further increase the power of the watermark without jeopardizing the perceptual quality of the audio signal.

Let $a(n)$ and $b(n)$ be the coefficients obtained from temporal and frequency analysis blocks respectively, $x(n)$ be the original audio, and $w(n)$ be the spread watermark in the time domain. Then the watermarked audio $y(n)$ is denoted by:

$$y(n) = x(n) + a(n)b(n)w(n) \tag{17}$$

Figure 10 provides a functional diagram of the watermark detection algorithm, which is based on a cross-correlation calculation. The novelty of this detection lies in the mean-removed cross-correlation between the watermarked audio signal and the equalized *m*-sequence. The watermarked audio is first processed by a high pass equalization filter, which filters out the strong low pass components, increases the correlation value, and enhances the detection result.

An adaptive filter, instead of a fixed coefficient one, is employed as the equalization filter to better accommodate the input audio. The output from the correlation calculation block is sampled by the detection and sampling block. The result is forwarded to the threshold detection block, which uses a majority rule to decide the value of the embedded watermark bit.

Spread Spectrum Watermarking with Psychoacoustic Model and 2-D Interleaving Array by Garcia (1999)

Garcia (1999) introduced a digital audio watermarking that uses a psychoacoustic model and spread spectrum theory. In this method, a psychoacoustic model similar to MPEG I Layer 1 is used to derive the masking thresholds from the original host audio signal. The watermark is generated with direct sequence spread spectrum and transformed into the frequency domain. Noise shaping is applied on the watermark to ensure transparent embedding.

The innovation part of this method lies in the two-dimension matrix used to enhance the watermark robustness. Here the watermark is first repeated for *M* times (Kirovski et al., 2003) and then interleaved by a *2-D* array. The purpose of the *2-D* array interleaving is to enhance the

Figure 10. Watermarking detection algorithm proposed in Cvejic (2002)

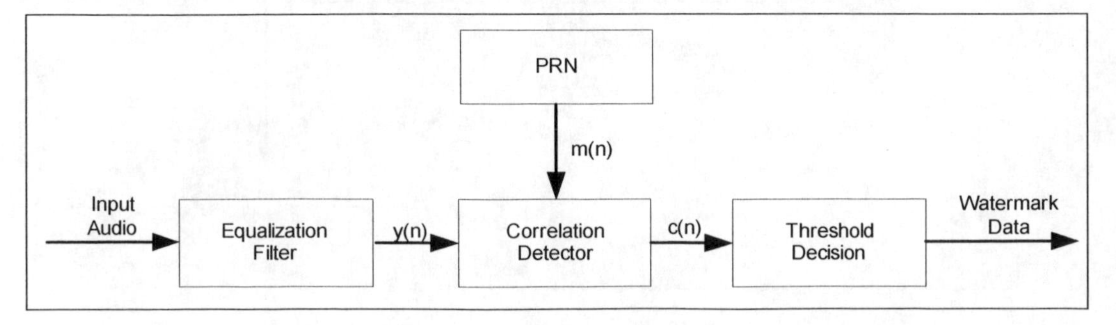

robustness against additive noise, especially pulse-like noise.

Suppose the data to be embedded comprise N samples denoted by $\{m_i\}$ consisting of '1's and '-1's ('0's in the watermark are transformed into '-1' to enhance the robustness). First every sample in $\{m_i\}$ is locally repeated M times resulting in the sequence $\{m_1 m_1 m_1 ... m_2 m_2 m_2 ... m_N m_N ... m_N\}$.

The new data sequence is interleaved by a *2-D* array with the size $N*M$ (N rows by M columns), the above sequence is used to form an array row-by-row and read out column-by-column to form a new data sequence as:

$$\{m_1 m_2 ... m_N m_1 m_2 ... m_N ... m_1 m_2 ... m_N\}.$$

Suppose more than $M/2$ samples of the same symbol in the uninterleaved watermark are corrupted by noise during transmission or attack. Then the symbol cannot be recovered at the decoder end. However, if the watermark is interleaved before embedding, there will be a better chance to recover it later.

As an illustrative example, suppose the data $\{m\}$ to be embed is a '*1001*', and since '0's are transformed into '*-1*'s, the watermark data $\{w\}$ become '*1 -1 -1 1*'. Then each symbol repeats M times (let $M = 5$) and the sequence now becomes '*1 1 1 1 1 -1 -1 -1 -1 -1 -1 -1 -1 -1 -1 1 1 1 1 1*'.

Table 2. Data interleaving example

1	1	1	1	1
-1	-1	-1	-1	-1
-1	-1	-1	-1	-1
1	1	1	1	1

Table 3. Data de-interleaving example

X	X	1	1	1
X	X	-1	-1	-1
X	X	-1	-1	-1
X	X	1	1	1

The repeated data is written into a 5*4 *2-D* array in rows as shown in Table 2.

The data is now read out column-by-column into before embedding resulting in a new sequence: '*1 -1 -1 1 1 -1 -1 1 1 -1 -1 1 1 -1 -1 1 1 -1 -1 1* '.

Suppose that during transmission, the first eight symbols become corrupted, and the decoder receives instead the sequence '*x x x x x x x x 1 1 -1 -1 1 1 -1 -1 1 1 -1 -1 1*' where 'x' denotes unknown symbol ('1' or '-1'). Deinterleaving is performed on the received sequence by writing it into the same *2-D* array by column resulting in the matrix of Table 3.

The data in the deinterleaving array are read out by row and we have '*x x 1 1 1 x x -1 -1 -1 x x -1 -1 -1 x x 1 1 1*'.

The decision rule for the k-th symbol of the recovered watermark is:

$$w_k = \begin{cases} 1 & if\ r_k > 0 \\ -1 & if\ r_k \le 0 \end{cases} \tag{18}$$

where

$$r_k = \sum_{i=(k-1)*M+1}^{k*M} m_i \tag{19}$$

M is the repetition number ($M = 5$ in this example), m_i is the *i*-th symbol in the above sequence.

According to the decision rule:

$$r_1 = r_4 = x + x + 1 + 1 + 1 = 3 + 2 * x > 0$$

$$r_2 = r_3 = x + x - 1 - 1 - 1 = -3 + 2 * x < 0$$

so $w_1 = w_4 = 1$ and $w_2 = w_3 = -1$ and the recovered watermark is '1 -1 -1 1', which is identical to the embedded.

By interleaving the embedded watermark, the possible additive noise is averaged by factor equal to the number of rows in the *2-D* array. On the other hand, if no interleave is involved during the embedding process, the sequence that the decoder received will be '*x x x x x x x x -1 -1 -1 -1 -1 -1 -1 1 1 1 1 1*'. We can easily see that the decoder can only recover the last two symbols correctly while the first two symbols are lost.

An Improved Spread Spectrum Method by Malvar and Florencio (2003)

An improved spread spectrum (ISS) method, which uses a new modulation technique for increased watermark robustness, was proposed by Malvar and Florencio (2003). In spread spectrum approaches, the host signal is considered as interference to the embedded watermark and affects the detection in the negative way. In ISS, the knowledge of the projection of the host signal (x_u in equation (8)) on the watermark is used to modulate the energy of the watermark, thus improving the detection performance. The new embedding rule is defined as:

$$y = x + \mu(x_u, b)u \tag{20}$$

where

$$x_u = \frac{<x,u>}{\sigma_u^2} \tag{21}$$

The traditional SS is a special case of ISS where the function μ is independent of x_u. A simple linear function is used in this method and the embedding function now becomes:

$$y = x + (\alpha b - \lambda x_u)u \tag{22}$$

The parameter α controls the distortion level caused by watermark embedding and λ controls the removal of host signal distortion. It can be noted that traditional spread spectrum is a special case of ISS with $\alpha = 1$ and $\lambda = 0$.

The normalized correlation now becomes:

$$r = \frac{<y,u>}{\|u\|} = \alpha b + (1-\lambda)x_u + x_n \tag{23}$$

As we can see from above equation, the more λ approaches 1, the less the influence of the host signal distortion on the watermark detection. Parameter α is chosen in a way that the distortion in the ISS is equal to the distortion in the SS.

The expected distortion in ISS is:

$$E[D] = E[\| y - x \|]$$

$$= E[| \alpha b - \lambda x_u |^2 \sigma_u^2] = (\alpha^2 + \frac{\lambda^2 \sigma_x^2}{N\sigma_u^2})\sigma_u^2 \tag{24}$$

Comparing this to equation (11) we have:

$$\alpha = \sqrt{\frac{N\sigma_u^2 - \lambda^2\sigma_x^2}{N\sigma_u^2}} \tag{25}$$

Equation 28.

$$p = \Pr\{\hat{m} < 0 \mid m = 1\} = \Pr\{\hat{m} > 0 \mid m = -1\} = \frac{1}{2} erfc\left(\frac{m_r}{\sigma_r \sqrt{2}}\right)$$

$$= \frac{1}{2} erfc\left(\sqrt{\frac{N\sigma_u^2 - \lambda^2\sigma_x^2}{2(\sigma_n^2 + (1-\lambda)^2\sigma_x^2)}}\right) = \frac{1}{2} erfc\left(\frac{1}{\sqrt{2}}\sqrt{\frac{\frac{N\sigma_u^2}{\sigma_x^2} - \lambda^2}{\frac{\sigma_u^2}{\sigma_x^2} + (1-\lambda)^2)}}\right)$$

The mean and variance of r are given by:

$$m_r = \alpha b \tag{26}$$

and

$$\sigma_r^2 = \frac{\sigma_n^2 + (1-\lambda)^2 \sigma_x^2}{N\sigma_u^2} \tag{27}$$

The error probability of ISS is calculated as, (see equation (28)).

The host signal is considered noise for watermark detection. The resulting watermark-to-noise ratio (WNR) is $\frac{\sigma_u^2}{\sigma_x^2}$.

By selecting an appropriate λ, the ISS can lead to an error rate several orders lower than traditional SS method. The optimal parameter of λ is given by:

$$\lambda_{opt} = \frac{1}{2}((1 + \frac{\sigma_n^2}{\sigma_x^2} + \frac{N\sigma_u^2}{\sigma_x^2})$$

$$- \sqrt{(1 + \frac{\sigma_n^2}{\sigma_x^2} + \frac{N\sigma_u^2}{\sigma_x^2})^2 - 4\frac{N\sigma_u^2}{\sigma_x^2}} \tag{29}$$

Novel Spread Spectrum Approach by Kirovski et al. (2001, 2002, 2003)

Kirovski et al. gradually and systematically developed a novel spread spectrum approach for audio signals. The key features of this method include:

a. Block repetition coding for prevention against desynchronization attacks.
b. Psycho-acoustic frequency masking (PAFM).
c. A modified covariance test to compensate the imbalance in the number of positive and negative watermark chips in the audible part of the frequency spectrum.
d. Cepstral filtering and chess watermarks are used to reduce the variance of the correlation test, thus improving reliability of watermark detection.
e. A special procedure used to identify the audibility of spread spectrum watermarks.
f. A technique which enables reliable covert communication over public audio channel.
g. A secure spread spectrum watermark that survives watermark estimation attacks.

Feature e is particularly worth noting and it will further be expanded upon in a later section. Detailed illustration of other features can be found in Kirovski and Attias (2002).

In order to embed spread spectrum WM imperceptibly, blocks with dynamic content are detected prior to watermark embedding. These blocks have some quiet parts and other parts rich in audio energy. Embedding watermarks into the quiet parts will result in audible distortion even is the amplitude of watermarks is low.

The procedure of detecting the problematic audio data blocks is as follows:

a. The signal is divided into blocks of K samples.
b. Each block is further partitioned into P interleaved subintervals.
c. The interval energy level is computed as:

$$E(i) = \sum_{j=1+K(i-1)/(2P)}^{iK/P} y_j, i = 1...P,$$

for each of the P subintervals of the tested signal y in the time domain.

d. If $\min_{1 \le j \le P} (E(j) / \sum_{i=1}^{P} E(i)) \le x_0$, then the watermark is audible in this block where x_0 is an experimentally obtained threshold.

Those problematic blocks are not used for watermark embedding or detection.

Enhanced Spread Spectrum Watermarking for AAC Audio from Cheng, Yu, and Xiong (2002)

While the above spread spectrum watermarking methods can successfully embed watermarks into uncompressed audio, none of them can deal effectively with audio in compressed domains. Cheng et al. proposed enhanced spread spectrum watermarking for compressed audio in MPEG-2 AAC (advanced audio coding) format.

The novelty of their methods lies in two contributions:

a. An enhanced spread spectrum technique, which reduces the variance of the host signal and enhances watermark detection. This contribution will be further analyzed in a later section.
b. Direct embedding of watermarks into the quantization indices instead of the MDCT coefficients, which saves processing time since no dequantization or requantization is necessary. Here a heuristic approach is used to select the indices and estimate the strength that those indices should be scaled to. Three steps are involved for the heuristic approach:
    1. Pick indices located at frequencies where the HAS is more sensitive and therefore they are perceptually important. This enhances the watermark robustness by preventing frequency truncation attack.
    2. Indices with zero values are not selected to avoid introducing the audible distortion of embedding watermark into silent areas.
    3. Quantization step is set to 1 to minimize distortion.

This method is low in both structural and computational complexity because only Huffman encoding and decoding are required to embed and estimate the watermark. The heuristic estimation on the perceptual weighting of the indices, which provides acceptable perceptual quality, is much simpler than incorporating a full psychoacoustic model.

Even though the most popular spread spectrum is DSSS and the key techniques have been presented, frequency hopping spread spectrum (FHSS) can also used for watermarking.

Frequency Hopping Spread Spectrum by Cvejic and Seppänen (2004)

Cvejic and Seppänen (2004) proposed an audio watermarking based on FHSS and so called "attack characterization" to battle against some limitations of spread spectrum based on direct sequence. Based on the observation that if the watermarked signal undergoes fading-like attacks, such as MPEG compressing or low-pass (LP) filtering for instance, the correlation between the host signal and the watermark can result in ineffective detection because those attacks cannot be modeled as additive white Gaussian noise (AWGN). It is claimed that a far more appropriate model is the frequency selective model, which takes into account more precisely the distortion introduced by those fading attacks.

In their method, characterization of the MPEG compression watermarking attack is analyzed first to find out the subset of fast Fourier transform coefficients that are least distorted by the attack. Frequency hopping spread spectrum is used to select two FFT coefficients from those least vulnerable subsets. The real watermark embedding is performed by the patchwork algorithm. If bit 1 is to be embedded, the magnitude of the lower frequency coefficient is increased by K decibels and the magnitude of higher frequency coefficient is decreased by the same amount. If bit 0 is to be embedded, the opposite operation is performed. An example of the frequency hopping method used during watermark embedding is illustrated in Figure 11.

Figure 11. Example of frequency hopping used during watermark embedding (Cvejic & Seppänen, 2004)

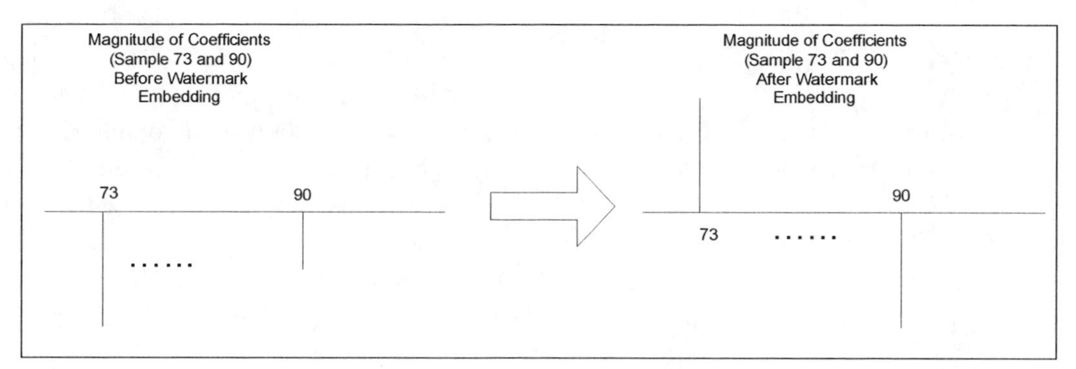

By using the frequency hopping method different FFT coefficient pairs are selected to embed watermarks in each block, making the watermark location very tough to guess by opponents.

EFFECTIVE SPREAD SPECTRUM WATERMARK DETECTION AND ATTACK RESISTANCE

In this section, techniques for improved spread spectrum detection are reviewed first followed by countermeasures to desynchronization attacks. A new fast and efficient synchronization method is highlighted later in this section.

Techniques to Improve Spread Spectrum Detection

A typical detection method for spread spectrum is the computation of the cross-correlation between the received stego-signal and the actual watermark. If the computed output exceeds a threshold, then a watermark is considered to have been detected in the received signal.

Matched Filter Approach

Some watermarking system employs a pseudo random sequence (PN sequence) for synchronization purpose. Matched filter is usually used in such cases to detect the existence of the PN sequence and to precisely locate the starting sample of the PN sequence.

In the watermarking system, the PN sequence is considered as noise added to the host signal. Since the PN sequence only lasts a very short period of time, it could be treated as transient noise pulses and detected by a filter whose impulse response is matched to the PN sequence. Such is the matched filter whose frequency response is defined as (Vaseghi, 2000):

$$H(f) = K\frac{PN^*(f)}{PSx(f)} \tag{30}$$

where K is a scaling factor, $PN^*(f)$ is the complex conjugate of the spectrum of PN sequence $\{u\}$, and $PSx(f)$ is the power spectrum of the host signal x.

In real world applications, the host signal is very close to a zero mean process with variance σ_x^2 and is uncorrelated to the PN sequence. Then, equation (30) becomes:

$$H(f) = \frac{K}{\sigma_x^2}PN^*(f) \tag{31}$$

with impulse response

$$h(n) = \frac{K}{\sigma_x^2}u(-n) \tag{32}$$

When the received signal contains the PN sequence, it is defined as:

$$y(n) = x(n) + \alpha u(n) \tag{33}$$

where α is a strength control factor. Then the output of the matched filter is:

$$\begin{aligned} o(n) &= \frac{K}{\sigma_x^2} u(-n) * (x(j) + \alpha u(j)) \\ &= \frac{K}{\sigma_x^2} u(-n) * x(j) + \frac{\alpha K}{\sigma_x^2} u(-n) * u(j) \end{aligned} \tag{34}$$

Since the host signal and PN sequence are uncorrelated, $d_{\mu x}$ is expected to be zero. The PN sequence itself is orthogonal to the host signal, and so d_{μ} will reach its maximum if the sequence is perfectly aligned, otherwise, it will be zero.

Therefore, the detected starting location of the PN sequence in a block will be:

$$pn_{loc} = \arg\max_i (o(n)) \tag{35}$$

Figure 12 shows a typical output of such matched filter. Note that the peak location marks the beginning of the embedded PN sequence.

The described matched filter detection is optimal for additive white Gaussian noise (Cvejic, 2004). However, the host audio signal is usually far from being white Gaussian noise. Adjacent audio sample are usually highly correlated with large variance, which increases the detection error for spread spectrum. Several techniques have been developed to decrease such correlation.

Savitzky-Golay Smoothing Filters

Cvejic (2004) applied least squares Savitzky-Golay smoothing filters in the time domain to smooth out a wideband noise audio signal. By doing so, the variance of the host audio signal is greatly reduced.

Let us recall the equal error extraction probability p as given by, (see equation (36)).

It is clear that a reduced variance σ_x will result in a decreased detection error rate.

Another way to decrease the audio signal variance is so called whitening process, which

Figure 12. Typical result of the correlation output

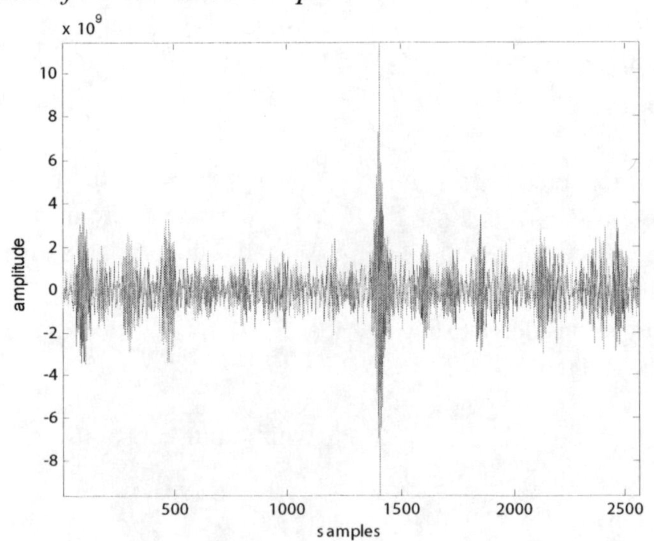

Equation 36.

$$p = \Pr\{\hat{m} < 0 \mid m = 1\} = \Pr\{\hat{m} > 0 \mid m = -1\} = \frac{1}{2} erfc\left(\frac{m_r}{\sigma_r \sqrt{2}}\right) = \frac{1}{2} erfc\left(\sqrt{\frac{N\sigma_u^2}{2(\sigma_x^2 + \sigma_n^2)}}\right)$$

subtracts a moving average from the frequency spectrum of the received signal right before correlation. This method, although it removes part of the correlation in the audio signal, it also removes part of the watermark as well since the watermarks are spread all over the frequency spectrum.

Cepstrum Filtering

Kirovski et al. (2001) proposed a cepstrum filtering method to reduce the variance which works as follows:

a. The received signal is a modulated complex lapped transform (MCLT) vector y.
b. Compute the cepstrum of the dB magnitude MCLT vector y by the discrete cosine transform (DCT). $z = DCT(y)$
c. Filter out the first K cepstral coefficients. $z_i = 0, i = 1 \ldots K$
d. Inverse DCT is applied to z and the output is used by the correlation detector.

The idea behind this method is that large variations in the received signal y can only come from large variations in the host signal x, considering that the watermark magnitude is low. By removing large variations in y, the variance of the host signal x is greatly reduced resulting a better detection performance.

Spectral Envelop Filtering

Another way to improve the detection is to minimize the effect of the original signal on the correlation. During detection, the host signal is considered as noise which degrades the detection performance due to its strong energy. This effect is much worse in audio watermarking compared to image watermarking since the human auditory system is more sensitive than human visual system. Audible distortion is prevented by keeping the magnitude of the watermark low which however leads to small watermark to signal ratio (SNR) at the detection phase.

In order to blindly remove the effects of host signal for better audio watermark extraction, Jung, Seok, and Hong (2003) proposed a method they called spectral envelop filtering.

This method removes the spectral envelop of the received stego signal and reduces the noise variance of the correlation value at the detector. Since most of the spectral features in the spectral envelope of the watermarked signal come from host signal, filtering the spectral envelope will reduce the effects of the host signal and improve watermark detection. The steps of this method are:

a. Low pass filter (LPF) the input watermarked signal spectrum X to get the spectral envelop vector Y in dB.

$$Y = LPF(\log(X)) \tag{37}$$

b. Remove Y from $log(X)$.
c. Convert the filtered data into linear scale.

$$\hat{X} = e^{(\log(X) - Y)} \tag{38}$$

According to their experiments, the spectral envelop filtering method improves the detection performance by one to two orders.

The Linear Prediction Method

Seok, Hong, and Kim (2002) introduced a method base on linear prediction (LP) as a way to remove the correlation in the audio signal. Using LP, the original audio signal $x(n)$ is estimated by the past p audio samples as:

$$\hat{x}(n) = a_1x(n-1) + a_2x(n-2) + ... + a_px(n-p) \tag{39}$$

where the LP coefficients $a_1, a_2 ... a_p$ are fixed over the audio analysis frame.

If the error between $x(n)$ and its estimation is denoted as $ex(n)$ then:

$$x(n) = \hat{x}(n) + ex(n) \tag{40}$$

the watermarked signal $y(n)$ becomes,

$$y(n) = x(n) + w(n) = \hat{x}(n) + ex(n) + w(n)$$

$$= \sum_{i=1}^{p} a_i x(n-i) + ex(n) + w(n) \tag{41}$$

Applying linear prediction to $y(n)$ we have:

$$y(n) = \hat{y}(n) + e(n) = \sum_{i=1}^{p} a_i y(n-i) + e(n) \tag{42}$$

where $\hat{y}(n) = \sum_{i=1}^{p} a_i y(n-i)$ is the estimation of $y(n)$ and $e(n) = y(n) - \hat{y}(n)$ is the residual signal of $y(n)$, which characterizes both the watermark $w(n)$ and the residual signal of $x(n)$.

An example of this method is illustrated in Figure 13 where (a) is the probability density function (pdf) of watermarked audio and (b) is the pdf of its residual signal. It is evident that the watermarked signal has large variance while its residual signal is smoothed out by the LP operation and has a much smaller variance. The residual signal and the PN sequence are used as input to a matched filter. The watermark recovery bit error rate with LP is greatly reduced compared to the non-LP detection.

Desynchronization Attacks and Traditional Solutions

While spread spectrum is robust to many attacks including noise addition, it remains susceptible to desynchronization attacks. In this section, we

Figure 13. Example probability density functions of the LP method for one audio frame (Seok et al., 2002)

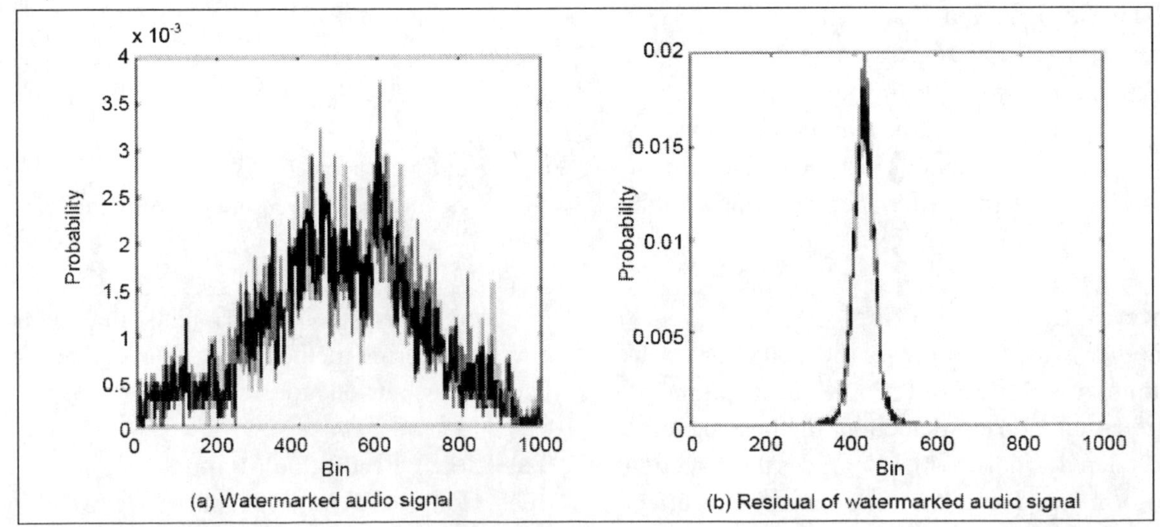

review several techniques aimed at improving spread spectrum detection under synchronization attacks. We also emphasize and illustrate in detail the features of a method developed by the authors.

Exhaustive Search

An intuitive solution for synchronization is the exhaustive search in the received signal, which takes into account all the possible attacks the signal might have undergone during transmission, inverses those attack effects followed by the watermark detection. Although this solution works for few applications where the search space is limited, it is usually time and computationally prohibitive and impossible to carry out for most real-time applications. Moreover, false positive errors increase as the search space increases making this approach useless for most applications where a low false positive error rate is usually required (Lichtenauer, Setyawan, Kalker, & Lagendijk, 2003; Licks & Jordan, 2005).

Redundant Watermark Eembedding

If watermarks are repeated N times before embedding in the audio it will have a better chance to achieve synchronization at the detection side. The idea was presented by Kirovski et al. (2001) and it is illustrated as shown in Figure 14.

In Figure 14, the leftmost subfigure shows perfect synchronization between the watermark and the received signal during detection where the normalized correlation value is $Q = 1$. The middle subfigure depicts the situation where the watermark is shifted by one sample due to desynchronization attacks and the normalized correlation is now $Q = -1/3$, which is a large change from the leftmost subfigure. In rightmost subfigure, the watermark is repeated three times before embedding. The normalized correlation is performed only at the center sample of each region and result a high value of $Q = 1$ despite the one chip shift desynchronization. It is clear that the correlation is correct as long as the linear shift is less than *floor(N/2)* samples. The drawback of the method is that the watermark capacity is reduced by N times.

Invariant Domain Embedding

Another method for synchronization is to embed the watermark in a special domain that is invariant to possible attacks that the signal may encounter. Such idea was first implemented in image watermarking. In Ruandaidh and Pun (1998) and Lin, Bloom, Cox, Miller, and Liu (2000), the authors proposed several watermarking systems using the Fourier-Mellin domain and claimed robustness to rotation, scale and translation.

The process of performing Fourier-Mellin transform includes three steps (Licks & Jordan, 2005). First the discrete Fourier transform (DFT) is performed on the signal. Then the DFT magnitude spectrum is transformed into a log-polar mapping followed by the second DFT transform.

One of the earliest attempts in introducing Fourier-Mellin transform into audio watermarking/fingerprinting was reported in Seok et al. (2002) where a linear speed-change resilient audio fingerprinting based on the Fourier-Mellin transform was proposed. They claimed their scale

Figure 14. Example given in Kirovski et al. (2001) to illustrate the redundant chip embedding method

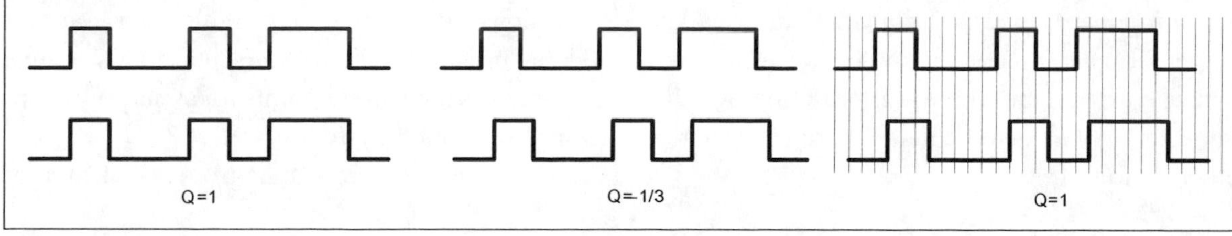

invariance audio fingerprinting algorithm was robust against speed changes of up to ±10%.

Using Synchronization Marks

Synchronization masks are codes with special features known both to the encoder and decoder. They usually do not carry any watermark information and are used only for synchronization purposes. The detector can use the known feature of those marks to perform an exhaustive search to achieve synchronization and keep the false positive error rate low. Pseudo random sequences are widely used as the synchronization marks for their well know good autocorrelation property (Bender et al., 1996; Cox et al., 1997).

Several authors (Garcia, 1999; Jung et al., 2003) have used the PN sequence as synchronization mark for audio watermarking during decoding and to achieve robustness against the de-synchronization attack. Synchronization marks, however, have their own limitations. They compromise watermark capacity since they do not carry any watermark information. They also exacerbate the watermark security problem since the opponents can guess the PN sequence and perform template estimation attacks (Gomes, 2001).

Use of Self-Synchronized Watermarks

In order to solve the problems brought by synchronization marks, self-synchronized watermarks may be used for watermarking. Those watermarks are specially designed so that the autocorrelation of the watermarks have one or several peaks, thus achieving the synchronization (Licks & Jordan, 2005). Although successful synchronization has been reported using self-synchronized watermarks (Kutter, 1999; Delannay & Macq, 2000), this method is also susceptible to estimation attacks. Opponent, once they know the watermark, can perform an autocorrelation between the watermarked signal and the watermark, remove all the periodic peaks, and cause future watermark detection to fail.

Feature Points or Content Synchronization

Feature points synchronization, which is also mentioned as salient point or content based synchronization, is regarded as a second generation synchronization method (Bas, Chassery, & Macq, 2002). Feature points are the stable sections of the host signal that are invariant or nearly invariant against common signal processing or attacks. The feature points are extracted both at the encoder and decoder and used to achieve synchronization (Bas et al., 2002).

Wu, Su, and Kuo (1999, 2000) presented robust audio watermarking using audio content analysis. The features points in their methods are defined as the fast energy climbing area, which is perceptually important and stable during the normal signal processing. Those feature points can survive many malicious attacks as long as signal perceptually transparency is maintained.

Kirovski et al. (2002) proposed an audio watermarking system robust to desynchronization via beat detection. The feature points used in this method are the beats, which are the most robust events in musical signals. The mean beat period of the music is estimated during the embedding process and watermarks are inserted at onset of the beat. The detector first extracts the mean beat period and locates the onset of the beat, thus synchronizing with the watermark.

According to Li and Xue (2003), the signal is first decomposed by discrete wavelet transform (DWT) and the statistical features in the wavelet domain are employed as feature points. The mean of the wavelet coefficient values at the coarsest approximation sub band provides information about the low frequency distribution of the audio signal (Tzanetakis, Essl, & Cook, 2001), which represents the mostly perceptual components of the signal and is relatively stable after common signal processing (Li & Xue, 2003). This mean is adopted as the feature points for synchronization purpose in their system.

The feature points synchronization method

does not need to embed extra marks for synchronization, thus saving watermark capacity and introducing less distortion to the host signal. However, it is very hard for this method to achieve accurate sample-to-sample synchronization.

An Improved Synchronization Method

Here we briefly present a novel synchronization algorithm that met the following goals:

a. Fast, making real-time synchronization possible.
b. Robust, making synchronization reliable under common signal processing or attacks
c. Precise, which enables sample-to-sample synchronization.
d. Secure, which deters the common estimation attack against guessing the location of PN sequence.

Feature Points Extraction

This method also employs feature points as synchronization features. In order to achieve fast synchronization to meet real-time requirement, the feature points should be chosen such that they can be extracted easily with low computational cost. In our method, we use the distribution of high energy areas as our main feature points. The high energy areas are the perceptually important components of the audio and remain very stable after common signal processing, including MP3 compression. If those areas change much, obvious audible distortion will be introduced to the audio and it will make it sound annoying.

Considering an audio signal $x(n)$ of N samples, the procedure of high energy area extraction proceeds as follows.

The average signal energy is calculated using the following equation:

$$x_{avg} = \frac{\sum_{i=1}^{N} x^2(i)}{N} \quad (43)$$

The signal is segmented into nonoverlapping blocks each contains L samples.

Let $M = floor(\frac{N}{L})$, where $floor(x)$ is the function that rounds x towards minus infinity, then M is the number of blocks in audio signal $x(n)$.

The average signal energy of each block is then calculated. Let $x(i,j)$ denote the j-th sample in the i-th block, then the average signal energy of i-th block is calculated as:

$$avg_i = \frac{\sum_{j=1}^{L} x^2(i,j)}{L} \quad (44)$$

The energy threshold is obtained by multiplying the average signal energy with a secret factor α, which is known both to encoder and decoder. Let T denotes the energy threshold, then:

$$T = \alpha x_{avg} \quad (45)$$

Record the location and length of each high energy section by the following pseudo code:

```
HEngLoc = zeros(1,M)
LenHEngLoc = zeros(1,M)
index = 1;
Eng_Ch_Inx = 1;
for i = 1 to M
 if avg_i ≥ T
     if Eng_Ch_Inx = 1
             HEngLoc(index) = i;
             Eng_Ch_Inx = 0;
                     end
     LenHEngLoc(index)++;
 end
 else
     index++;
     Eng_Ch_Inx = 1;
 end
end
```

In the above code, HEngLoc records the start location of each high energy areas and LengHEngLoc records the length of each high energy area. Index denotes the number of high energy areas and the Eng_Ch_Inx denotes the change between high and low energy areas.

The longest K high energy areas are chosen as the features points for synchronization. The factor K is defined according to the watermarking application requirements.

In the decoder side, the same procedure is performed to locate all the high energy areas (as well as their neighbor blocks) which are later correlated with the PN sequence to achieve synchronization purpose.

Figures 15 and 16 list the high energy block distribution before and after MP3 compression of a sample audio file. The x axis is the high energy block location starting frame and the y axis is the length of each high energy block. As we can see from those figures, the locations of high energy areas are very stable under MP3 compression. Further experiments showed that the high energy block distribution is also robust to normal signal processing as well as under malicious attacks including white noise addition, cropping, equalization, and so forth.

The secret factor α in equation (45) controls the high energy blocks distribution. Different α results in different high energy blocks distribution. A lower value for α decreases the thresholds for high energy areas, thus more areas are treated as high energy areas. Although this provides more room to embed the watermarks, it reduces the system robustness since some high energy areas may become unstable after common signal processing or attacks. A higher α, on the other hand, increases such threshold, thus decreasing the number of high energy areas. Better robustness is achieved at the cost of lower capacity for watermark embedding. By keeping this factor secret, the system prevents the attackers from guessing the locations of high energy distribution used between encoder and decoder, thus making the synchronization method more secure.

PN Sequence Scrambling and Multiple PN Sequences Embedding

PN sequence is used to perform the synchronization in our method. However, the PN sequence employed is carefully designed to meet the following goals:

- The length of the PN sequence should be as short as possible to avoid introducing audible noise, but long enough to achieve robust correlation (Gomes, 2001).
- The strength of PN sequence should be adaptive to the host signal energy so a stronger PN sequence could be used to achieve better robustness while keeping the introduced distortion under perceptible levels (Cox et al., 2002).
- PN sequence should be hard to guess to mitigate the estimation attacks.

In order to meet the above goals, the PN sequence employed in our system has the some special features.

The PN sequence is 2560 samples long, or about 58 ms for 44.1 kHz audio sampling rate. It is long enough to achieve robust synchronization and short enough to avoid introducing audible noise, as well as hard to estimate for attackers.

To further improve the security of the PN sequence and mitigate the estimation attacks, the PN sequence is scrambled before adding to the host signal. The purpose of scrambling is to make guessing the PN sequence very difficult while scrambled marks may still maintain the good autocorrelation property of the PN sequence.

In order to take the advantage of temporal masking phenomena in the psychoacoustic model, the PN sequence is embedded in the high energy area, starting from the beginning of the high energy blocks with some fixed offset. Let $\mu(n)$ be the PN sequence, $x(n)$ be the host signal, and $y(n)$ be the host signal with embedded PN sequence, then (Kim, 2003):

$$y(n) = x(n) + c\mu(n) \tag{46}$$

where c is a factor adaptively controls the strength

of PN sequence to meet both the robustness and inaudibility requirements.

For some watermarking applications, multiple watermarks are embedded in the host signal to protect different information. For instance, the system may embed one watermark representing the author's name and use another watermark to embed the author's signature which may be a

Figure 15. Original high energy block distribution

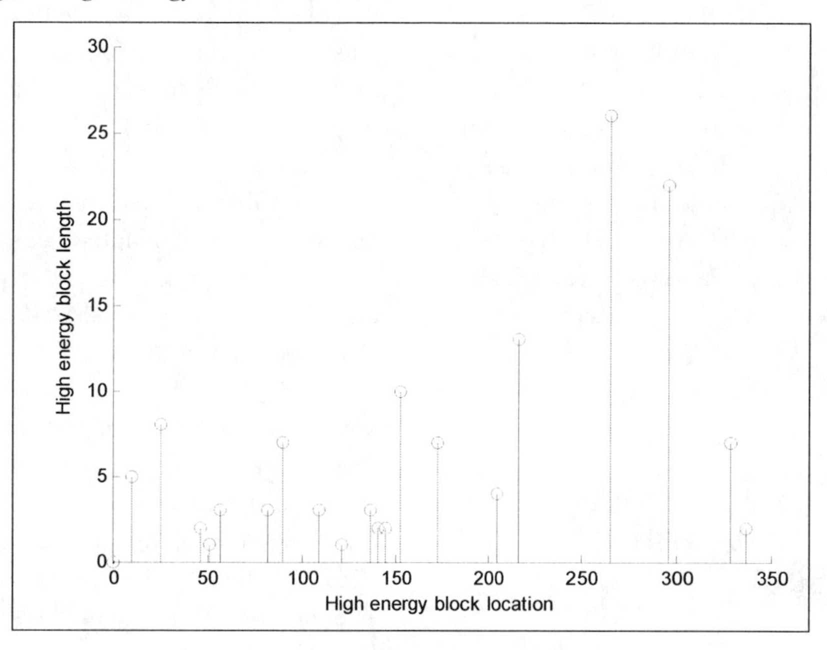

Figure 16. High energy block distribution after mp3 compression

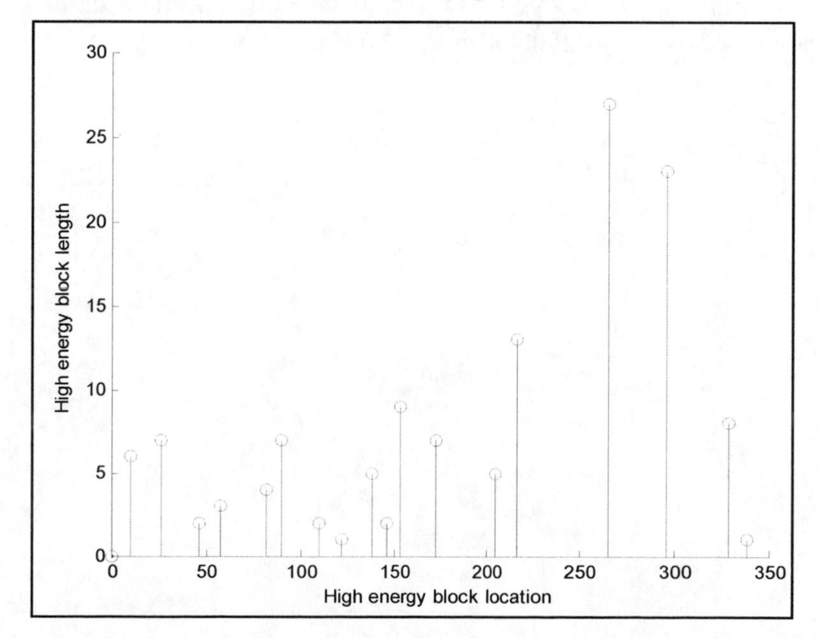

small graph (Cox et al., 2002). This brings up the multiple watermarking synchronization problems. Fortunately, due to the orthogonality of the PN sequence (Cox et al., 1997), multiple different PN sequences can be embedded using the proposed method to solve this problem. Matched filter is used in our method to successfully detect the existence of the PN sequence and precisely locate the starting sample of the PN sequence.

The whole synchronization system is illustrated in Figure 17.

Our experimental results show that this synchronization system is robust against normal signal processing including random cropping, white noise addition (snr = 20dB, 10dB and 5dB), color noise addition, digital to analog (D/A) and analog to digital (A/D) conversion, and mp3 compression at 64 bits/s rate.

PSYCHOACOUSTIC MODELING FOR SPREAD SPECTRUM WATERMARKING

Psychoacoustic modeling has made important contributions in the development of recent high quality audio compression methods (ISO/IEC 11172-3, 1993; Painter & Spanias, 2000; Pan, 1995), and it has enabled the introduction of effective audio watermarking techniques (Cox et al., 2002; Liu, 2004; Swanson et al., 1998). In audio analysis and coding, psychoacoustic modeling strives to reduce the signal information rate in lossy signal compression, while maintaining transparent quality. This is achieved by accounting for auditory masking effects, which makes it possible to keep quantization and processing noises inaudible. In speech and audio watermarking, the inclusion of auditory masking has made possible the addition of information that is unrelated to the signal in a manner that keeps it imperceptible and can be effectively recovered during the identification process.

Brief Review of Previous Psychoacoustic Models

Most psychoacoustic models used in audio compression or watermarking have so far utilized the short-time Fourier transform (STFT) to construct a time-varying spectral representation of the signal (Bosi & Goldberg, 2003; Johnston, 1998; Painter & Spanias, 2000). A window sequence of fixed length is used to capture a signal section, resulting in a fixed spectral resolution. The STFT is applied on windowed sections of the signal thus providing an analysis profile at regular time instances.

Figure 17. Synchronization system illustration

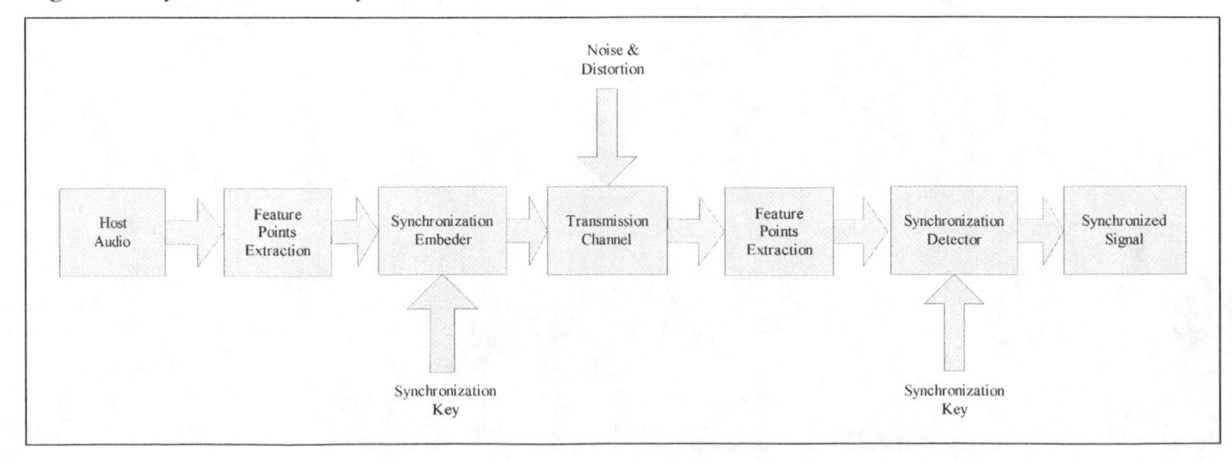

However, the STFT can provide only averaged frequency information of the signal and it lacks the flexibility of arbitrary time-frequency localization (Polikar, 2006). Such a rigid analysis regime is in striking contrast with the unpredictably dynamic spectral-temporal profile of information-carrying audio signals. Instead, signal characteristics would be analyzed and represented more accurately by a more versatile description providing a time-frequency multiresolution pertinent to the signal dynamics.

The approaches included in the MPEG 1 standard and elsewhere allow the switching between two different analysis window sizes depending on the value of the signal entropy (ISO/IEC 11172-3, 1993) or the changes in the estimated signal variance (Lincoln, 1998). Greater flexibility, however, is needed. The wavelet transform presents an attractive alternative by providing frequency-dependent resolution, which can better match the hearing mechanism (Polikar, 2006). Specifically, long windows analyze low frequency components and achieve high frequency resolution while progressively shorter windows analyze higher frequency components to achieve better time resolution. Wavelet analysis has found numerous signal processing applications including video and image compression (Abbate, Decusatis, & Das, 2002; Jaffard, Meyer, & Ryan, 2001), perceptual audio coding (Veldhuis, Breeuwer, & van der Wall, 1998), high quality audio compression, and psychoacoustic model approximation (Sinha & Tewfik, 1993).

Wavelet-based approaches have been previously proposed for perceptual audio coding. Sinha and Tewfik (1993) used the masking model proposed in Veldhuis et al. (1998) to first calculate masking thresholds in the frequency domain by using the fast Fourier transform. Those thresholds were used to compute a reconstruction error constraint caused either by quantization or by the approximation of the wavelet coefficients used in the analysis. If the reconstruction errors were kept below those thresholds, then no perceptual distortion was introduced. The constraints were then translated into the wavelet domain to ensure transparent wavelet audio coding.

Black and Zeytinoglu (1995) mimicked the critical bands distribution with a wavelet packet tree structure and directly calculated the signal energy and hearing threshold in the wavelet domain. This information was in turn used to compute masking profiles. Since the N-point FFT was no longer needed, the computational complexity was greatly reduced. Specifically, it was reported that the new method only requires 1/3 of the computational effort when compared to the MPEG 1 Layer 1 encoder.

Zurera, Ferreras, Amores, Bascon, and Reyes (2001) presented a new algorithm to effectively translate psychoacoustic model information into the wavelet domain, even when low-selectivity filters were used to implement the wavelet transform or wavelet packet decomposition. They first calculated the masking and auditory thresholds in the frequency domain by using the Fourier transform. Based on several hypotheses (orthogonality of sub band signals and white noise-like quantization noise in each sub band), those thresholds were divided by the equivalent filter frequency response magnitude of the corresponding filter bank branch, forming the overall masking threshold in wavelet domain.

Carnero and Drygajlo (1999) constructed a wavelet domain psychoacoustic model representation using a frame-synchronized fast wavelet packet transform algorithm. Masking thresholds due to simultaneous frequency masking were estimated in a manner similar to Johnston (1998). The energy in each sub band was calculated as the sum of the square of the wavelet coefficients, scaled by the estimated tonality, and finally extended by a spreading function. Masking thresholds due to temporal masking were found by further considering the energy within each sub band. Final masking thresholds were obtained by considering both simultaneous and temporal masking as well as the band thresholds in absolute quiet. However, this model was tailored specifically for speech signals and its effectiveness on wideband audio remains untested.

The above psychoacoustic modeling methods are either computationally expensive (Reyes, Zurera, Ferreras, & Amores, 2003; Sinha & Tewfik, 1993; Veldhuis et al., 1998; Zurera et al., 2001) by relying on the Fourier transforms for the computation of the psychoacoustic model, or approximate the critical bands sub-optimally (Black & Zeytinoglu, 1995; Carnero & Drygajlo, 1999), which may result in objectionable audible distortion in the reconstructed signal. In contrast, it has been shown (Wu, Huang, Huang, & Shi, 2005) that the discrete wavelet transform (DWT) requires a significantly lower computational load when compared both to the DCT or the DFT while supporting psychoacoustically-based models, and therefore it still presents an attractive alternative.

Figure 18. Illustration of the psychoacoustic model computation

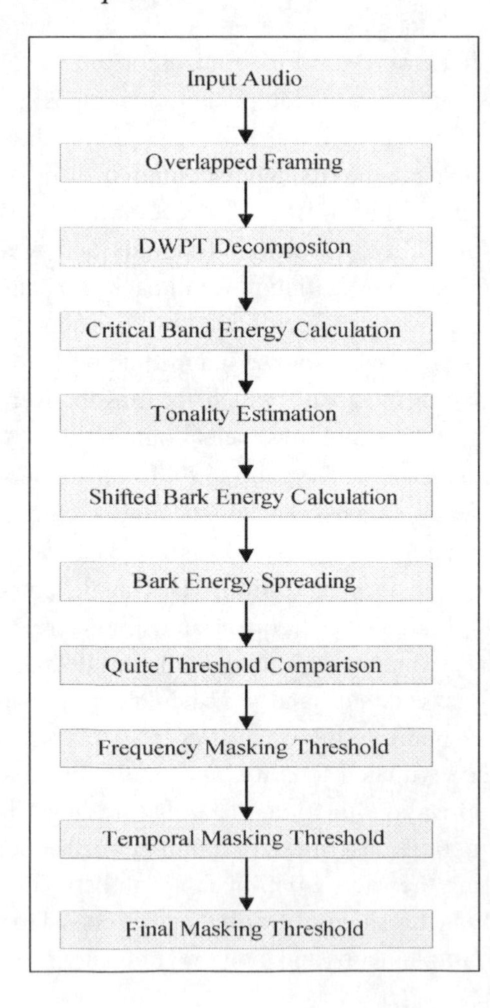

In this section, we present a new psychoacoustic model in the wavelet domain. Wavelet analysis results are incorporated in effective simultaneous and temporal masking. Furthermore, the proposed model introduces a wavelet packet-based decomposition that better approximates critical bands distribution. The proposed model maintains perceptual transparency and provides an attractive alternative appropriate for audio compression and watermarking.

A Novel Psychoacoustic Model

This section includes the process of building the psychoacoustic model, and it presents an improved decomposition of the signal into 25 bands using discrete wavelet packet transform (DWPT) to closely approximate the critical bands, followed by the implementation of temporal masking.

The computation of the psychoacoustic model has a general structure similar to several related techniques (Black, 1995; Carnero & Drygajlo, 1999; Sinha & Tewfik, 1993; Zurera et al., 2001), and it is illustrated in Figure 19.

Signal analysis depicted in Figure 18 proceeds as follows:

a. The input audio is segmented into overlapping frames.
b. Each frame is decomposed by the DWPT into 25 sub bands that approximate the auditory critical bands as illustrated in Figure 19.
c. Signal energy in each band is computed in the wavelet domain to provide the Bark spectrum energy.
d. Tonality is estimated in each band to determine the extent the signal in that band

is noise-like or tone-like.

e. Shifted bark energy is calculated by scaling the bark energy according to the tonality factor.
f. The energy spreading effects on neighboring bands is computed by convolving the shifted bark energy with the spreading function to provide the effective masking threshold in each critical band.
g. The masking threshold in each critical band is normalized by the band width and then compared with the threshold in absolute quiet. The maximum of the two is selected as the masking threshold.
h. Temporal masking threshold is calculated within each band. The frequency masking and temporal masking thresholds are compared. The final masking threshold for the particular signal frame is the maximum of the two.

As it can be seen in Figure 15, both temporal and spectral masking information are considered in the computation of the final masking threshold. In the introduced model, temporal masking effects include both pre- and post-echo. However, in contrast to other approaches (e.g., Sinha & Tewfik, 1993), here the computation uses more precise critical bank approximations as well as simultaneous masking results obtained in the wavelet domain.

Signal Decomposition with the DWPT

As mentioned, the discrete wavelet packet transform can conveniently decompose the signal into a critical band-like partition (Sinha & Tewfik, 1993; Carnero & Drygajlo, 1999; Zurera et al., 2001). The standard critical bands are included in the ISO/IEC 11172-3 (1993). The critical bands approximation used in alternative approaches can be found in Liu (2004) and Carnero and Drygajlo (1999).

In this work, we divided the input audio signal into 25 sub bands using DWPT in the manner shown in Figure 19, where the band index is enumerated from 1 to 25 to cover the complete audible spectrum (frequencies up to 22 kHz). The corresponding frequencies obtained by this DWPT decomposition are listed in He and Scordilis (2006).

This decomposition achieves a closer approximation of the standard critical bands than that used elsewhere (Carnero & Drygajlo, 1999; Liu, 2004), thus providing a more accurate psychoacoustic model computation.

The signal decomposition resulting from wavelet analysis needs to satisfy the spectral resolution requirements of the human auditory system, which match the critical bands distribution. On the other hand, the selection of the wavelet basis is critical for meeting the required temporal resolution, which is less than 10 ms at high frequency areas and up to 100 ms at low frequency areas (Bosi & Goldberg, 2003). In order to have the appropriate analysis window size to accommodate that temporal resolution for wideband audio, we choose wavelet base of order 8 (length $L = 16$ samples) whose specific properties are as follows:

The frame length (in samples) at level j ($2 \leq j \leq 8$) is given by:

$$F_j = 2^j \qquad (47)$$

The duration of the analysis window (in samples) at level j is (Carnero & Drygajlo, 1999; Liu, 2004):

$$W_j = (L-1)(F_j - 1) + 1 \qquad (48)$$

where L is the length of Daubechies filter coefficients ($L = 16$ in this case). The Daubechies wavelet is selected because it is the most compactly supported wavelet (finer frequency resolution) compared to other wavelet bases with the same number of vanishing moments.

For signal bandwidth of 22 kHz, the maximum frame length is $F_{max} = 2^8 = 256$ samples, which provides resolution of $\frac{22kHz}{256} = 86Hz$. The minimum frame length is $F_{\min} = 2^2 = 4$ samples with frequency resolution $\frac{22kHz}{4} = 5.5kHz$. The maximum duration of the analysis window is $W_{\max} = 15*(256-1)+1 = 3826$ samples, which at sampling rate of 44.1 kHz corresponds to 87 ms and it applies to the low frequency end, while the minimum duration of the analysis window is $W_{\min} = 15*(4-1)+1 = 46$ samples, or about 1 ms, which applies to the high frequency end.

Wavelet Decomposition Evaluation

While techniques based on the Fourier transform dominate in the implementation of psychoacoustic models, wavelet-based approaches are relatively new. In Carnero and Drygajlo (1999), frame-synchronized fast wavelet packet transform algorithms were used to decompose wideband speech into 21 sub bands, which approximate the critical bands. The spreading function used was optimized to speech listening. For wideband audio, Liu (2004) has extended that work and appropriately altered the spreading function in order to ensure transparency and inaudibility in audio watermarking applications. A similar critical bands partition was implemented, which in that case spanned the entire audible spectrum

Figure 19. DWPT-based signal decomposition

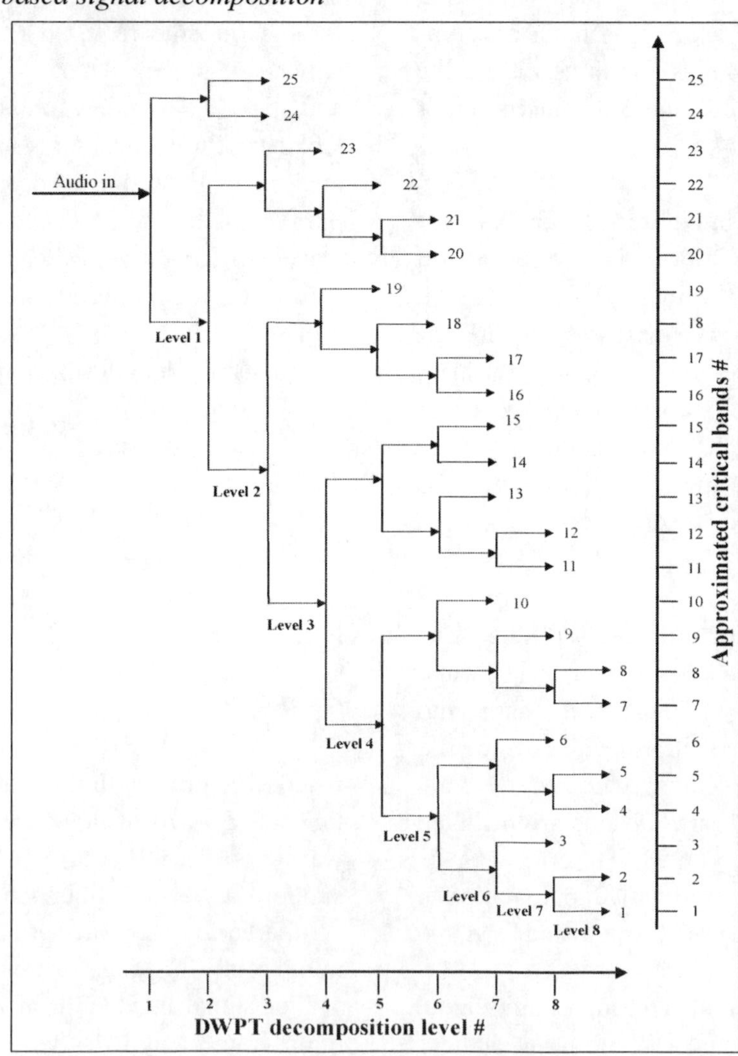

and consisted of 26 bands.

The degree to which the two different approaches approximate the standard critical bands partition can be examined by plotting the critical bands starting frequencies, as shown on Figure 20. When the differences in starting frequency are plotted, as shown in Figure 21, it is readily observed that the proposed band partition is substantially closer to the standard, particularly beyond the 16th critical band (frequencies of 2800 Hz and higher). The differences between the two approaches are more striking when critical bands center frequency differences are examined, as depicted on Figure 22, where it can be seen that the proposed approach is considerably closer to the standard. A better approximation to the standard critical bands can provide a more accurate computation of the psychoacoustic model. While this wavelet approach yields a spectral partition that is much closer to the standard critical bands frequencies, the inherent continuous subdivision of the spectrum by a factor of 2 prevents an exact match. However, the overall analysis features of this approach outlined elsewhere in this discussion uphold its overall appeal over competing techniques.

Window Size Switching and Temporal Masking

Temporal masking is present before and after a strong signal (masker) has been switched on and off abruptly. If a weak signal (maskee) is present in the vicinity of the masker, temporal masking may cause it to become inaudible even before the masker onset (premasking) or after the masker vanishes (postmasking). Typically, the duration of premasking is less than one tenth that of the postmasking, which is in the order of 50 to 100 ms (Lincoln, 1998), depending on the masker amplitude.

Although premasking is relatively a short-time phenomenon, it has important implications and has had to be addressed in audio analysis and compression, particularly in minimizing the so-called pre-echo problem (Bosi & Goldberg, 2003; Lincoln, 1998; Painter & Spanias, 2000). A good

Figure 20. Starting frequencies (lower edge) of each critical band

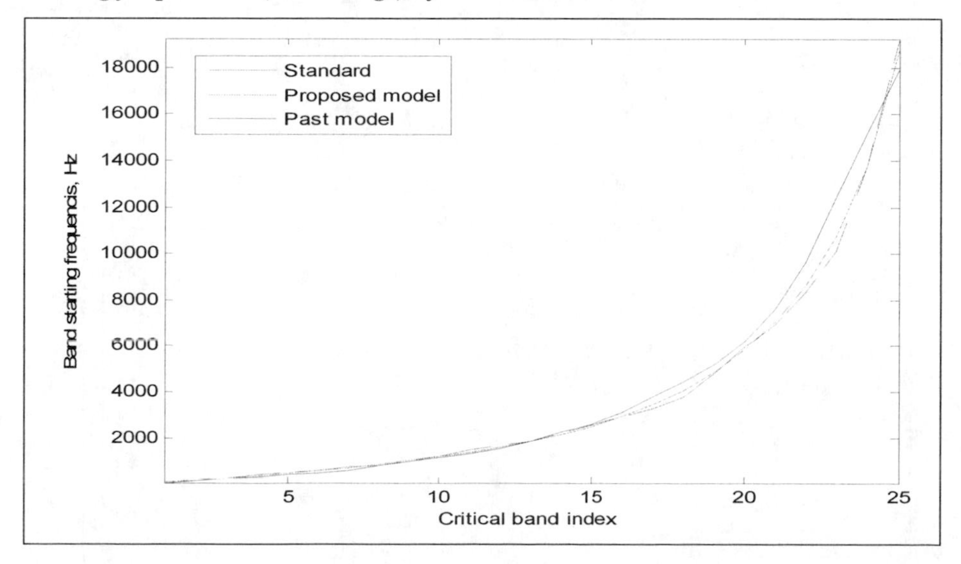

way to mitigate this problem has been the use of adaptive analysis frame size. Analysis using long frame size requires less computation than shorter frame size, which is a critical aspect in real time applications and it can also yield lower bit rate in compression applications or higher watermarking capacity but at the risk of permitting audible distortion caused by large variations in the signal stationarity, such as the presence of an attack in only part of the frame (Sinha & Tewfik, 1993). Therefore, several approaches have been proposed for seamlessly switching between different frame sizes.

The proposed psychoacoustic model was tested on CD-quality audio (sampling rate of 44.1 kHz) with nominal analysis frame duration of 46 ms (2048 samples). While this frame size is adequate

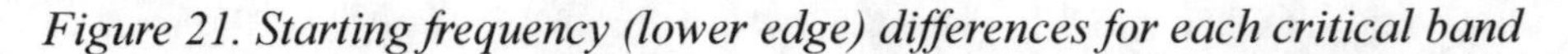

Figure 21. Starting frequency (lower edge) differences for each critical band

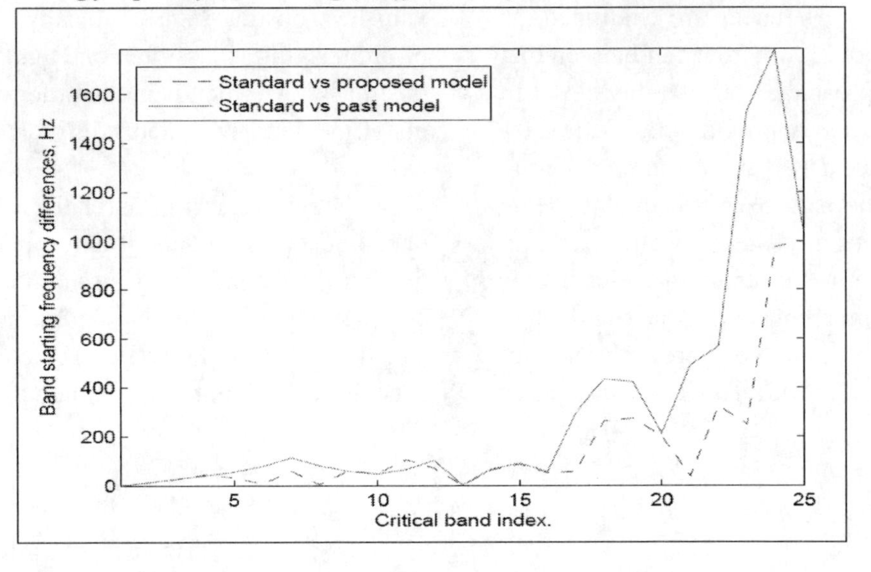

Figure 22. Center frequency differences for each critical band

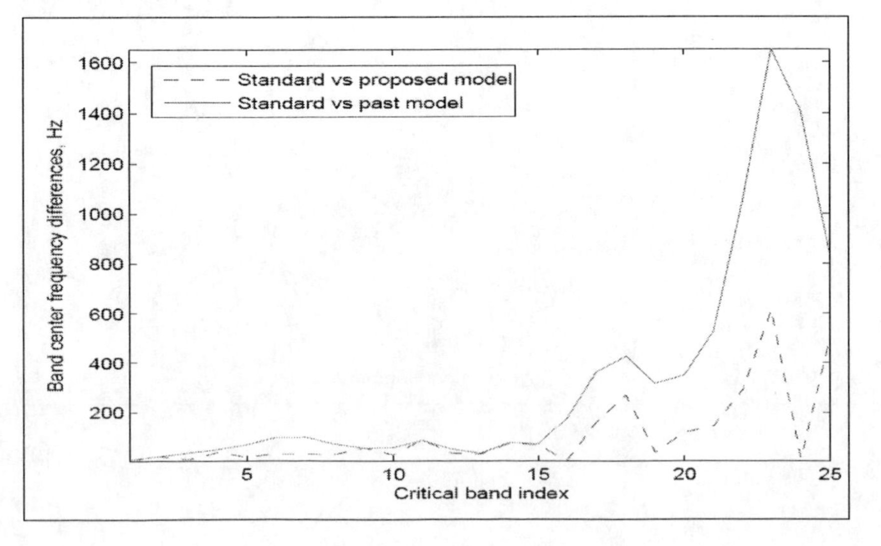

for slowly changing signal characteristics, it may be too long for rapidly changing signals. Consider for example the case that a silent period is followed by a percussive sound, such as from castanet or triangles within the same analysis block. In this case, quantization noise during the coding process will spread over the entire block and it will be audible in the portion before the signal attack. Premasking effects will be unable to cover the offending pre-echo noise in its entirety, which will therefore become audible (Sinha & Tewfik, 1993).

Window size switching is used to address pre-echo effects and in the proposed model it proceeds as follows:

a. The input signal is divided into frames of 2048 samples with 50% overlap.
b. Each frame is decomposed into 25 sub bands using the DWPT method outlined before.
c. The wavelet coefficients x_i are normalized by dividing each coefficient with the maximum absolute coefficient in the frame:

$$x_i = \frac{x_i}{\max(|x_i|)} \quad (49)$$

d. The normalized wavelet energy entropy of the frame is calculated:

$$E = -\sum_{i=a}^{b} x_i^2 \log_2 x_i^2 \quad (50)$$

where a and b are the starting and ending locations of the analysis frame containing the wavelet coefficients.
e. If the change of E from the previous frame to the present frame exceeds a certain threshold, then the frame is switched to a frame of half the size. In this approach, the initial 2048 samples frame is switched to three 1024 samples long frames with 50% overlap.
f. The change of the energy entropy of the new 1024 samples frames is calculated using equations (47) and (48). Exceeding a threshold will again result in a further switching step, which divides each 1024 samples frame into three 512 samples frames with 50% overlap.
g. Further frame switching to 256 samples is permissible and it may occur in some extreme transient signals. Beyond that point no subsequent switching is performed and the switching process is terminated.

The postmasking effect is also considered in this psychoacoustic model by implementing an approach similar to Lincoln (1998). However, this time the entire algorithm operates in the wavelet domain.

$$T^p(i) = \lambda T^p(i-1) + (1-\lambda)T^s(i) \quad (51)$$

where i is the frame (time) index, λ is a control factor with value between 0 and 1, $T^p(i)$ and $T^p(i-1)$ are the temporal masking thresholds for the current and previous frame respectively, and $T^s(i)$ is the frequency (simultaneous) masking threshold derived by the proposed DWPT as described in a later section. The derived masking threshold is provided by:

$$T = \max(T^s(i), T^p(i)) \quad (52)$$

Note that in equation (7), factor λ controls how much the previous frame temporal masking threshold and the current frequency masking threshold contribute to the current temporal masking threshold. A greater λ denotes stronger temporal masking effects and therefore a slower masker decay slope. A smaller λ on the other hand, denotes weaker temporal masking effects and a steeper masker decay slope.

The experiments by He and Scordilis (2006) showed that the new psychoacoustic model accurately calculates the masking and auditory thresholds directly in the wavelet domain by closely approximating the critical bands, thus making the watermarking process (He & Scordilis, 2006) transparent.

The proposed method also provides broader masking capabilities compared to the DFT-based psychoacoustic model proposed in Johnston (1998), thus revealing that larger signal regions are in fact inaudible and therefore can provide more space for watermark embedding without noticeable effect. Furthermore, the signal-to-masking ratio is further reduced indicating that in audio watermarking applications this approach can lead to increased watermark robustness by embedding relative higher energy watermarks without audible quality degradation.

This new psychoacoustic model is incorporated into an improved spread spectrum audio watermarking in He and Scordilis (2006), and the experiments there showed that compared to the perceptual entropy psychoacoustic model-based spread spectrum watermarking system, the proposed watermarking system provides improved system performance by offering better watermark robustness, higher watermark payload, and shorter watermark length.

CONCLUSION

This chapter provides a comprehensive exposure to the principles and methods of spread spectrum audio watermarking and a discussion of related issues. It includes a brief introduction of the communication model for watermarking, and fore spread spectrum technology. It reviews the current state-of-the-art spread spectrum audio watermarking schemes proposed by researchers over the last decade as well as techniques to improve spread spectrum detection. Desynchronization and psychoacoustic modeling are two of the most important components in spread spectrum audio watermarking. Solutions for desynchronization of spread spectrum were presented, followed by a brief introduction of traditional psychoacoustic models used in various spread spectrum audio watermarking. We also presented some of our work in solving those problems (He & Scordilis, 2006). Future spread spectrum audio watermarking while maintaining the current basic structure will probably include better synchronization and more elaborate psychoacoustic modeling. These will remain critical for effective watermarking detection and security. In addition, increasing payload, surviving new attacks, and real time embedding of robust and secure watermarks into compressed audio will continue to be the focus of future efforts.

FUTURE RESEARCH DIRECTIONS

Spread spectrum will continue to play an important role in audio watermarking while the basic concepts, structure, and process will remain unchanged in the near future. However, some critical enhancements may be made on the current spread spectrum-based audio watermarking, which could be along the following directions:

a. **Psychoacoustic modeling:** Psychoacoustic research is an active field affecting a broad range of application areas. Although a variety of psychoacoustic models have been proposed, the mechanism of the human auditory system is still not fully understood. Better psychoacoustic modeling will definitely contribute to better audio watermarking by revealing more room in the auditory space for embedding stronger watermarks while maintaining high quality audio.
b. **Watermark payload:** Compared to some other watermarking schemes such as phase coding, spread spectrum-based audio watermarking provides relatively small watermark payload, which means that for the same length of audio a much shorter secret message can be accommodate. Nevertheless, the resulting watermark is much more secure than in competing techniques. However, this limitation makes spread spectrum unsuitable

for some applications where large amount of data hiding is desired; for example when audio enhancing metadata information needs to be added. Due to the nature of spread spectrum technology, where the hidden data is spread over a much large bandwidth spectrum, high capacity watermark payload remains a challenge to be overcome in future research efforts.

c. **Desynchronization attacks:** It is widely acknowledged that spread spectrum-based audio watermarking is vulnerable to many attacks, but especially desynchronization, which aims at scrambling the alignment between the encoder and decoder, making watermark detection hard to perform if not impossible. Mild desynchronization attacks on the watermarked audio, such as time or frequency scaling of the order of 5% to 10%, introduce little distortion to the audio quality, but could be totally debilitating for spread spectrum based watermarking. Some enhancements have been proposed, but how to solve this problem effectively is an interesting and challenging topic for future research.

d. **Other attacks and signal transformations:** Besides desynchronization, other attacks could also cause trouble for spread spectrum watermarking systems. Those attacks include, but are not limited to, MP3 compression, digital to analog and analog to digital conversion, white or color noise addition, multiple bands equalization, echo addition, cropping, quantization, resampling, low pass, high pass, or band pass filtering. Such attacks can cause watermark detection to fail even if changes to the audio signal are subtle and imperceptible. Immunity to attacks will remain a challenge as more types attacks, new channels, signal coding, and transmission methods are continuously introduced.

e. **Watermarking in the compressed domain:** The proliferation of the Internet, which has enabled audio material (copyrighted or not) to be widely disseminated has also made wider the use of many compressed audio formats (MP3, AAC, WMA, etc.) at a global scale. A good example is the Windows Media Audio (WMA) format from Microsoft, which enables both real time watermarking and audio compression in an effective way. This scheme is used by many online music vendors including Napster. However, once the user downloads the music with the valid key, the embedded watermark can be easily removed by some software and infringe on the purpose of audio watermarking. The development of effective watermarking schemes in the compressed domain will remain an important research area in the future.

REFERENCES

Abbate, A., Decusatis, C. M., & Das, P. K. (2002). *Wavelets and subbands, fundamentals and applications.* Boston: Birkhauser.

Bas, P., Chassery, J. M., & Macq, B. (2002). Geometrically invariant watermarking using feature points. *IEEE Transactions on Image Processing, 11*(9), 1014-1028.

Bender, W., Gruhl, D., Morimoto, N., & Lu, A. (1996). Techniques for data hiding. *IBM Systems Journal, 35*(3/4), 313-336.

Black, M., & Zeytinoglu, M. (1995). Computationally efficient wavelet packet coding of wideband stereo signals. In *Proceedings of the IEEE International Conference on Acoustics, Speech, and Signal Processing* (pp. 3075-3078).

Bosi, M., & Goldberg, R. E. (2003). *Introduction to digital audio coding and standards.* Kluwer

Academic Publishers.

Carnero, B., & Drygajlo, A. (1999). Perceptual speech coding and enhancement using frame-synchronized fast wavelet packet transform algorithms. *IEEE Transactions on Signal Processing, 47*(6), 1622-1635.

Cheng, S., & Yu, H., & Xiong, Z. (2002). Enhanced spread spectrum watermarking of MPEG-2 AAC audio. In *Proceeding of the IEEE International Conference on Acoustics, Speech, and Signal Processing* (pp. 3728-3731).

Chipcenter. (2006). *Tutorial on spread spectrum.* Retrieved March 18, 2007, from http://archive.chipcenter.com/knowledge_centers/digital/features/showArticle.jhtml?articleID=9901240

Cox, I. J., Kilian, J., Leighton, T., & Shamoon, T. (1997). Secure spread spectrum watermarking for multimedia. *IEEE Transactions on Image Processing, 6*(12), 1673-1687.

Cox, J., Miller, M. L., & Bloom, J. A. (2002). *Digital watermarking.* Academic Press.

Cvejic, N. (2004). *Algorithms for audio watermarking and steganography.* Unpublished doctoral thesis, University of Oulu, Department of Electrical and Information Engineering.

Cvejic, N., Keskinarkaus, A., & Seppänen, T. (2001, October). Audio watermarking using m sequences and temporal masking. In *Proceedings of the IEEE Workshop on Applications of Signal Processing to Audio and Acoustics* (pp. 227-230). New York.

Cvejic, N., & Seppänen, T. (2004). Spread spectrum audio watermarking using frequency hopping and attack characterization. *Signal Processing, 84*, 207-213.

Delannay, D., & Macq, B. (2000). Generalized 2D cyclic patterns for secret watermark generation. In *Proceedings of the IEEE International Conference of Image Processing* (pp. 72-79).

Garcia, R. A. (1999, September). Digital watermarking of audio signals using a psychoacoustic model and spread spectrum theory. In *Proceedings of the Audio Engineering Society Meeting* (Preprint 5073). New York.

Gomes, L. de C. T. (2001). Resynchronization methods for audio watermarking. In *Proceedings of the 111th AES Convention* (Preprint 5441). New York.

Gruhl, D., & Lu, A., & Bender, W. (1996). Echo hiding. In *Proceedings of the Information Hiding Workshop* (pp. 295-315). University of Cambridge.

He, X., Iliev, A., & Scordilis, M., (2004). A high capacity watermarking technique for stereo audio. In *Proceedings of the IEEE International Conference on Acoustics, Speech, and Signal Processing* (Vol. 5, pp. 393-396).

He, X., & Scordilis, M. (2005). Improved spread spectrum digital audio watermarking based on a modified perceptual entropy psychoacoustic model. In *Proceedings of the IEEE Southeast Conference* (pp. 283-286).

He, X., & Scordilis, M. (2006). A psychoacoustic model based on the discrete wavelet packet transform. *Journal of Franklin Institute, 343*(7), 738-755

ISO/IEC 11172-3. (1993). *Coding of moving picture and associated audio for digital storage media at up to about 1.5 Mbits - part 3* (Audio Recording).

Jaffard, S., Meyer, Y., & Ryan, R. D. (2001). *Wavelets tools for science and technology.* SIAM.

Johnston, J. D. (1998). Transfrom coding of audio signals using perceptual noise criteria. *IEEE Journal on Selected Areas in Communications, 6*(2), 314-323.

Jung, S., Seok, J., & Hong, J. (2003). An improved

detection technique for spread spectrum audio watermarking with a spectral envelope filter. *ETRI Journal, 25*(1), 52-54.

Kim, H. J. (2003). Audio watermarking techniques. In *Proceedings of the Pacific Rim Workshop on Digital Steganography*, Kyushu Institute of Technology, Kitakyushu, Japan.

Kirovski, D., & Attias, H. (2002). Audio watermark robustness to de-synchronization via beat detection. *Information Hiding Lecture Notes in Computer Science*, *2578*, 160-175.

Kirovski, D., & Malvar, H. (2001). Robust spread-spectrum audio watermarking. In *Proceeding of the IEEE International Conference on Acoustics, Speech, and Signal Processing* (Vol. 3, pp. 1345-1348).

Kirovski, D., & Malvar, H. S. (2003). Spread-spectrum watermarking of audio signals. *IEEE Transaction on Signal Processing Special Issue on Data Hiding, 51*(4), 1020-1033.

Kuo, S. S., Johnston, J. D., Turin, W., & Quackenbush, S. R. (2002). Covert audio watermarking using perceptually tuned signal independent multiband phase modulations. In *Proceeding of the IEEE International Conference on Acoustics, Speech, and Signal Processing* (Vol. 2, pp. 1753-1756).

Kutter, M. (1999). Watermarking resisting to translation, rotation, and scaling. In *Proceedings of the SPIE Mutimedia Systems and Applications* (Vol. 3528, pp. 423-431). International Society for Optical Engineering.

Li, W., & Xue, X. (2003). An audio watermarking technique that is robust against random cropping. *Computer Music Journal*, *27*(4), 58-68.

Lichtenauer, J., Setyawan, I., Kalker, T., & Lagendijk, R. (2003). Exhaustive geometrical search and false positive watermark detection probability. In *Proceedings of the SPIE Electronic Imaging on Security and Watermarking of Multimedia Contents* (Vol. 5020, pp. 203-214).

Licks, V., & Jordan, R. (2005). Geometric attacks on image watermarking systems. *IEEE Transactions on Multimedia*, *12*(3), 68-78.

Lin, C. Y., Bloom, J. A., Cox, I. J. Miller, M. L., & Liu, Y. M. (2000). Rotation, scale and translation-resilient public watermarking for images. *Proceedings of SPIE*, *3971*, 90-98.

Lincoln, B. (1998). An experimental high fidelity perceptual audio coder. *Project in MUS420 Win97.* Retrieved March 18, 2007, from http://www-ccrma.stanford.edu/jos/bosse/

Liu, Q. (2004). *Digital audio watermarking utilizing discrete wavelet packet transform.* Unpublished master's thesis, Chaoyang University of Technology, Institute of Networking and Communication, Taiwan.

Malvar, H. S., & Florencio, D. A. (2003). Improved spread spectrum: A new modulation technique for robust watermarking. *IEEE Transactions on Signal Processing*, *51*(4), 898-905.

Meel, I. J. (1999). Spread spectrum (SS) introduction. *Sirius Communications*, *2*, 1-33.

Painter, T., & Spanias, A. (2000). Perceptual coding of digital audio. *Proceedings of the IEEE*, *88*(4), 451-513.

Pan, D. (1995). A tutorial on mpeg/audio compression. *IEEE Multimedia*, *2*(2), 60-74.

Polikar, R. (2006). *The wavelet tutorial.* Retrieved March 18, 2007, from http://users.rowan.edu/polikar/Wavelets/ wtpart1.html

Reyes, N. R., Zurera, M. R., Ferreras, F. L., & Amores, P. J. (2003). Adaptive wavelet-packet analysis for audio coding purposes. *Signal Processing*, *83*, 919-929.

Ruandaidh, J. J. K. O., & Pun, T. (1998). Rotation,

scale and translation invariant spread spectrum digital image watermarking. *Signal Processing, 66*(3), 303-317.

Seo, J. S., & Haitsma, J., & Kalker, T. (2002). Linear speed-change resilient audio fingerprinting. In *Proceedings of the First IEEE Benelus Workshop on Model based Processing and Coding of Audio.*

Seok, J., Hong, J., & Kim, J. (2002). A novel audio watermarking algorithm for copyright protection of digital audio. *ETRI Journal, 24*(3), 181-189.

Sinha, D., & Tewfik, A. (1993). Low bit rate transparent audio compression using adapted wavelets. *IEEE Transactions on Signal Processing, 41*(12), 3463-3479.

Swanson, M. D., Zhu, B., Tewfik, A. H., & Boney, L. (1998). Robust audio watermarking using perceptual masking. *Elsevier Signal Processing, Special Issue on Copyright Protection and Access Control, 66*(3), 337-355.

Tzanetakis, G., Essl, G., & Cook, P. (2001). Audio analysis using the discrete wavelet transform. In *Proceedings of the 2001 International Conference of Acoustics and Music: Theory and Applications,* Skiathos, Greece. Retrieved March 23, 2007, from http://www.cs.princeton.edu/~gessl/papers/amta2001.pdf

Vaseghi, S. V. (2000). *Advanced digital signal processing and noise reduction.* John Wiley & Sons, Ltd.

Veldhuis, R. N. J., Breeuwer, M., & van der Wall, R. G. (1998). Subband coding of digital audio signals. *Philips Res. Rep., 44*(2-3), 329-343.

Wu, S., Huang, J., Huang, D., & Shi, Y. Q. (2005). Efficiently self-synchronized audio watermarking for assured audio data transmission. *IEEE Transactions on Broadcasting, 51*(1), 69-76.

Wu, C. P., Su, P. C., & Kuo, C. C. J. (1999). Robust audio watermarking for copyright protection. *Proceeding of SPIE, 3807*, 387-397.

Wu, C. P., Su, P. C., & Kuo, C. C. J. (2000). Robust and efficient digital audio watermarking using audio content analysis. *Proceedings of SPIE, 3971*, 382-392.

Zurera, M. R., Ferreras, F. L., Amores, M. P. J., Bascon, S. M. & Reyes, N. R. (2001). A new algorithm for translating psychoacoustic information to the wavelet domain. *Signal Processing, 81*, 519-531.

ADDITIONAL READING

Altun, O., Sharma, G., Celik, M., Sterling, M., Titlebaum, E., Bocko, M. (2005). *Morphological steganalysis of audio signals and the principle of diminishing marginal distortions.* Paper presented at the IEEE International Conference on Acoustics, Speech, and Signal Processing (IEEE Cat. No.05CH37625, Part 2, Vol. 2, pp. ii/21-4).

Audio box, Digital audio watermarking. Retrieved March 22, 2007, from http://www.ece.uvic.ca/499/2003a/group09/w/watermarking.htm

Augustine, R.J., Devassia, V.P. (2006). Watermarking of audio signals in sub-band domain wavelet transform. *WSEAS Transactions on Signal Processing, 2*(2), 238-44.

Baras, C., Moreau, N. (2005). *An audio spread-spectrum data hiding system with an informed embedding strategy adapted to a Wiener filtering based receiver.* Paper presented at the IEEE International Conference on Multimedia and Expo, p 4.

Erfani, Y., & Ghaemmaghami, S. (2004). Increasing robustness of one bit embedding spread spectrum audio watermarking against malicious attack. In *Proceedings of Circuits, Signals, and Systems* (p. 449).

Fang, Yanmei, Gu, Limin, Huang, Jiwu (2005).

Performance analysis of CDMA-based watermarking with quantization scheme. *Lecture Notes in Computer Science*, v 3439, *Information Security Practice and Experience - First International Conference, ISPEC 2005, Proceedings* (pp. 350-361).

George, T., & Essl, G., & Cook, P. (2001). *Audio analysis using the discrete wavelet transform.* Paper presented at the International Conference on Acoustics and Music: Theory and Applications.

Gunsel, B., Ulker, Y., & Kirbiz, S. (2006). A statistical framework for audio watermark detection and decoding. *Lecture Notes in Computer Science (including subseries Lecture Notes in Artificial Intelligence and Lecture Notes in Bioinformatics)*, v 4105 LNCS, *Multimedia Content Representation, Classification and Security - International Workshop, MRCS 2006. Proceedings*, (pp. 241-248).

Horvatic, P., Zhao, J., & Thorwirth, N. (2000). Robust audio watermarking based on secure spread spectrum and auditory perception model. In *Proceedings of the IFIP TC11 Fifteenth Annual Working Conference on Information Security for Global Information Infrastructures* (Vol. 175, pp. 181-190).

Kim, S., & Bae, K. (2005). Robust estimation of amplitude modification for scalar Costa scheme based audio watermark detection. *Lecture Notes in Computer Science*, v 3304, *Digital Watermarking - Third International Workshop, IWDW 2004* (pp. 101-114).

Kim, Y. H., Kang, H. I., Kim, K. I., & Han, S. (2002). A digital audio watermarking using two masking effects. In *Proceedings of the Third IEEE Pacific Rim Conference on Multimedia* (pp. 655-662).

Ko, C-S., Kim, K-Y., Hwang, R-W., Kim, Y., & Rhee, S-B. (2005). *Robust audio watermarking in wavelet domain using pseudorandom sequences. Proceedings.* Paper presented at the Fourth Annual ACIS International Conference on Computer and Information Science (pp. 397-401).

Kraetzer, C., Dittmann, J., & Lang, A. (2006). Transparency benchmarking on audio watermarks and steganography. *Proceedings of SPIE - The International Society for Optical Engineering*, v 6072, *Security, Steganography, and Watermarking of Multimedia Contents VIII - Proceedings of SPIE-IS and T Electronic Imaging*, 2006, p 60721L

Lee, S., Lee, S-K., Seo, Y.-H., & Yoo, C.D. (2006). *Capturing-resistant audio watermarking based on discrete wavelet transform.* Paper presented at the IEEE International Conference on Multimedia and Expo (IEEE Cat. No. 06TH8883C) p 4 pp.

Lemma, A., Katzenbeisser, S., Celik, M., & van der Veen, M. (2006). Secure watermark embedding through partial encryption. *Lecture Notes in Computer Science (including subseries Lecture Notes in Artificial Intelligence and Lecture Notes in Bioinformatics)*, v 4283 LNCS, *Digital Watermarking - 5th International Workshop, IWDW 2006, Proceedings*, 2006, p 433-445

Li, Z., Sun, Q., & Lian, Y. (2006, August). Design and analysis of a scalable watermarking scheme for the scalable audio coder. *IEEE Transactions on Signal Processing, 54*(8), 3064-77.

Lie, W-N., & Chang, L-C. (2006, February). Robust and high-quality time-domain audio watermarking based on low-frequency amplitude modification. *IEEE Transactions on Multimedia, 8*(1), 46-59.

Liu, Q., Jiang, X., & Zhou, Z. (2005, October). A unified digital watermark algorithm based on singular value decomposition and spread spectrum technology. *Chinese Journal of Electronics, 14*(4), 621-624.

Liu, Z., Kobayashi, Y., Sawato, S., & Inoue, A. (2003). Robust audio watermark method using sinusoid patterns based on pseudo-random sequences. *Proceedings of the SPIE, 5020*, 21-31.

Ma, X.F., Jiang, T. (2006). The research on wavelet audio watermark based on independent component analysis. *Journal of Physics: Conference Series*, 48(1), 442-6.

Malik, H., Khokhar, A., & Ansari, R. (2004, May). Robust audio watermarking using frequency selective spread spectrum theory. In *Proceedings of the ICASSP'04*, Canada.

Malik, H., Khokhar, A., & Ansari, R. (2005). Improved watermark detection for spread-spectrum based watermarking using independent component analysis. In *Proceedings of the Fifth ACM Workshop on Digital Rights Management, DRM'05* (pp. 102-111).

Megias, D., Herrera-Joancomarti, J., & Minguillon, J. (2005). Robust frequency domain audio watermarking: A tuning analysis. *Lecture Notes in Computer Science*, v 3304, *Digital Watermarking - Third International Workshop, IWDW 2004*, 2005, p 244-258

Microsoft Audio Watermarking Tool. Retrieved March 23, 2007, from http://research.microsoft.com/research/downloads/Details/885bb5c4-ae6d-418b-97f9-adc9da8d48bd/Details.aspx?CategoryID

Mitrea, M., Duta, S., & Preteux, F. (2006). Informed audio watermarking in the wavelet domain. In *Proceedings of SPIE: The International Society for Optical Engineering*, v 6383, *Wavelet Applications in Industrial Processing IV*, p 63830P

Muntean, T., vGrivel, E., Nafornita, I., & Najim, M. (2002). Audio digital watermarking based on hybrid spread spectrum IEEE Wedel Music, 2002.

O'Donovan, Finbarr B., Hurley, Neil J., Silvestre, Guenole C.M. (2005). An investigation of robustness in non-linear audio watermarking. In *Proceedings of SPIE - The International Society for Optical Engineering*, v 5681, *Proceedings of SPIE-IS and T Electronic Imaging - Security, Steganography, and Watermarking of Multimedia Contents VII*, 2005, p 769-778

Petrovic, R. (2005). *Digital watermarks for audio integrity verification.* Paper presented at the 7th International Conference on Telecommunications in Modern Satellite Cable and Broadcasting Services (IEEE Cat. No. 05EX1072, Part 1, Vol. 1, pp, 215-220).

Sedghi, S., Khademi, M., & Cvejic, N. (2006). Channel capacity analysis of spread spectrum audio watermarking for noisy environments. In *Proceedings of the 2006 International Conference on Intelligent Information Hiding and Multimedia* (pp. 33-36).

Sriyingyong, N., & Attakitmongcol, K. (2006). *Wavelet-based audio watermarking using adaptive tabu search.* Paper presented at the 1st International Symposium on Wireless Pervasive Computing (Vol. 200, pp. 1-5).

Sterling, M., Titlebaum, E.L., Xiaoxiao Dong, Bocko, M.F. (2005). *An adaptive spread-spectrum data hiding technique for digital audio.* Paper presented at the IEEE International Conference on Acoustics, Speech, and Signal Processing (IEEE Cat. No.05CH37625, Part 5, Vol. 5, pp. v/685-688).

Wang, R-D., Xu, D-W., & Li, Q. (2005). Multiple audio watermarks based on lifting wavelet transform. In *Proceedings of the 2005 International Conference on Machine Learning and Cybernetics* (IEEE Cat. No. 05EX1059, Part 4, Vol. 4, pp. 1959-64).

Wang, X-Y., & Zhao, H. (2006, December). A novel synchronization invariant audio watermarking scheme based on DWT and DCT. *IEEE Transactions on Signal Processing, 54*(2), pp. 4835-40.

Wu, Y.J., & Shimamoto, S. (2006). A study on

DWT-based digital audio watermarking for mobile ad hoc network. *Proceedings - IEEE International Conference on Sensor Networks, Ubiquitous, and Trustworthy Computing*, v 2006 I, *Proceedings - Thirteenth International Symposium on Temporal Representation and Reasoning, TIME 2006*, 2006, p 247-251

Zaidi, A., Boyer, R., Duhamel, & P. (2006, February). Audio watermarking under desynchronization and additive noise attacks. *IEEE Transactions on Signal Processing*, 54(2), 570-584.

Zhang, L., Chen, L-m., Qian, G-b. (2006). *Self-synchronization adaptive blind audio watermarking*. Paper presented at The 12th International Multi-Media Modelling Conference Proceedings, p. 4.

Zhang, L., Qian, G-b., Chen, L-m., Li, X. (2005). *The design of optimum blind audio watermark decoder using higher order spectrum.* Paper presented at the IEEE 2005 International Symposium on Microwave, Antenna, Propagation and EMC Technologies for Wireless Communications (IEEE Cat. No. 05EX1069, Part 2, Vol. 2, pp. 1237-1240).

Chapter III
Audio Watermarking Through Parametric Synthesis Models

Yi-Wen Liu
Boys Town National Research Hospital, USA

ABSTRACT

This chapter promotes the use of parametric synthesis models in digital audio watermarking. It argues that, because human auditory perception is not a linear process, the optimal hiding of binary data in digital audio signals should consider parametric transforms that are generally nonlinear. To support this argument, an audio watermarking algorithm based on aligning frequencies of spectral peaks to grid points is presented as a case study; its robustness is evaluated and benefits are discussed. Toward the end, research directions are suggested, including watermark-aided sound source segregation, cocktail watermarking, and counter-measure against arithmetic collusive attacks.

BACKGROUND

Frequency-domain masking is often regarded as the standard perceptual model in audio watermarking. Below the masking threshold, a spread-spectrum watermark (e.g., Kirovski & Malvar, 2003; Swanson, Zhu, & Tewfik, 1998) distributes its energy; the same threshold also sets a limit to the step size of quantization in informed watermarking (e.g., Chou, Ramchandran, & Ortega, 2001). However, ways to manipulate sounds can go far beyond masking to deceive the human auditory system. This has been explored by a group of attackers; in 2001, participants invited by Secure Digital Music Initiative (SDMI) successfully defeated audio watermarking schemes that were then state of the art (Wu, Craver, Felten, & Liu, 2001). Typically, successful attacks exploited human insensitivity to small changes in phase, pitch, and time. Since then, many have wondered how these aspects may also be used for improved watermarking. A plethora of audio watermarking schemes thus came into existence.

Without pursuing details of all the schemes and their applications, this section aims to provide a unifying view. First, the concept of parametric sound synthesis is introduced. Then, general design principles for watermarking with synthesis models will be stated. Finally, existing watermarking methods will be categorized according to the signal models.

Parametric Modeling in Digital Audio: A Brief Tutorial

Simply put, "parametric modeling" means to represent a signal comprised of a large number of samples with fewer variables called *parameters*. Whenever such a model can be constructed for a signal, it immediately leads to significant data compression. Historically, data compression through parametric modeling has worked very well in speech through *linear prediction* (Markel & Gray, 1976). Human vocalization has been studied extensively, and speech signals are nowadays routinely encoded in parameters derived from estimated motion of the vocal tract and vibration rates of the vocal cord (Shroeder & Atal, 1985).

Nevertheless, speech is not the only type of audio signals that can be parameterized, and benefits of parameterization go beyond saving the data storage space. In mid 1980s, *sinusoidal modeling* was independently proposed by two groups; McAulay and Quatieri (1986) sought to parameterize speech by tracking spectral peaks, and Smith and Serra (1987) devised a similar scheme to generate musically expressive effects such as stretching a signal in time without dropping its pitch.

Predictive modeling and sinusoidal modeling are not mutually exclusive; the residual component in sinusoidal modeling can be parameterized by linear prediction (Serra & Smith, 1990). The hybrid system is referred to as "deterministic plus stochastic," because the residual component lacks tonal quality and thus is modeled as filtered noise. The deterministic plus stochastic model was refined by Levine (1998) by further decomposing the stochastic component into a quasi-stationary "noise" part and a rapidly changing "transient" part. Thus, Levine's model led to a versatile audio coding scheme that is both efficient and expressive. Development in sinusoidal modeling has culminated in its adoption by MPEG-4 as an extension to the audio coding standard (Purnhagen & Meine, 2000). F-QIM, the main watermarking algorithm to be described in later sections, is heavily based on Smith, Serra, and Levine's work.

In a futuristic fashion, parametric signal modeling has been used in network-based musical interaction, and languages and protocols are still being designed (e.g., Vercoe, Gardner, & Scheirer, 1998; Wessel & Wright, 2004). Often, sounds are encoded by physically meaningful parameters and synthesized through wave equations in object-oriented manners (Cook & Scavone, 1999; Scavone & Cook, 2005). As new synthesis models and music formats emerge, their impact on the audio watermarking is yet to be seen.

Watermarking Through Parametric Modeling

For the convenient of discussion, hereafter, a parametric model is defined as a mapping from a parameter space to a Euclidean signal space R^N (or C^N). The mapping is also referred to as a "synthesis." Conversely, a recorded signal $\mathbf{s}$ can be decomposed into a perfectly parameterized part $\mathbf{s}_{|\theta\rangle}$ and a residual $\mathbf{r}$ orthogonal to $\mathbf{s}_{|\theta\rangle}$. Demonstrated in Figure 1, the triangle-shaped region $\Xi(\Theta)$ on the left represents the image of the parameter space Θ in the signal space R^N. For an arbitrary signal $\mathbf{s}$ in R^N, its closest neighbor $\mathbf{s}_{|\theta\rangle}$ is searched in $\Xi(\Theta)$, and an inverse mapping takes $\mathbf{s}_{|\theta\rangle}$ back to its parametric representation $|\theta\rangle$. The inverse mapping is also referred to as an "analysis," or "parameter estimation". Mathematical notations used in this chapter are listed in Table 5.

The diagram in Figure 2 shows a generic watermarking scheme based on parametric analysis and synthesis. To embed of a binary watermark W, the cover signal $\mathbf{s}$ is first decomposed into $\mathbf{s}_{|\theta\rangle} + \mathbf{r}$

Figure 1. Geometric interpretation of signal decomposition and parameter estimation

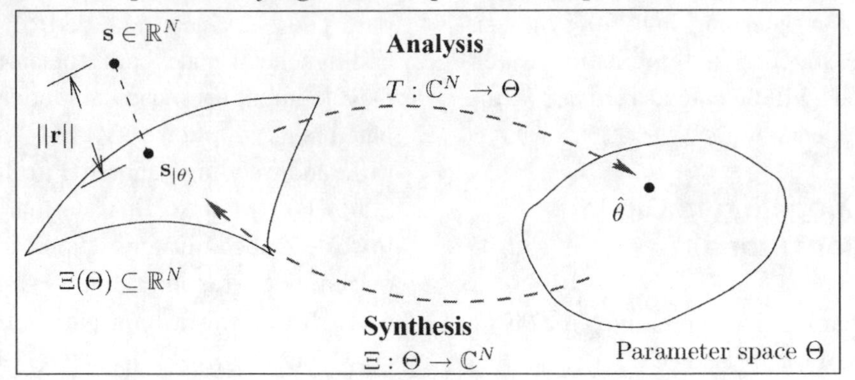

Figure 2. Watermark embedding and decoding through parametric models

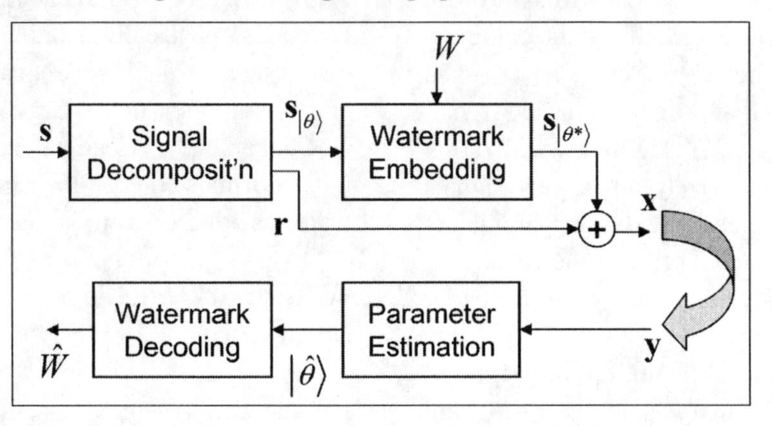

as depicted in Figure 1. Then, $|\theta\rangle$ is modified to $|\theta *\rangle$ to carry W, and $\mathbf{s}_{|\theta*\rangle}$ is superposed with **r** to form a *watermarked signal* **x**, which is subjected to unknown attacks. Upon the reception of a corrupted copy **y**, parameters are estimated so as to decode W. The attempt of watermarking is considered successful if the estimated parameters $|\hat{\theta}\rangle$ are close enough to $|\theta *\rangle$ such that the decoded binary message $\hat{W}$ is identical to W.

Two criteria ought to be considered regarding how θ is modified to θ *. First, the modification should be small enough not to introduce perceptible distortion. Second, this modification should be as big as possible so that all choices of θ * are distinguishable after attacks.

Intuitively, if the magnitude of the attack increases, the uncertainty in parameter estimation also increases. As a result, the chance that $\hat{W}$ is successfully decoded decreases. Thus, it is crucial to predict how much uncertainty an attack causes in parameter estimation. Interested readers should refer to the Appendix for a quantitative definition of uncertainty based on the Cramér-Rao inequality.

With this mathematical framework in mind, audio watermarking schemes can be classified into the following categories.

Transform-Domain Amplitude Manipulation

These schemes typically utilize the masked portion in a spectrum to embed watermarks. "Masking" refers to the capability of a loud stimulus to make softer sounds inaudible at its immediate time-frequency vicinity. For example, near a railroad crossing, a person can hardly talk on the cell phone if a train passes by.

The masking curve can be computed in a many similar ways (see, e.g., Bosi & Goldberg, 2003; Cox, Miller, & Bloom, 2002; Painter & Spanias, 2000). Using the knowledge about masking, at first, it was thought that information can be hidden as inaudible components below the threshold (Boney, Tewfik, & Hamdy, 1996). However, inaudible components are prone to the removal by perceptual audio coders (Cox, Kilian, Leighton, & Shamoon, 1999). Hence, the spread spectrum (SS) method emerged and watermarks were embedded into significant time-frequency components. Theories predict that significant components hide information with higher security, because a watermark cannot be removed by denoising if its power spectrum matches with the cover signal's (Hartung, Su, & Girod, 1999). Also, the SS approach is by its nature invariant to amplitude scaling, and its robustness against desynchronization has been reported (He, Iliev, & Scordilis, 2004; Kirovski, 2003). However, it cannot achieve high data hiding rates as promised by the theory of "writing on dirty paper" (Chen & Wornell, 2001; Costa, 1983).

To achieve the high rates, Chou et al. (2003) designed an audio watermarking scheme using Chen and Wornell's (2001) quantization index modulation (QIM). The quantization level is set at the masking threshold given by the MPEG-1 psychoacoustic model (see ISO/IEC 11172-3), which was also adopted by Silvestre, Hurley, Hanau, and Dowling (2001), and Lee, Kim, Kim, and Moon (2003). Over 100 kbits/sec of data payload was reported to sustain additive noise at SNR = 15 dB (Chou et al., 2003). However, the simplest form of QIM is not robust against amplitude scaling, which is unfortunately a common operation in music processing. To mend this obvious shortcoming of QIM, Lee et al. (2003) proposed to estimate the amplitude scaling factor by an expectation-maximization (EM) algorithm. Based on the EM estimation, the quantization step size was corrected at the watermark decoder. Alternatively, watermark encoding schemes could introduce redundancy to achieve robustness against scaling (e.g., Miller, Doërr, & Cox, 2002), but this inevitably reduces the data hiding payload.

Phase Manipulation

Phase quantization for watermarking was first proposed by Bender, Gruhl, Morimoto, and Lu (1996). For each long segment of sound to be watermarked, the phase at 32 to 128 frequency bins of the first short frame was replaced by $\pi / 2$, representing the binary 1, or $-\pi / 2$, representing the binary 0. In all the frames to follow, the relative phase relation was kept unchanged. It was reported that such phase manipulation was inaudible to "normal people" but might cause audible dispersion to "professionals". Therefore, a heuristic procedure was outlined to reduce the perceptibility of the phase change. Kuo, Johnston, Turin, and Quackenbush (2002) refined the procedure and empirically set a constraint in the phase spectrum $\Phi(z)$:

$$\left|\frac{d\Phi}{dz}\right| < 30^\circ / \text{Bark}, \qquad (1.1)$$

where z is the frequency in terms of critical band rate (see Table 4 in the Appendix for a list of critical band locations). It was reported that perceptual audio compression alters the phase spectrum significantly, and a Viterbi search algorithm was developed for the recovery of the quantized phase. Twenty-eight bits/sec of data hiding was attempted, sustaining MPEG AAC compression at 64k bits/sec/channel with a decoding bit error rate (BER) of < 1%.

A more recent phase quantization scheme assumes harmonic structure of the cover signal, and the absolute phase of each harmonic is modified by QIM (Dong, Bocko, & Ignjatovic, 2004). About 80 bits/sec of data hiding was reported robust to MP3 compression at 64 kbits/sec/channel with a BER of 1% to 5% depending on the phase quantization step size, which was set at $\pi/2$, $\pi/4$, or $\pi/8$.

These works indeed proved that phase quantization is robust to perceptual audio compression. However, the human hearing is *not highly* sensitive to phase distortion, argued originally in Bender et al. (1996). Thus, an attacker can use imperceptible frequency modulation to drive the absolute phase of a component arbitrarily. This would defeat the watermark decoding of phase quantization schemes.

Frequency Manipulation

Audio watermarking by frequency manipulation dates back to "replica modulation" (Petrovic, 2001). Inspired by echo-hiding (Bender et al., 1996), Petrovic observed that an echo is a "replica" of the cover signal placed at a delay and the echo becomes transparent if sufficiently attenuated. Petrovic then envisioned that an attenuated replica can also be placed at a shifted frequency to encode hidden information. Following Petrovic's vision, Shin, Kim, Kim, and Choil (2002) proposed to utilize pitch scaling of up to 5% at mid-frequency (3 to 4 kHz) for watermark embedding. Data hiding of 25 bits/sec robust to 64 kbits/sec/channel of audio compression was reported with BER < 5%.

Succeeding Petrovic's and Shin's work, the main thrust of this chapter departs in a few ways. First, the cover signal is *replaced by* instead of *superposed with* the replica. This is done in a few steps: signal decomposition, frequency QIM, and sinusoidal resynthesis. Second, the amount of frequency shift in QIM is based on studies of pitch sensitivity in human hearing. The just noticeable difference (JND) in frequency is approximately constant at low frequency (e.g., below 500 Hz) and linearly increases above. Thus, the step size of QIM is set accordingly. Finally, the frequency shift incurred by QIM is about an order of magnitude smaller than the pitch shift described by Shin et al. (2002). The watermark decoding therefore requires high accuracy of frequency estimation. To this end, a frequency estimation algorithm that approaches the Cramér-Rao bound (CRB) is adopted.

Independently, Girin and Marchand (2004) studied frequency modulation for audio watermarking in speech signals. A harmonic structure of the signal is assumed, and frequency modulation in the sixth harmonic or above is found, surprisingly, not noticeable up to a deviation of 0.5 times the fundamental frequency. Based on this observation, transparent watermarking at 150 b/sec was achieved by coding 0 and 1 with a positive and negative frequency deviation, respectively. However, it remained as future work to test the robustness of this algorithm under attacks.

Modifying Speech Models

Possibly inspired by methods in speech recognition and speech compression, watermarking schemes specifically for speech signals have been invented. For example, Li and Yu (2000) devised a quantization scheme on cepstral parameters, and achieved robust audio watermarking that sustained several types of operations. More recently, Gurijala, Deller, Seadle, and Hansen (2002) attempted to watermark *The National Gallery of the Spoken Word* through linear predictive (LP) modeling. From each short segment of speech, the encoder derives LP coefficients and modifies them to embed a watermark. The decoder estimates LP coefficients and compared to their original values; a private-decoding scenario is assumed. Robustness to random cropping, MP3 compression, filtering, and additive noise was reported.

Salient Point Extraction

It was observed that, if a signal has clear onsets in its amplitude envelope, the locations of such onsets, called *salient points*, are invariant to a few common signal processing operations (Wu, Su, & Kuo, 2000). Salient points can be identified by wavelet decomposition and quantized in time to embed watermarks. Mansour and Tewfik (2003) reported robustness to MPEG compression and low-pass filtering, and their system sustained up to 4% of time-scaling modification with a probability of error less than 7%. Repetition codes were applied to achieve reliable data hiding at 5 bps.

Schemes mentioned above are listed in Table 1 in ascending data payload. Note that the payload spans across many orders of magnitude, and the degree of robustness varies significantly. Therefore, it is difficult to sort out one single best method for all applications. Instead, this chapter shall focus on a specific method next, and then suggest the *cocktail* usage of multiple methods in the end.

METHODS: IMPLEMENTING THE F-QIM SCHEME

The following two sections will focus on a watermarking scheme first proposed by Liu and Smith (2003, 2004a) based on QIM of the frequency parameters in sinusoidal models, thus to be called F-QIM.

To begin, F-QIM decomposes a cover signal into sines + noise + transient (SNT) (Levine, 1998). Each sinusoid is subject to QIM in frequency using codebooks $\{C_0, C_1\}$ that consist of binary signaling grid points. As illustrated in Figure 2, $|\theta\rangle$ now denotes the frequencies $\left| f_1, f_2, f_3, \ldots, f_J \right\rangle$ of J prominent sinusoids detected at a frame; $\mathbf{s}_{|\theta^*\rangle}$ denotes the sum of sinusoids synthesized with quantized frequencies; $\mathbf{r}$ denotes Noise + Transient, and the sum $\mathbf{x} = \mathbf{s}_{|\theta^*\rangle} + \mathbf{r}$ is referred to as

*Table 1. Performance of audio watermarking schemes, sorted by attempted data payload (*Performance of F-QIM will be described later in this chapter)*

Reference	Method	Payload	Robustness
Kirovski & Malvar (2003)	Spread spectrum	0.5-1 bps	BER < 10^{-6} to additive noise, bandpass filtering, MP3 compression, time-scaling 4%, wowwing, echo of 100 ms
Mansour & Tewfik (2003)	Salient point	5 bps	MP3 compression, low-pass filtering, time-scaling 4%
Li & Yu (2000)	Cepstrum	22 bps	1% jittering, reverberation, time-warping, low-pass filtering, MP3 compression
Gurijala et al. (2002)	Linear prediction	24 bps	White and colored noise, low- and high-pass filtering, random cropping, MP3 compression, CELP45 speech coding. Cover signal is required for decoding the watermark.
Shin et al. (2002)	Pitch scaling	25 bps	BER < 5% to AAC/MP3 compression
Kuo et al. (2002)	Phase quantization	28 bps	BER < 1% to AAC compression
Liu (2005)	F-QIM	28-43 bps	*.
Dong et al. (2004)	Phase quantization	80 bps	BER < 5% to MP3; harmonic signal structure assumed.
Girin & Marchant (2004)	Frequency shift	150 bps	Robustness not reported; harmonic signal structure assumed.
Van der Veen et al. (2003)	Low-bit modulation	~ 40k bps	Not aimed for robust applications.
Chou et al. (2001)	QIM	>100k bps	Robust to white noise at SNR = 15 dB.

the *watermarked signal*. The decoder, receiving a corrupted copy **y**, estimates the frequencies and decodes a binary message by rounding off the frequencies to the grid points in $\{C_0, C_1\}$.

Figure 3 shows a closer look at watermark embedding. Initially, the spectrum of the cover signal is computed by the short-time Fourier transform (STFT). If the current frame contains a sudden rise of energy and the sine-to-residual ratio (SRR) is low, it is labeled a *transient* and fed to the output unaltered. Otherwise, prominent peaks are detected and represented by sinusoidal parameters. The *residual* component is computed by removing all the prominent peaks from the spectrum, transforming the spectrum back to the time domain through inverse FFT (I-FFT), and then overlap-adding (OLA) the frames in time. Parallel to residual processing, a peak tracking unit memorizes sinusoidal parameters from the past and connects peaks across frames to form trajectories. The watermark is embedded in the trajectories via QIM in frequency. The signal that takes quantized trajectories to synthesize is called the *watermarked sinusoids*. A watermarked signal is comprised of the watermarked sinusoids, the residual, and the transients. Details of each building block are described next.

Spectral Decomposition

Window Selection

To compute STFT, the Blackman window (Harris, 1978) of length $L = 2N + 1$ is adopted.[2] The Blackman window has close to 60 dB of sidelobe rejection, and a roll-off rate at 18 dB per octave; both make it an appealing choice for sinusoidal modeling.

Calculating the Masking Curve

Only unmasked peaks are used for watermark embedding in F-QIM. The masking curve is computed via a spreading function that approximates the pure-tone excitation pattern on the human basilar membrane (Bosi, 1997), (see equation (1.2)).

The slope 27 dB/Bark on the low frequency side is fixed while the slope on the high frequency side changes as a function of Λ, the sound pressure level (SPL). Note that SPL is a physically measurable quantity. To align it with digital signals, a pure-tone of maximum level is arbitrarily set equal to 100 dB SPL. The masking level $M(z)$ is given by the following equation:

Figure 3. SNT decomposition and watermark embedding

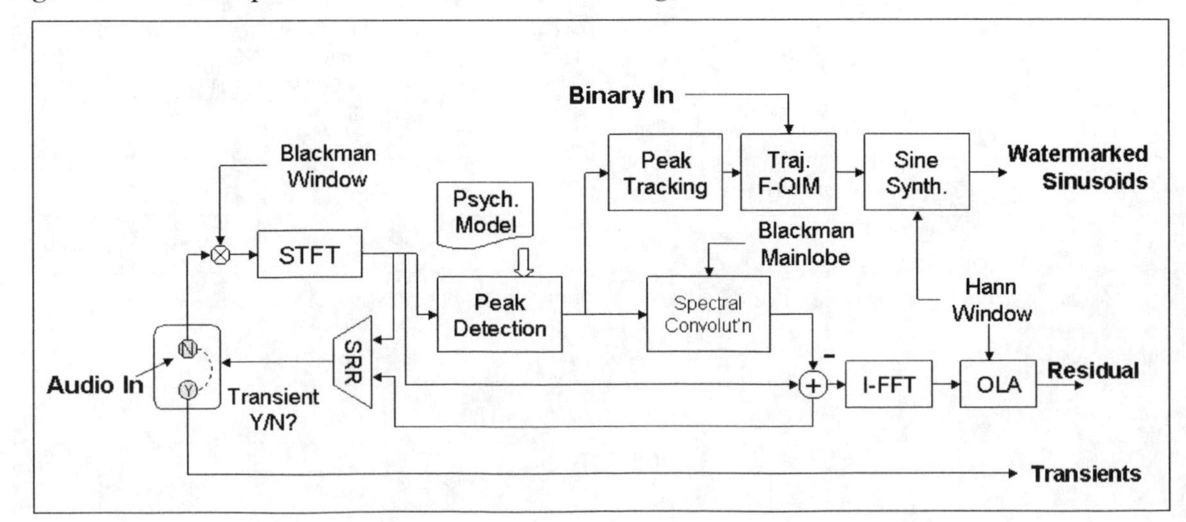

Equation 1.2.

$$\begin{cases} d\psi / dz = 0, & |z - z_0| \le .5 \\ d\psi / dz = 27, & z < z_0 - 0.5 \\ d\psi / dz = \begin{cases} -27, & \text{if } \Lambda \le 40 \\ -27 + 0.37(\Lambda - 40), & \text{if } \Lambda > 40 \end{cases}, & z > z_0 + 0.5 \end{cases}.$$

$$M(z) = \Lambda - \Delta(z_0) + \psi(z) \qquad (1.3)$$

where the offset $\Delta(z_0)$ can be approximated by the following (Jayant, Johnston, & Safranek, 1993)

$$\Delta = (14.5 + z_0) \text{ dB}. \qquad (1.4)$$

To express $M(z)$ in the unit of power per frequency bin, the following normalization is necessary (Cox et al., 2002):

$$M_k^2 = 10^{M(z)/10} / N(z) \qquad (1.5)$$

where $N(z)$ is the equivalent number of FFT bins within a critical bandwidth (CBW) centered at $z = z(k\Omega_s)$, $k\Omega_s$ being the frequency of the kth bin. The CBW as a function of its center frequency f can be approximated by the following formula (Zwicker & Fastl, 1990):

$$\text{CBW (in Hz)} = 25 + 75[\,1 + 1.4(f/\text{kHz})^2\,]^{0.69}. \qquad (1.6)$$

When more than one tone is present, the overall masking curve $\sigma^2(k\Omega_s)$ is set as the maximum of the spreading functions and the threshold in quiet (TIQ) denoted as $I_0(f)$ below:

$$\sigma^2(k\Omega_s) = \max\{\, M_{1,k}^2,\ M_{2,k}^2, \ldots, M_{J,k}^2\ \ 10^{I_0(k\Omega_s)/10} \} \qquad (1.7)$$

where $M_{j,k}$ denote the masking level at frequency bin k due to the presence of tone j, and $I_0(f)$ is calculated using Terhardt's (1979) approximation, (see equation (1.8)).

Equation (1.7) is called a rule of *pooling*. More sophisticated pooling rules involve taking higher-order mean among the masking levels (Bosi & Goldberg, 2003).

In this chapter, a peak is considered "prominent" if its intensity is higher than the masking curve. To carry a watermark, the prominent peaks shall be subtracted from the spectrum and then added back at quantized frequencies.

Note that noise maskers are ignored in this psychoacoustic model. Consequently, some of the prominent peaks that stay above it could actually be inaudible. It is left as future work to lump up noise components in a critical band, to define that as a noise masker, and to update the masking curve accordingly.[3]

Spectral Interpolation

Abe and Smith (2004) showed that, at sufficiently high SNR, frequency estimation by quadratic interpolation of the log-magnitude spectrum (QI-FFT) approaches the CRB if the signal is sufficiently zero-padded in the time domain. In the F-QIM system, the frequency and amplitude of each prominent peak are estimated by QI-FFT,

Equation 1.8.

$$I_0(f)/\text{dB} = 3.64(f/\text{kHz})^{-0.8} - 6.5e^{-0.6(f/\text{kHz}-3.3)^2} + 10^{-3}(f/\text{kHz})^4$$

and the phase of the peak is estimated afterwards by linear interpolation.

To understand intuitively why QI-FFT works well, let us assume that, contrary to the fact, a continuous-time Gaussian window $w(t)$ is applied to a continuous-time signal:

$$w(t) = e^{-pt^2},\ p > 0 \tag{1.9}$$

It was derived in Papoulis (1977) that the Fourier transform of $w(t)$ is:

$$W(\omega) = \sqrt{\frac{\pi}{p}}\, e^{-\frac{\omega^2}{4p}} \tag{1.10}$$

From equation (1.10), observe that the log-magnitude spectrum of the continuous-time Gaussian window is exactly parabolic:

$$\log |W(\omega)| = \log\sqrt{\frac{\pi}{p}} - \frac{\omega^2}{4p} \tag{1.11}$$

By the shift theorem of Fourier transform, a Gaussian-windowed stationary sinusoid $\mathbf{s}_{|A,\omega_0,\phi\rangle}$ should have a quadratic log-magnitude spectrum:

$$\log |S_w(\omega)| = \log\sqrt{\frac{\pi}{p}} + \log A - \frac{(\omega-\omega_0)^2}{4p} \tag{1.12}$$

In practice, a discrete spectrum is computed. Denote $S_k = S_w(k\Omega_s)$. For any given peak at bin number k (such that $|S_k| > |S_{k+1}|$ and $|S_k| > |S_{k-1}|$), denote $a^- = \log |S_{k-1}|$, $a^+ = \log |S_{k+1}|$, and $a = \log |S_k|$. The frequency and amplitude estimates are given by the following equations:

$$\hat{\omega} = \left(k + \frac{1}{2}\frac{a^- - a^+}{a^- - 2a + a^+}\right)\Omega_s; \tag{1.13}$$

$$\log \hat{A} = a - \frac{1}{4}\frac{\hat{\omega} - k\Omega_s}{\Omega_s}(a^- - a^+) - C, \tag{1.14}$$

where the constant $C = \log\left(\sum_{n=-N}^{N} w[n]\right)$ is a normalization factor. Denote $q = \frac{\hat{\omega}}{\Omega_s} - k$. The phase estimate is given by linear interpolation:

$$\hat{\phi} = \angle S_k + q(\angle S_{k+1} - \angle S_k). \tag{1.15}$$

Spectral Subtraction

Having estimated the parameters $|\hat{A},\hat{\omega}_0,\hat{\phi}\rangle$, the corresponding sinusoid can be removed efficiently in the frequency domain because it is concentrated in a few bins. Described below is a sinusoid removal algorithm, to be called *spectral subtraction*.

- **Step 0:** Initialize the sum spectrum $\hat{S}(\omega) = 0$. Denote $\hat{S}_k = \hat{S}(k\Omega_s)$.
- **Step 1:** For each peak, fit the main lobe of the Blackman window transform $W(\omega)$ at $\hat{\omega}$, and scale it by $\hat{A}\exp(j\hat{\phi})$. For the convenience of discussion, assume the normalization factor $\sum_n w[n] = 1$. Denote the scaled and shifted main lobe of the window as $\hat{W}(\omega)$:

$$\hat{W}(\omega) = \begin{cases} \hat{A}e^{j\hat{\phi}}W(\omega - \hat{\omega}), & \text{if } |\omega - \hat{\omega}| \le 3\frac{2\pi}{L}, \\ 0, & \text{otherwise} \end{cases}. \tag{1.16}$$

- **Step 2:** Denote $\hat{W}_k = \hat{W}(k\Omega_s)$. Update $\hat{S}_k$ by $\hat{S}_k + \hat{W}_k$.
- **Step 3:** Take the next prominent peak and repeat Step 1 and Step 2 until all prominent peaks are processed. Toward the end, $\hat{S}_k$ becomes the spectrum to be subtracted.
- **Step 4:** Define the residual spectrum R_k as the following:

$$R_k = \begin{cases} S_k - \hat{S}_k, & \text{if } \left|S_k - \hat{S}_k\right| < \left|S_k\right|, \\ S_k, & \text{otherwise.} \end{cases} \tag{1.17}$$

The reasoning behind the if-condition in equation (1.17) is such that the residual spectrum is smaller than the signal spectrum everywhere, in terms of its magnitude. Figure 4 shows results of particularly thorough spectral subtraction; nominally, the residual is almost entirely masked for this very clean recording of the trumpet.

Residual Postprocessing

Inaudible portion of the residual is removed by setting R_k to zero if $|R_k|^2$ is below the masking curve. Then, inverse FFT is applied to obtain a residual signal **r** of length N_{FFT}. Due to concerns that will be discussed later regarding perfect reconstruction, **r** is shaped in the time-domain:

$$r^{\text{sh}}[n] = r[n]\, w^{\text{H}}[n] / w[n] \qquad (1.18)$$

where $w^{\text{H}}[n]$ denotes Hann window of length $N + 1$. Then, across frames, $r^{\text{sh}}[n]$ is overlap-added with a hopsize of $N/2$ to form the final residual signal $r^{\text{OLA}}[n]$:

$$r^{\text{OLA}}[n] = \sum_m r_m^{\text{sh}}[n - mN/2], \qquad (1.19)$$

where the subscript m is an index pointing to the frame centered around time $n = mN/2$.

Transient Detection

Regions of rapid transients need to be treated with caution so as to avoid *pre-echoes*. A pre-echo refers to an energy spread in time when the short-time spectrum of a rapid onset is modified. If the rapid onset is preceded by a period of silence originally, a pre-echo is not premasked (Zwicker & Fastl, 1990) in the time domain and thus can be heard.

To avoid pre-echoes, in the F-QIM system, regions of rapid onsets are conservatively kept unaltered. The following variables are defined for the convenience of discussion:

1. Sine-to-residual ratio (SRR)
$$= \frac{\sum_{k=1}^{N_{\text{FFT}}/2-1} |S_k|^2}{\sum_{k=1}^{N_{\text{FFT}}/2-1} |R_k|^2} - 1.$$

2. Current-to-previous energy ratio (CPR): The ratio of $\sum_{k=1}^{N_{\text{FFT}}/2-1} |S_k|^2$ in the current frame to the previous frame.

A frame is labeled "transient" if SRR is below a threshold ρ_s, CPR is above a threshold ρ_c, and there is at least a peak greater than K dB SPL in

Figure 4. The residual is almost masked by a dominant signal

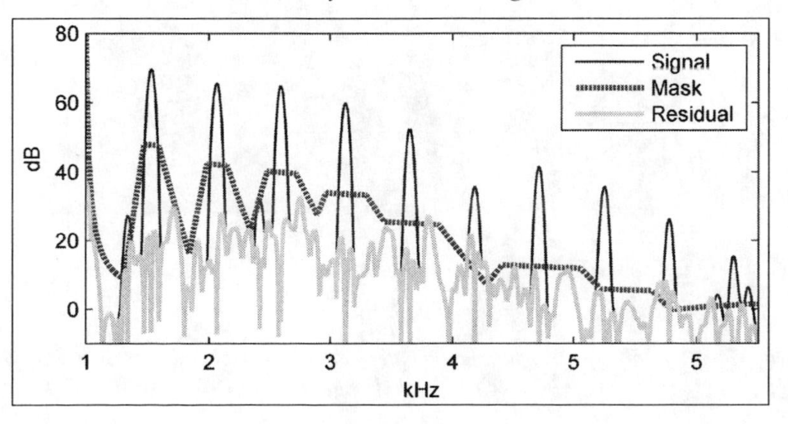

a selected frequency range $[f_l, f_h]$. The following configuration is chosen empirically: for speech signals, $\rho_s = 5.0$; for signals with sharp onsets, such as glockenspiel or the harpsichord recordings, $\rho_s = 20.0$. For all signals, $\rho_s = 1.5$, $[f_l, f_h] = [2000, 8000]$ Hz, and $K = 30$. When all three criteria are met, the Hann window of length $N + 1$ is applied, and 5 consecutive frames are marked transient. Inside the transient region, spectral subtraction and watermark embedding are disabled.

Transient detection is itself a rich research topic. Methods based on Bayesian statistics are worth exploration, and interested readers can refer to Thornburg (2005).

Sinusoidal Synthesis

Peak Tracking

Let J_m denote the number of prominent peaks at frame m. Also, denote the estimated frequencies as $|\omega'_j\rangle$ for the previous frame $m - 1$, and $|\omega_j\rangle$ for the current frame m, respectively. The following procedure connects peaks from frame $m - 1$ to frame m:

- **Step 1—(Find closest link):** For each peak j in the current frame, find its closest neighbor $i(j)$ from the previous frame; $i(j) = \arg\min_k |\omega'_k - \omega_j|$. Connect peak $i(j)$ of the previous frame to peak j of the current frame.
- **Step 2—(Forbid large jumps):** If a connection has a frequency jump $|\omega'_{i(j)} - \omega_j|$ that is greater than a threshold, then break the connection and label peak j of the current frame as an onset to a new trajectory.
- **Step 3—(Resolve splits):** For each peak i_0 in the previous frame, if it is connected to more than one peaks of the current frame, then keep only the link with the smallest frequency jump and break all the other ones. Mark all the other peaks j such that $i(j) = i_0$ as onsets to new trajectories.

Figure 5. Frequency trajectories from a German female speech recording.

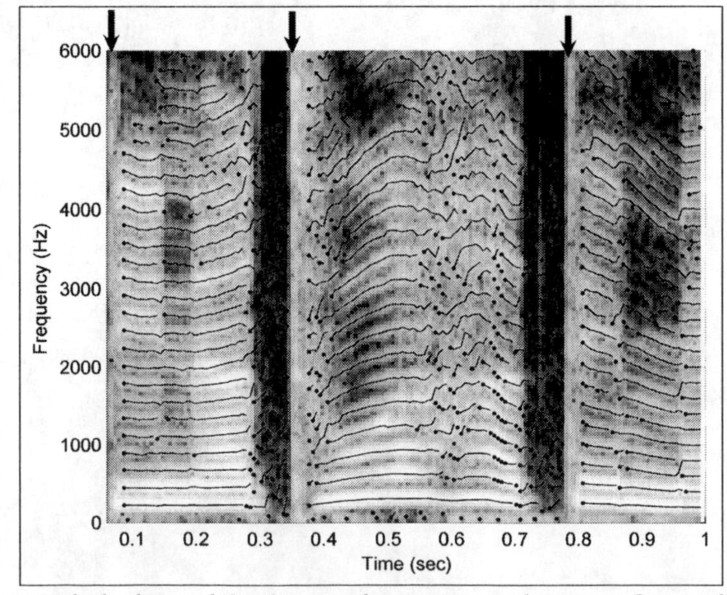

Note: Spectrogram is shown in the background; brightness indicates acoustical intensity. Onsets of trajectories are marked with dots. Arrows on the top point to transient regions, where peak detection is temporarily disabled.

A trajectory starts at an onset and ends whenever the connection can not continue. Figure 5 shows trajectories extracted from one second of female speech. The maximal allowed frequency jump is 1.0 Bark.

Overlap-Add and Phase Continuation

Smith and Serra (1987) stated that "it is customary in computer music to *linearly interpolate* the amplitude and frequency trajectories from one hop to the next." Presented here is an alternative signal synthesis approach by superposing windowed stationary sinusoids. This approach, if implemented in the frequency domain, is computationally more efficient.

For each trajectory k, let $\phi_0^{(k)}$ denote the initial phase, $|A_{km}\rangle$ denote its amplitude envelope, and $|\omega_{km}\rangle$ denote its frequency envelope. A window-based synthesis can be written as the following:

$$s_{\text{total}}[n] = \sum_k \sum_m A_{km} w[n-mh] \cos(\phi_m^{(k)} + \omega_{km}(n-mh)) \tag{1.20}$$

where the phase $\phi_m^{(k)}$ updates by the following equation-

$$\phi_{km} = \phi_{k,m-1} + \left(\frac{\omega_{k,m-1} + \omega_{km}}{2} \right) h \tag{1.21}$$

In equation (1.20), the window $w[n]$ needs to satisfy a perfect reconstruction condition:

$$\sum_{m=-\infty}^{\infty} w[n-mh] = 1 \quad \text{at all } n \tag{1.22}$$

To be consistent with residual post-processing in equation (1.18), the Hann window is adopted in equation (1.20).

Figure 6. Frequency JND as a function of frequency

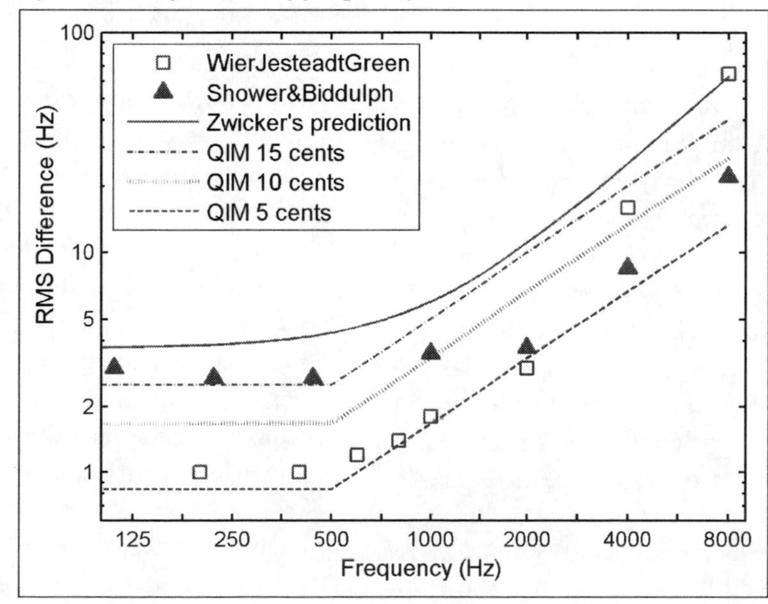

Note. The data in Wier, Jesteadt, and Green (1977) were obtained using pure tone stimuli; the data in by Shower and Biddulph (1931) were obtained using frequency-modulated tones. The RMS frequency difference introduced by QIM at step size 5 to 15 cents is comparable to the JND.

Entering and Exiting Transient Regions

Having described the core algorithm for SNT decomposition, now, continuity problems can occur due to the switching between transient and non-transient regions. When leaving a transient region, it is important to consider every prominent peak as an onset to a trajectory; that is, during synthesis, its phase should be estimated using equation (1.15) to ensure phase matching. For sinusoidal trajectories that last to the next transient region, their entrance there needs to be phase-matched, too. Otherwise, the discontinuity there can be heard as a wide-band click. Phase preservation at the entrance to transient regions[4], can be achieved by cubic-phase synthesis (McAulay & Quatieri, 1986), or by inserting break points in frequency trajectories (Smith & Serra, 1987).

Watermark Embedding at the Encoder

Designing Frequency Quantization Codebooks

In the F-QIM system, frequency parameters $\left|\omega_{km}\right\rangle$ in equation (1.20) are quantized to embed a watermark. To design the quantization codebook properly, the JND in frequency, or *frequency limen* (FL), should be considered. Figure 6 shows existing measurements of the FL of human (Shower & Biddulph, 1931; Wier et al., 1977). The general trend of the FL data agrees well with Zwicker's (1956) prediction[5], which implies that the FL is approximately a fixed proportion of the CBW. Levine (1998) reported that a sufficiently small frequency quantization at approximately a fixed fraction of CBW did not introduce audible distortion. The step size in Levine (1998) is 3 Hz if $f < 500$ Hz and linearly increases above 500 Hz.[6] This design is adopted in F-QIM, and the root-mean-square (RMS) frequency difference introduced by F-QIM is plotted in Figure 6 for comparison. The step size, to be determined based on informal listening, will vary from 5 to 15 cents (at above 500 Hz) depending on the complexity of the timbre of cover signals.

Watermark Encoding

Earlier in this section, a peak is regarded prominent if it is not masked. To generalize from this, each peak's watermarking validity can be characterized by an error probability $P_{m,j}$, where m and j denote the frame index and the peak number, respectively. $P_{m,j}$ is determined by how well the decoder can estimate the frequencies $f_{m,j}$ under attacks. While a method to estimate $P_{m,j}$ will be presented later, the bottom line is that they all differ. Consequently, sinusoidal trajectories can be regarded as communication channels whose capacities vary with $P_{m,j}$. Proposed next are a few heuristic rules to utilize these fading channels for information encoding:

1. **High-rate encoding.** Every prominent peak j in every frame m encodes one bit $b_{m,j}$. Liu and Smith (2004a) reported near 500 b/sec of data hiding that sustained perceptual audio compression with a small error probability. However, the sound example used there was synthesized with a constant number of well-separated trajectories. For audio signals in general, trajectories emerge and disappear, therefore the decoder may not be able to recover the order of the peaks (m, j) if some peaks are deleted by attacks. This makes high-rate encoding improper for reliable data hiding.
2. **Constant-rate encoding.** Assume peaks (m, j) distribute in time and frequency randomly with arbitrary $P_{m,j}$. A conservative data encoding strategy is to parse the signal into long blocks to guarantee that many peaks exist in every block. Therefore, robustness can be achieved even if some of $\{b_{m,j}\}$ is decoded wrong. In the evaluation section of this chapter, it will be shown that, at 0.5

bit/sec, probabilities of false detection and false rejection can be kept low simultaneously.

3. **Per-frame encoding**. At each frame, all peaks encode the same binary value $b_1 = b_2 = \ldots = b_J$. Compared to high-rate encoding, the data hiding payload is cut by $1/J$ to trade for higher robustness. In the evaluation section of this chapter, the performance of per-frame encoding is evaluated under various types of common operation.
4. **Per-trajectory encoding**. The same binary value $b_1 = b_2 = \ldots = b_M$ is embedded in all peaks of the same trajectory, while peaks at the same frame encode independently. Although this scheme is not proper for robust watermarking due to the presence of spurious trajectories, the Future Trends section presents a way to embed information that transcribes certain properties of the cover signal. Particularly, it will be demonstrated that per-trajectory encoding helps identify fundamental frequencies of mixed sound sources.

Watermark Retrieval at the Decoder

To decode a watermark from a signal, frequencies of prominent peaks are estimated using windows of length *h*, the hop size. It is desired that the frequency estimation is not biased and that the error is minimized. Abe and Smith (2004) showed that QI-FFT efficiently achieves both goals if, first, the spectrum is sufficiently interpolated, second, the peaks are sufficiently well-separated, and third, the SNR is sufficiently high.

When only one peak is present, a zero-padding factor of 5.0 confines frequency estimation bias to $0.0001F_s/M$ (M=Hann window length). If multiple peaks are present but are separated by at least $2.28F_s/M$, the frequency estimation bias is bounded below $0.042F_s/M$. If peaks are well-separated and SNR is sufficiently high, for example greater than 20 dB, then the mean-square frequency estimation error decreases as SNR increases. The error either approaches the CRB (at moderate SNR) or is negligible compared to the bias (at very high SNR).

In all experiments to be reported in the next section, QI-FFT is adopted as the frequency estimator at the decoder. Nevertheless, if the above criteria are not met, the bias can be corrected by iterative methods, such as in Liu and Smith (2004a).

Maximum-Likelihood Combination of "Opinions"

Assuming per-frame data encoding, b_js in a frame are either all 0s or all 1s at the watermark encoder. When the watermark decoder receives a signal and identifies peaks at frequencies $\left| \hat{f}_1, \hat{f}_2, \ldots, \hat{f}_J \right\rangle$, these frequencies are decoded to a binary vector **b**=($\hat{b}_1, \hat{b}_2, \ldots, \hat{b}_J$) with error probability $\{P_j\}$. A question arises when some $\hat{b}_j$s are zeros and some are ones: how is the hidden bit best determined?

If the prior distribution is such that $p(b=0)=p(b=1)=0.5$, the optimal way to combine $\hat{b}_j$s the "opinions" is given by the following hypothesis test:

$$b^{\text{opt}} = \begin{cases} 1, & \text{if} \sum_{j=1}^{J} \left[\log\left(\frac{1-P_j}{P_j} \right) \right] \left(\hat{b}_j - \frac{1}{2} \right) > 0, \\ 0, & \text{otherwise} \end{cases} \quad (1.23)$$

Equation (1.23) is a maximum-likelihood (ML) estimator if bit error occurs independently. However, the error probabilities $\{P_j\}$ are not known *a priori*. In the F-QIM decoder, they are estimated by the following:

$$P_j \approx 2Q\left(\frac{\Delta f_j / 2}{J_{ff}^{-1/2}} \right) \quad (1.24)$$

where $Q(x) = (1/\sqrt{2\pi}) \int_x^{\infty} e^{-x^2/2} dx$, Δf_j is the QIM step size near f_j, and $J_{ff}^{-1/2}$ is the CRB of frequency estimation. Equation (1.24) assumes that frequency estimation is normally distributed, is not biased, and approaches the CRB. Note that the CRB depends on how the attack is modeled. Assume that the attack is additive Gaussian noise, then:

$$J_{ff} = \left(\frac{\partial \mathbf{S}}{\partial f_j}\right)^{\dagger} \mathbf{\Sigma}^{-1} \left(\frac{\partial \mathbf{S}}{\partial f_j}\right) \quad (1.25)$$

where **S** represents the DFT of the signal $s_{\text{total}}[n]$ defined in equation (1.20), and $\mathbf{\Sigma}$ is power spectral density of the attack. In the experiments to be reported next, the noise spectrum $\mathbf{\Sigma}$, unknown to the decoder *a priori*, is taken as the maximum of the masking curve in equation (1.7) and the residual magnitude in equation (1.17).[7]

EVALUATION

In this section, the robustness of F-QIM against masked noise is evaluated at five different quantization step sizes. Based the result of robustness evaluation and informal listening, a quantization step size is chosen in for each sound clip in a test bench. F-QIM is further tested against common types of signal processing. Discussion of the results is given in the end.

Performance Against Colored Noise

Two types of noise are considered, namely additive colored Gaussian noise (ACGN) and coding noise (CN), respectively. For each type of noise, the ML decoding scheme introduced in the previous section is compared with a reference scheme called "intensity weighting" (IW), defined as the following (Liu & Smith, 2004c):

$$b = \begin{cases} 1, & \text{if} \sum_{j=1}^{J} A_j^2 \left(\hat{b}_j - \frac{1}{2}\right) > 0 \\ 0, & \text{otherwise,} \end{cases} \quad (1.26)$$

where A_j is the amplitude of peak j. To test the robustness of F-QIM against both types of noise, watermarks are embedded in selected soundtracks from a Web posting of the European Broadcast Union's *sound quality assessment materials* (EBU SQAM).[8] All the soundtracks are 10 seconds long, and their data hiding payload varies between 333 and 468 bits, depending on how many frames are not transient and contain prominent peaks.

The magnitude of ACGN is set at the masking threshold, and the CN is imposed by 9 to 12:1 compression[9] using *Ogg Vorbis*, a "completely open, patent-free, professional audio encoding and streaming technology."[10]

Shown in Figure 7, the solid lines present ML weighting in watermark decoding, and the dotted lines present watermark decoding by IW. The horizontal axis shows the step size Δf in cents (at f> 500 Hz), and the vertical axis shows the decoding success rate, defined as 100% minus BER. In almost all cases, ML weighting performs better than IW. Therefore, IW is not considered further. Also, there is a clear trend that the watermark decoding accuracy increases as a function of the quantization step size.

Choosing the F-QIM Step Size

Given the performance shown in Figure 7, it becomes crucial to find the frequency QIM step size that has an acceptable BER and does not introduce objectionable artifacts. Does this step size have to be smaller than the JND at every frequency? Informal listening tests by the author and friends suggested that human tolerance to frequency QIM depends on the timbral complexity of the cover signal. Among the selected soundtracks in EBU SQAM, 10 cents of QIM is quite noticeable in the trumpet soundtrack, but not so much in vocal recordings (bass and quartet). It remains a future work to choose adaptively the F-QIM step size according to the timbral complexity.

Hereafter, the F-QIM step size is empirically chosen at {5, 10, 15, 15} cents for {trumpet, cello, bass, quartet (S.A.T.B.)} soundtracks, respectively. This gives BER of {12%, 5%, 9%, 7%} against ACGN, and BER of {15%, 6%, 11%, 9%} against Ogg Vorbis compression.

Figure 7. Decoding success rate as a function of quantization step size Δf

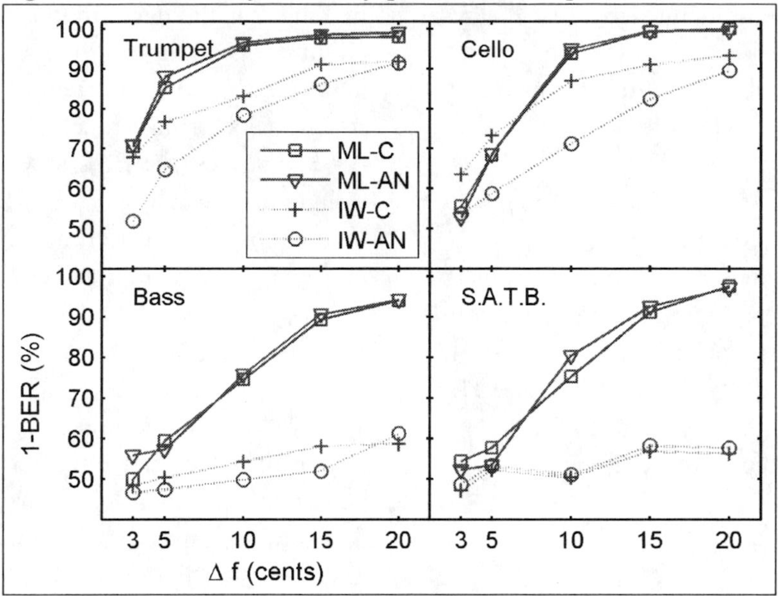

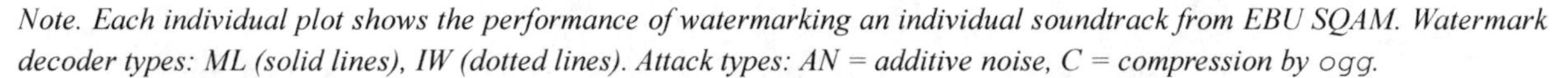
Note. Each individual plot shows the performance of watermarking an individual soundtrack from EBU SQAM. Watermark decoder types: ML (solid lines), IW (dotted lines). Attack types: AN = additive noise, C = compression by ogg.

Figure 8. Decoding performance against LPF and RVB

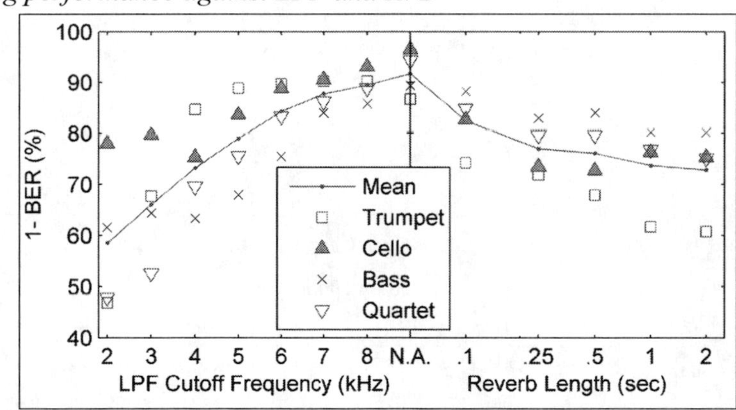

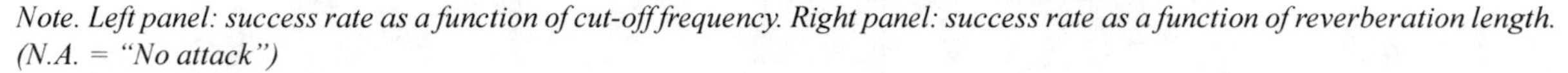
Note. Left panel: success rate as a function of cut-off frequency. Right panel: success rate as a function of reverberation length. (N.A. = "No attack")

Resilience to Common Operations

For each type of the following attacks, the performance of the F-QIM system is evaluated in terms of BER at various attack levels.

Low-Pass Filtering (LPF)

Low-pass FIR filters are obtained by length-512 windowing of the ideal low-pass response. The cut-off frequency f_c is controlled between 2 to 8 kHz. As shown on the left panel of Figure 8, not surprisingly, the performance of F-QIM gets better when the bandwidth increases. 1 – BER stays above 70% at $f_c > 5$ kHz. Degradation below 5 kHz indicates that F-QIM requires to operate at a sampling rate of 10 kHz or higher.

Reverberation (RVB)

The RVB attack is conducted by using the reverb toolkit from the music editing software and running on Linux. The length of reverberation is controlled between 100 msec and 2.0 sec. The right panel in Figure 8 shows that 1 - BER stays above 70% when the reverberation is shorter than 500 ms.

Gain Fading (GF)

The GF attack simulates the situation when the volume of a watermarked signal is turned up and down. GF attack is implemented by amplitude modulation at a rate $f_M = 2$, 5, or 10 Hz:

$$y[n] = x[n] \{ 1 + \cos(2\pi f_M nT) \} / 2 \qquad (1.27)$$

where α is the magnitude of gain-fading. Listed in Table 2 is the performance of F-QIM against GF in terms of bit decoding success percentage; items (a), (b), (c), and (d) correspond to the four soundtracks {trumpet, cello, bass, and quartet}. As α increases, the BER stays below 20% until a slight drop when α reaches 1.0, the full-range modulation. The difference is not significant between $f_M = 2$, 5, and 10 Hz.

Playback Speed Wowing (WOW)

The WOW attack emulates an analog playback device that does not have a stable speed. The watermarked signals are partitioned into segments of 167 msec; the segments are stretched or compressed in time by digital resampling and then concatenated at the original sampling rate f_s. Define the speed change as a dimensionless quantity:

$$\Delta v = \frac{f_s}{f_{s,\text{new}}} - 1. \qquad (1.28)$$

Table 2. Decoding success rate of F-QIM against gain fading

		Gain Fading Factor α						
	f_M (Hz)	**0**	**0.1**	**0.2**	**0.3**	**0.5**	**0.8**	**1**
(a)	**2**	86.67	84.18	84.75	84.14	83.81	84.33	79.24
	5	--	83.94	85.07	85.71	85.59	84.79	81.32
	10	--	83.33	84.70	83.90	84.83	83.90	73.01
(b)	**2**	96.44	92.87	92.64	92.64	92.64	91.69	87.50
	5	--	92.87	92.87	92.64	92.16	91.69	87.62
	10	--	93.11	92.40	93.35	93.35	92.14	85.95
(c)	**2**	89.32	85.41	85.05	85.05	83.63	80.78	79.34
	5	--	85.77	86.12	86.12	84.34	82.50	82.37
	10	--	86.83	86.12	85.77	85.77	80.78	77.94
(d)	**2**	94.30	91.69	91.92	91.92	91.45	89.31	87.10
	5	--	92.64	92.87	92.64	91.21	90.26	85.00
	10	--	91.45	90.26	91.21	91.90	90.95	85.00

Table 3. Performance of F-QIM against WOW attack

	Maximum Speed Change r (%)						
	0	**0.1**	**0.2**	**0.3**	**0.5**	**0.75**	**1**
(a)	86.67	85.83	79.17	60.56	--	--	--
(b)	96.44	95.25	93.11	92.16	75.30	60.57	--
(c)	89.32	88.61	90.39	90.39	87.19	76.16	62.28
(d)	94.30	92.87	93.35	92.87	88.60	74.35	64.13

Figure 9. Performance against pitch scaling

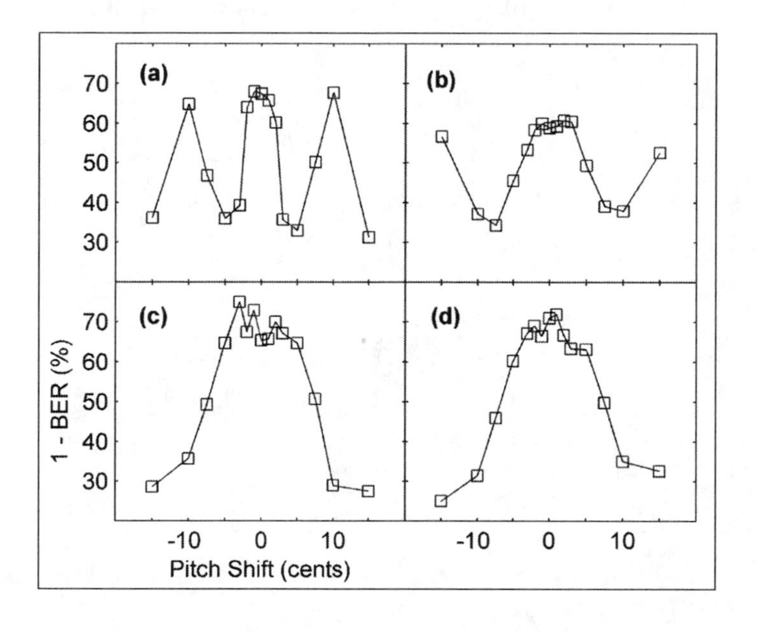

For every 12 consecutive segments in 2.0 seconds, Δv is set at {1, .67, .33, -.33, -.67, -1, -1, -.67, -.33, .33, .67, 1} times a fixed unit r, respectively. Table 3 shows the performance of the F-QIM system against WOW as a function of r. Nominally, degradation happens at the point where the speed change causes a frequency shift by half of the QIM step size. As r increases, the decoding success rate in soundtrack (a) falls below 75% first, near $r = 0.25\%$, (b) near $r = 0.5\%$, and (c) and (d) near $r = 0.75\%$.

Pitch Scaling (PSC)

The PSC attack is implemented by sinusoidal reanalysis of the watermarked signals followed by sinusoidal resynthesis with a constant frequency shift Δp in logarithmic scale. Performance of F-QIM at various level of Δp is shown in Figure 9. Recall that the F-QIM step size was set at {5, 10, 15, 15} cents for soundtracks (a), (b), (c), and (d), respectively. It is clear that the BER is close to 50% when Δp equals to half of the step size, and is significantly *above* 50% when Δp equals to the step size. The current implementation of F-QIM is sensitive to pitch scaling for it relies on the ability to align quantization grid points at the encoder and the decoder.

As a subtle observation, the sinusoidal re-modeling degrades the watermark performance significantly even without pitch scaling; in Figure 9, BER reaches 30% to 40% when $\Delta p = 0$. This is because sinusoidal modeling tends to smooth the

Table 4. Critical bandwidths as a function of frequency (Source: Zwicker, 1990; Bosi & Goldberg, 2003)

Band No.	Frequency Range (Hz)	CBW (Hz)	Band No.	Frequency Range (Hz)	CBW (Hz)
0	0-100	100	13	2000-2320	320
1	100-200	100	14	2320-2700	380
2	200-300	100	15	2700-3150	450
3	300-400	100	16	3150-3700	550
4	400-510	110	17	3700-4400	700
5	510-630	120	18	4400-5300	900
6	630-770	140	19	5300-6400	1100
7	770-920	150	20	6400-7700	1300
8	920-1080	160	21	7700-9500	1800
9	1080-1270	190	22	9500-12000	2500
10	1270-1480	210	23	12000-15500	3500
11	1480-1720	240	24	15500-*	
12	1720-2000	280	* Upper limit at approximately 20kHz		

Table 5. Mathematical notations

$s[n]$	Scalar-valued deterministic time-domain signal.
$s[k:l]$ or $s_{k:l}$	Column vector $(s[k], s[k+1], \ldots, s[l])^t$.
$\lvert\theta\rangle = \lvert\theta_1, \theta_2, \ldots, \theta_K\rangle$	**(Dirac's notation)** Vector enumeration of parameters; a ket is a column vector by convention and its transpose is denoted as $\lvert\theta\rangle$.
U_n, X_n, Y_n	Scalar random processes with time-index n.
$U_{k:l}$	Vector random variable $(U_k, U_{k+1}, \ldots, U_l)^t$.
u_n, x_n, y_n	Specific time-sequences generated by corresponding uppercased random processes. In other words, "stochastic signals".
$u_{k:l}$	Specific time-sequence generated by $U_{k:l}$.
s, **x**, **y**	**(Bold lowercase)** Finite-dimensional vectors in the signal space; can be deterministic or stochastic.
$\mathbf{s}_{\lvert\theta\rangle}$	Deterministic signal with a parametric representation $\lvert\theta\rangle$.
$S(\cdot), X(\cdot), \Phi(\cdot)$	**(Uppercase)** Functions in the z- or the Fourier transform domain.
$\hat{\theta}, \hat{S}$	**(Hat)** Estimated parameters or spectral lines.
$\tilde{\theta}$	**(Tilde)** Estimation error.
J, **Σ**, **Q**	**(Bold uppercase)** Matrices, could be in the signal space or the parameter space.

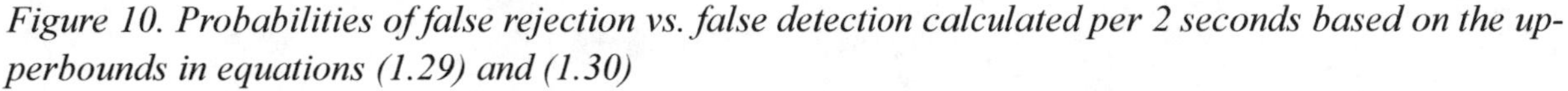

Figure 10. Probabilities of false rejection vs. false detection calculated per 2 seconds based on the upperbounds in equations (1.29) and (1.30)

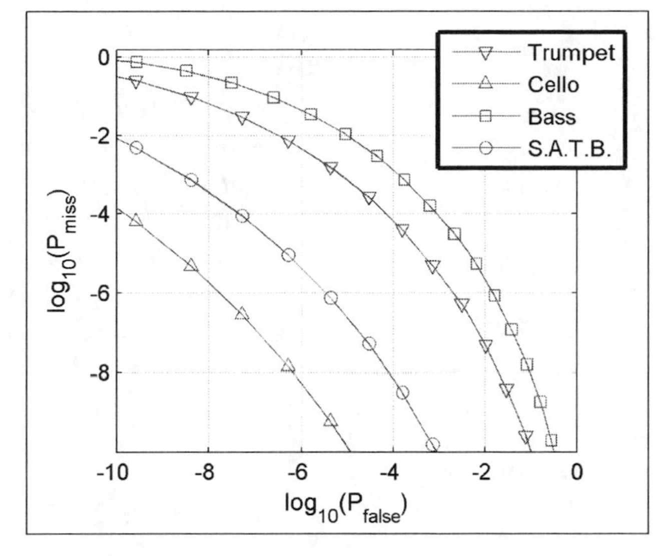

frequency envelopes of the watermark-embedded trajectories. Designing a trajectory quantization scheme that endures smoothing could be a future investigation.

SUMMARY AND DISCUSSION

The current implementation of F-QIM demonstrates BER < 15% against 12:1 compression, comparable to its performance without attack (see Figure 8 or Table 2 and 3); BER < 25% against LPF at cutoff frequency of 6 kHz, BER < 30% against 250 msec of reverberation, and BER < 20% at 5 Hz of full-range amplitude modulation. BER stays below 25% against WOW only if the change of speed does not cause misalignment in frequency or desynchronization in time. Similarly, F-QIM is sensitive to pitch scaling.

To improve from the current implementation of F-QIM, first, derived features from the trajectories can be chosen as watermark embedding parameters so as to provide immunity to pitch scaling. Second, by estimation and removal of WOW (Wang, 1995), F-QIM could possibly be made immune to playback speed variation.

Sensitivity of F-QIM to RVB and PSC indicates that the current implementation is not robust against attackers that have the knowledge of trajectory smoothing. Higher-dimensional quantization lattices, such as the spread-transform scalar Costa scheme (ST-SCS) in Eggers, Bäuml, Tzschoppe, and Girod (2003) and vector QIM codes in Moulin and Koetter (2005), are worth investigation. Note that these schemes to boost up robustness will inevitably lower the data hiding payload, which is generally true. This section shall conclude with a simple scheme of introducing redundancy to lower BER.

Lowering the Probability of Error by Introducing Redundancy

Assuming low-rate watermark encoding, let N frames in a few seconds jointly encode one bit of information b. To this end, a threshold value $0.5 < \eta < (1\text{-BER})$ is chosen so that a signal is authenticated if and only if more than ηN bits are decoded to match b.

Denote the probability of false rejection and false detection as $\{P_{miss}; P_{false}\}$. The following upper bounds can be derived:

$$P_{\text{false}} \le \left(\frac{1}{1-\gamma}\right) 2^{-ND(\eta\|0.5)} \tag{1.29}$$

and

$$P_{\text{miss}} \le \left(\frac{1}{1-\alpha}\right) 2^{-ND(\eta\|q)} \tag{1.30}$$

where $q = 1 - \text{BER}$, $\gamma = \frac{1-\eta}{\eta} < 1$, and $\alpha = \frac{1-q}{1-\eta}\left(\frac{\eta}{q}\right) < 1$, and $D(\cdot\|\cdot)$. The Kullback-Leibler divergence between two Bernoulli processes (Cover & Thomas, 1991), is defined by the following equation:

$$D(p\|q) = p\log_2\left(\frac{p}{q}\right) + (1-p)\log_2\left(\frac{1-p}{1-q}\right). \tag{1.31}$$

It can be shown that $D(p\|q)$ is non-negative, and derivation of (1.29) and (1.30) can be found in Liu (2005).

Figure 10 shows *the receiver operating characteristic* (ROC) curves of F-QIM when encoding one bit every 2 seconds. The q-value of each soundtrack is taken as shown in Figure 7 with ML decoding against ACGN. For all the soundtracks, P_{miss} and P_{false} can be kept below 10^{-3} simultane-

Figure 11. Transcriptions of the sound sources

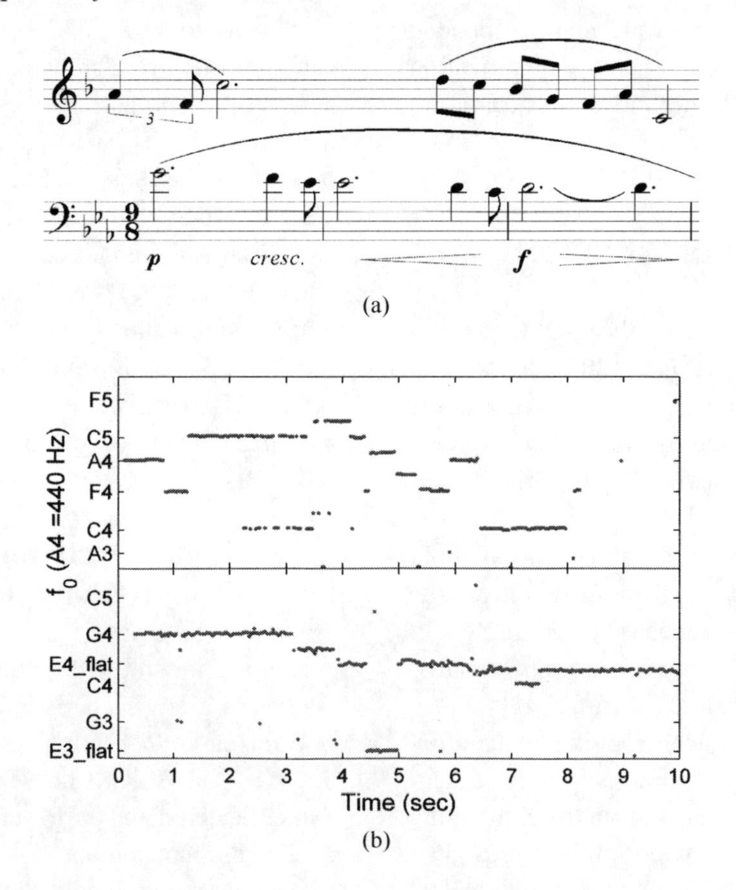

Note: (a) Manual transcription. Top trace: trumpet; bottom trace: cello. (b) Automatic transcription after watermark-assisted segregation

ously, a significant reduction from their per-frame encoding performance {12%, 5%, 9%, 7%}.

FUTURE RESEARCH TRENDS

The aim of this section is not to suggest incremental improvement of existing methods, but rather, to present unexplored directions that are unique to audio watermarking based on parametric models.

Watermark Assisted Sound Source Segregation

Computer segregation of multiple sound sources from a single-channel mixture has been a challenging problem. Its success largely relies on the ability to estimate multiple fundamental frequencies (e.g., Klapuri, 2003). Instead of solving the problem in traditional ways, what is proposed here might be regarded as cheating; using watermarks to assist in sound source segregation. Here, sound sources are assumed to be available for watermark embedding before they are mixed.

The F-QIM and other schemes based on sinusoidal models (e.g., Dong et al., 2004; Girin & Marchand, 2004) are highly *granular* in the sense that every peak in every short-time spectrum encodes one bit of information. The granularity makes them attractive for applications when watermarks are expected to encode side information about the cover signal itself. For sound source segregation, the side information would be the source labels.

In an experiment, two sound sources {A, B} are separately analyzed and, before they are mixed, watermarked using F-QIM codebooks C_0 and C_1, respectively. Thus, all the peaks in A represent binary 0 and peaks in B represent 1. To recognize A and B from the linear mixture M = A + B, M is analyzed, and its trajectories are classified according to a vote; if a frequency envelope more often aligns to C_0 than C_1, the corresponding trajectory is recognized as belonging to source A, and vice versa. Thus, sound source segregation is assisted by watermark decoding.

Figure 11 shows the results of demixing the trumpet and the cello soundtracks from EBU SQAM. The QIM step size is set at 10 cents. The melodies transcribed manually are shown for comparison with computerized monophonic pitch estimation after source segregation. Except occasional octave confusion, the computer-estimated pitch contours match well to the score even though the two sources have an overlapped pitch range (C4-C5).

Cocktail Audio Watermarking

Since existing methods in audio watermarking operate at widely different rates and each has its unique strength (see Table I), it would be advantageous to utilize all of them jointly. In principle, Mintzer and Braudaway (1999) have argued that two or more watermarks, when embedded in the correct order, can simultaneously achieve data hiding at multiple rates. In practice, Lu and Liao (2001) have demonstrated a time-sharing scheme that allows two watermarks to be embedded in an image for different purposes. Audio watermarking in this *cocktail* manner has also been reported (e.g., Lu, Liao, & Chen, 2000; van der Veen, Bruekers, Haitsma, Kalker, Lemma, & Oomen, 2001).

As a suggestion here, the F-QIM scheme can embed a watermark at low rates, the spread-spectrum (SS) approach can provide synchronization (e.g., Kirovski & Attias, 2003), and a quantization scheme can insert a fragile watermark on the top (e.g., Chou et al., 2001; van der Veen, Bruekers, van Leest, & Cavin, 2003). The F-QIM watermark should be embedded first, followed by the SS, then the fragile, based on the following reasons. First, the F-QIM watermark sees the subsequent watermarks as masked noise, whose presence should not degrade the F-QIM decoding any more than shown in Figure 7. Second, SS is robust to quantization, but the fragile watermark may not be robust to noise. Liu and Smith (2004b) showed that power sharing between SS and QIM watermarking provides better data hiding capacity than time sharing[11] only if SS is embedded first. Based on these reasons, the fragile watermark should be embedded last.

Counter-Measure to Collusion

For certain applications, including transaction tracking, it is desired to embed orthogonal watermarks in different copies of the same signal so as to bear identification numbers. However, orthogonal watermarks average to zero asymptotically at the rate of $O(N^{-1/2})$, where N is the number of different watermarked copies available to collusive attackers. This becomes a problem for watermarking schemes that operate on linear transform domains, because attackers can simply sum and divide to estimate the original signal (Kilian, Leighton, Matheson, Shamoon, Tarjan, & Zane, 1998).

If a watermarking scheme operates on nonlinear parametric models, instead, embedding a cover signal with orthogonal binary sequences produces nearly orthogonal watermarked signals. Therefore, when they are added up linearly, the mean signal does not give an estimate of the original; it contains audible artifacts, such as beating, which render the mean signal invalid. The attackers are forced to conduct signal decomposition, too. Thus their computation load becomes heavier than summing and averaging.

REFERENCES

Abe, M., & Smith, J.O. (2004). *Design criteria for simple sinusoidal parameter estimation based on quadratic interpolation of FFT magnitude peaks.* Paper presented at the AES 117th Convention, San Francisco.

Boney, L., Tewfik, A. H., & Hamdy, K.N. (1996). Digital watermarks for audio signals. In *Proceedings of the IEEE International Conference on Multimedia Computing and Systems* (pp. 473-480).

Bosi, M. (1997). Perceptual audio coding. *IEEE Signal Processing Magazine, 14*(5), 43-49.

Bosi, M., & Goldberg, R.E. (2003). *Introduction to digital audio coding and standards.* Boston: Kluwer Academic Publishers.

Bender, W., Gruhl, D., Morimoto, N., & Lu, A. (1996). Techniques for data hiding. *IBM Systems Journal, 35,* 313-336.

Chen, B., & Wornell, G. (2001). Quantization index modulation: A class of provably good methods for digital watermarking and information embedding. *IEEE Trans. Information Theory, 47,* 1423-1443.

Chou, J., Ramchandran, K., & Ortega, A. (2001). Next generation techniques for robust and imperceptible audio data hiding. In *Proceedings of the IEEE International Conference on Acoustics, Speech, and Signal Processing* (pp. 1349-1352), Salt Lake City, Nevada.

Cook, P., & Scavone, G. P. (1999). The synthesis toolKit (STK). In *Proceedings of the International Computer Music Conference*, Beijing, China. Retrieved March 19, 2007, from http://ccrma.stanford.edu/software/stk/papers/stkicmc99.pdf

Costa, M. (1983). Writing on dirty paper. *IEEE Trans. Information Theory, IT, 29,* 439-441.

Cover, T., & Thomas, J. A. (1991). *Elements of information theory.* New York: Wiley-Interscience.

Cox, I. J., Kilian, J., Leighton, F., & Shamoon, T. (1999). Secure spread spectrum watermarking for multimedia. *IEEE Trans. Image Processing, 6*, 1673-1687.

Cox, I. J., Miller, M. L., & Bloom, J. A. (2002). *Digital watermarking*. San Francisco: Morgan Kaufmann.

Dong, X., Bocko, M. F., & Ignjatovic, Z. (2004). Data hiding via phase manipulation of audio signals. In *Proceedings of the IEEE International Conference on Acoustics, Speech, and Signal Processing* (pp. 377-380). Montréal, Canada.

Eggers, J., Bäuml, R., Tzschoppe, R., & Girod, B. (2003). Scalar Costa scheme for information embedding. *IEEE Trans. Signal Processing, 51*, 1003-1019.

Girin, L., & Marchand, S. (2004). Watermarking of speech signals using the sinusoidal model and frequency modulation of the partials. In *Proceedings of the IEEE International Conference on Acoustics, Speech, and Signal Processing* (pp. 1633-636). Montréal, Canada.

Gurijala, A., Deller, D. D. R., Seadle, M. S., & Hansen, J. H. L. (2002). Speech watermarking through parametric modeling. In *Proceedings of the International Conference on Spoken Language Processing* (CD-ROM), Denver, Colorado.

Harris, F. J. (1978). On the use of windows for harmonic analysis with the discrete Fourier transform. *Proceedings IEEE, 66*, 51-83.

Hartung, F., Su, J. K., & Girod, B. (1999). Spread spectrum watermarking: Malicious attacks and counterattacks. In *Proceedings of the SPIE Security and Watermarking of Multimedia Contents* (pp. 147-158).

He, X., Iliev, A. I., & Scordilis, M. S. (2004). A high capacity watermarking technique for stereo audio. In *Proceedings of the IEEE International Conference on Acoustics, Speech, and Signal Processing* (pp. 393-396). Montréal, Canada.

Jayant, N., Johnston, J., & Safranek, R. (1993). Signal compression based on method of human perception. *Proceedings IEEE, 81*, 1385-1422.

Kilian, J., Leighton, F. T., Matheson, L. R., Shamoon, T. G., Tarjan, R. E., & Zane, F. (1998). Resistance of digital watermarks to collusive attacks. In *Proceedings of the 1998 International Symposium on Information Theory* (p. 271). Cambridge, UK.

Kirovski, D., & Attias, H. (2002). Audio watermark robustness to desynchronization via beat detection. *Information hiding 2002* (LNCS 2578, pp. 160-176). Berlin: Springer-Verlag.

Kirovski, D., & Malvar, H. S. (2003). Spread-spectrum watermarking of audio signals. *IEEE Trans. Signal Processing, 51*, 1020-1033.

Klapuri, A. P. (2003). Multiple fundamental frequency estimation based on harmonicity and spectral smoothness. *IEEE Trans. Speech and Audio Processing, 11*, 804-816.

Kuo, S. S., Johnston, J. D., Turin, W., & Quackenbush, S. R. (2002). Covert audio watermarking using perceptually tuned signal independent multiband phase modulation. In *Proceedings of the IEEE International Conference on Acoustics, Speech, and Signal Processing* (pp. 1753-1756). Orlando, Florida.

Lee, K., Kim, D.S., Kim, T., & Moon, K.A. (2003). EM estimation of scale factor for quantization-based audio watermarking. In *Proceedings of the 2nd International Workshop on Digital Watermarking* (pp. 316-327). Seoul, Korea.

Levine, S.N. (1998). *Audio representations for data compression and compressed domain processing*. Unpublished doctoral dissertation, Stanford University.

Li, X., & Yu, H.H. (2000). Transparent and robust audio data hiding in cepstrum domain. In *Proceedings of the IEEE International Conference on Multimedia* (pp. 397-400). New York.

Liu, Y.W., & Smith, J.O. (2003). Watermarking parametric representations for synthetic audio. In

Proceedings of the IEEE International Conference on Acoustics, Speech, and Signal Processing (pp. V660-663). Hong Kong, China.

Liu, Y.W., & Smith, J.O. (2004a). Watermarking sinusoidal audio representations by quantization index modulation in multiple frequencies. In *Proceedings of the IEEE International Conference on Acoustics, Speech, and Signal Processing* (pp. 373-376). Montréal, Canada.

Liu, Y.W., & Smith, J.O. (2004b). Multiple watermarking: Is power-sharing better than time-sharing? In *Proceedings of the IEEE International Conference Multimedia, Expo* (pp. 1939-1942). Taipei, Taiwan.

Liu, Y.W., & Smith, J.O. (2004c). Audio watermarking based on sinusoidal analysis and synthesis. In *Proceedings of the International Symposium on Musical Acoustics* (CD-ROM), Nara, Japan.

Liu, Y.W. (2005). *Audio watermarking through parametric signal representations.* Unpublished doctoral dissertation, Stanford University.

Lu, C.S., Liao, H.Y.M., & Chen, L.H. (2000). Multipurpose audio watermarking. In *Proceedings of the IEEE 15th International Conference on Pattern Recognition* (pp. 282-285). Barcelona, Spain.

Lu, C.S., & Liao, H.Y.M. (2001). Multipurpose watermarking for image authentication and protection. *IEEE Trans. Image Processing, 10*, 1579-1592.

Mansour, M.F., & Tewfik, A.H. (2003). Time-scale invariant audio data embedding. *EURASIP Journal on Applied Signal Processing, 10*, 993-1000.

Markel, J.D., & Gray, A.H. (1976). *Linear prediction of speech.* New York: Springer-Verlag.

McAulay, R.J., & Quatieri, T.F. (1986). Speech analysis/synthesis based on a sinusoidal representation. *IEEE Trans. Acoustics, Speech, Signal Processing, 34*, 744-754.

Miller, M.L., Doërr, G.J., & Cox, I.J. (2002). Dirty-paper trellis codes for watermarking. In *Proceedings of the 2002 IEEE International Conference on Image Processing* (pp. 129-132).

Mintzer, F., & Braudaway, G.W. (1999). If one watermark is good, are more better? In *Proceedings of the IEEE International Conference on Acoustics, Speech, and Signal Processing* (pp. 2067-2069), Phoenix, Arizona.

Moulin, P., & Koetter, R. (2005). Data-hiding codes. *Proceedings IEEE, 93*, 2083-2126.

Painter, T., & Spanias, A. (2000). Perceptual coding of digital audio. *Proceedings IEEE, 88*, 451-513.

Papoulis, A. (1977). *Signal analysis.* New York: McGraw-Hill.

Petrovic, R. (2001). Audio signal watermarking based on replica modulation. In *Proceedings of the IEEE TELSIKS* (pp. 227-234). Niš, Yugoslavia.

Purnhagen, H., & Meine, N. (2000) HILN: The MPEG-4 parametric audio coding tools. In *Proceedings of the IEEE International Symposium on Circuits and Systems* (pp. 201-204), Geneva, Switzerland.

Scavone, G.P., & Cook, P. (2005) RtMIDI, RtAudio, and synthesis tookKit (STK) update. In *Proceedings of the International Computer Music Conference,* Barcelona, Spain. Retrieved March 19, 2007, from http://ccrma.stanford.edu/software/stk/papers/stkupdate.pdf

Schroeder, M.R., & Atal, B.S. (1985). Code-excited linear prediction (CELP): High-quality speech at very low bit rates. In *Proceedings of the IEEE International Conference on Acoustics, Speech, and Signal Processing* (pp. 937-940). Tampa, Florida.

Serra, X., & Smith, J. (1990). *Spectral modeling synthesis: A sound analysis/synthesis based on a deterministic plus stochastic decomposition. Computer Music Journal, 14*(4), 12-24.

Shin, S., Kim, O., Kim, J., & Choil, J. (2002). A robust audio watermarking algorithm using pitch scaling. In *Proceedings of the IEEE 10th DSP Workshop* (pp. 701-704). Pine Mountain, Georgia.

Shower, E.G., & Biddulph, R. (1931). Differential pitch sensitivity of the ear. *Journal of the Acoustical Society of America, 3*, 275-287.

Silvestre, G.C.M., Hurley, N.J., Hanau, G.S., & Dowling, W.J. (2001). Informed audio watermarking scheme using digital chaotic signals. In *Proceedings of the IEEE International Conference on Acoustics, Speech, and Signal Processing* (pp. 1361-1364). Salt Lake City, Nevada.

Smith, J., & Serra, X. (1987). PARSHL: An analysis/synthesis program for non-harmonic sounds based on a sinusoidal representation. In *Proceedings of the 1987 International Computer Music Conference,* Urbana-Champaign, Illinois.

Swanson, M.D., Zhu, B., & Tewfik, A.H. (1998). Robust audio watermarking using perceptual masking. *Signal Processing, 66*, 337-355.

Terhardt, E. (1979). Calculating virtual pitch. *Hearing Research, 1*, 155-182.

Thornburg, H. (2005) *On the detection and modeling of transient audio signals with prior information.* Unpublished doctoral dissertation, Stanford University.

Van der Veen, M., Bruekers, F., Haitsma, J., Kalker, T., Lemma, A. N., & Oomen, W. (2001). *Robust, multi-functional and high-quality audio watermarking technology.* Paper presented at the AES 110th Convention, Amsterdam, Netherlands.

Van der Veen, M., Bruekers, F., van Leest, A., & Cavin, S. (2003). High capacity reversible watermarking for audio. In *Proceedings of the SPIE Security and Watermarking of Multimedia Contents* (pp. 1-11). San Jose, California.

Vercoe, B.L., Gardner, W.G., & Scheirer, E.D. (1998). Structured audio: Creation, transmission, and rendering of parametric sound representations. *Proceedings IEEE, 86*, 922-940.

Wang, A. (1995). Instantaneous and frequency-warped techniues for source separation and signal parameterization. In *Proceedings of the IEEE Workshop on Applications of Signal Processing to Audio and Acoustics,* New Paltz, New York.

Wessel, D., & Wright, M., (Eds.). (2004). In *Proceedings of the Open Sound Control Conference*. Berkeley, CA: Center For New Music and Audio Technology (CNMAT). Retrieved March 23, 2007, from http://www.opensoundcontrol.org/proceedings

Wier, C.C., Jesteadt, W., & Green, D.M. (1977). Frequency discrimination as a function of frequency and sensation level. *Journal of the Acoustical Society of America, 61*, 178-184.

Wu, M., Craver, S., Felten, E.W., & Liu, B. (2001). Analysis of attacks on SDMI audio watermarks. In *Proceedings of the IEEE International Conference on Acoustics, Speech, and Signal Processing* (pp. 1369-1372). Salt Lake City, Nevada.

Wu, C.P., Su, P.C., & Kuo, C.C.J. (2000). Robust and efficient digital audio watermarking using audio content analysis. In *Proceedings of the SPIE Security and Watermarking of Multimedia Contents II* (pp. 382-392). San Jose, California.

Zwicker, E. (1956). Die elementaren Grundlagen zur Bestimmung der Informationskapazität des Gehörs [The elementary bases for the determination of the information capacity of the hearing]. *Acustica, 6*, 365-381.

Zwicker, E., & Fastl, H. (1990). *Psychoacoustics, facts and models.* Berlin: Springer Verlag.

ADDITIONAL READINGS

On Parametric Audio Coding and Synthesis:

Chowning, J.M. (1980). Computer synthesis of the singing voice. *Sound generation in winds, strings, computers.* Stockholm: Royal Swedish Academy of Music.

Rodet, X., & Depalle, P. (1992). *Spectral envelopes and inverse FFT synthesis*. Paper presented at the AES 93rd Convention, San Francisco.

Chaigne, A. (1992). On the use of finite differences for the synthesis of musical transients: Application to plucked stringed instruments. *Journal d'Acoustique, 5*, 181-211.

Lu, H.L., & Smith, J.O. (1999). Joint estimation of vocal tract filter and glottal source waveform via convex optimization. In *Proceedings of the IEEE Workshop on Applied Signal Processing to Audio and Acoustics,* New Paltz, New York.

Cook, P. (2002). Modeling Bill's gait: Analysis and parametric synthesis of walking sounds. In *Proceedings of the AES 22nd International Conference on Virtual, Synthetic, and Entertainment Audio* (pp. 73-78). Espoo, Finland.

Bilbao, S. (2004). *Wave and scattering methods for the numerical integration of partial differential equations.* New York: Wiley.

Smith, J.O. (2006). *Physical audio signal processing: For virtual musical instruments and digital audio effects.* Retrieved March 19, 2007, from http://ccrma.stanford.edu/~jos/pasp/

Bilbao, S. (2007). Robust physical modeling sound synthesis for nonlinear systems. *IEEE Signal Processing Magazine, 24*(2), 32-41.

Rabenstein, R., Petrausch, S., Sarti, A., de Sanctis, G., Erkut, C., & Karjalainen, M. (2007). Block-based physical modeling for digital sound synthesis. *IEEE Signal Processing Magazine, 24*(2), 42-54.

Beauchamp, J. W. (Ed.). (2007). *Analysis, synthesis, and perception of musical sounds: Sound of music*. Berlin: Springer.

On Psychoacoustics and Auditory Scene Analysis:

Moore, B.C.J., & Glasberg, B.R. (1983). Suggested formulae for calculating auditory-filter bandwidths and excitation patterns. *J. Acoustical Soc. America, 74*, 750-753.

Bregman, A.S. (1994). *Auditory scene analysis.* Boston: MIT Press.

Moore, B.C.J. (1996). Masking in the human auditory system. *Collected papers on digital audio bit-rate reduction.* New York: Audio Engineering Society.

Wang, D., & Brown, G. J. (2006) *Computational auditory scene analysis: Principles, algorithms, and applications.* NJ: Wiley-IEEE Press.

On Fisher Information and Parameter Estimation theory:

Rife, D.C., & Boorstyn, R.R. (1974). Single tone parameter estimation from discrete-time observations. *IEEE Trans. Information Theory, IT20,* 591-598.

Rife, D.C., & Boorstyn, R.R. (1976). Multiple tone parameter estimation from discrete-time observations. *Bell System Tech. J., 55*, 1389-1410.

Kay, S. (1988). *Modern spectral estimation: Theory and application.* NJ: Prentice Hall.

Porat, B. (1994). *Digital processing of random signals: Theory and methods.* NJ: Prentice Hall.

Golden, S., & Friedlander, B. (1999). Maximum likelihood estimation, analysis, and applications of exponential polynomial signals. *IEEE Trans. Signal Processing, 47*, 1493-1501.

Amari, S., & Nagaoka, H. (2000). *Methods of information geometry.* Providence, RI: American Mathematical Society.

Van Trees, H.L. (2001). *Detection, estimation, and modulation theory: Part I.* New York: Wiley-Interscience.

On Watermarking of Other Parameterized Objects:

Hartung, F., Eisert, P., & Girod, B. (1998). Digital watermarking of 3D head model animation parameters. In *Proceedings of the 10th IMDSP Workshop* (pp. 119-122), Alpach, Austria.

Cayre, F., & Macq, B. (2003). Data hiding on 3-D triangle meshes. *IEEE Trans. Information Theory, 49*, 939-949.

ENDNOTES

1. The residual is referred to as "noise" in previous work (e.g., Serra, 1989; Levine, 1998). In this chapter, "residual" and "noise" are used interchangeably.
2. Assuming a sampling rate of 44.1 kHz, N = 1024 or 1536 is chosen empirically.
3. Interested readers can refer to MPEG psychoacoustic model 1.
4. Currently not implemented in the F-QIM system.
5. Zwicker predicted that JND is approximately 1/27 Bark at every frequency.
6. That is, about 10 cents at 500 Hz or above.
7. The conver signal remains unknown to the decoder: the masking curve is computed and the signal decomposition is conducted purely based on the signal received at the decoder.
8. Can be retrieved from http://www.tnt.uni-hannover.de/project/mpeg/audio/sqam/, as of June 20, 2006.
9. The ratio assumes 16-bit PCM format sampled at 44.1 kHz before compression.
10. Free download available from http://www.vorbis.com/ as of June 20, 2006.
11. Performance of time sharing does not depend on the order of embedding. The two watermarks can be seen as being embedded together.

APPENDIX: FISHER INFORMATION AND CRAMÉR-RAO BOUNDS

Probabilistic Attack Characterization

Assume that an attack can be characterized by a probability distribution function:

$$Y_{-N:N} \sim f(\mathbf{y} \mid \mathbf{s}_{|\theta\rangle}) = f(\mathbf{y};\theta) \tag{1.32}$$

where $\mathbf{s}_{|\theta\rangle}$ is the watermarked signal and $\mathbf{y}$ is a corrupted copy of it. The Fisher information matrix $\mathbf{J}(\theta)$ for parameter estimation is defined as:

$$[\mathbf{J}(\theta)]_{ij} = \mathrm{E}\left[\frac{\partial}{\partial\theta_i}\ln f(\mathbf{y};\theta)\frac{\partial}{\partial\theta_j}\ln f(\mathbf{y};\theta)\right]. \tag{1.33}$$

The Cramér-Rao matrix inequality (see e.g., Cover & Thomas, 1991) states that, for any unbiased parameter estimator $\theta = T(\mathbf{y})$, the following holds true:

$$\mathrm{Var}[T(\mathbf{y}) - \theta] \geq \mathbf{J}^{-1}(\theta). \tag{1.34}$$

Denote the estimation error $\hat{\theta} - \theta$ as $|\tilde{\theta}\rangle$. Then, (1.34) means that $\mathrm{E}|\tilde{\theta}\rangle\langle\tilde{\theta}| - \mathbf{J}^{-1}$ is a non-negative definite matrix. In particular, for any individual parameter θ_i, the following is true:

$$\mathrm{E}\,\tilde{\theta}_i^2 \geq (\mathbf{J}^{-1})_{ii}. \tag{1.35}$$

The right hand side of this inequality is the Cramér-Rao bound (CRB) for the estimation of θ_i. Next, let us calculate $\mathbf{J}(\theta)$ if $f(\mathbf{y};\theta)$ is Gaussian.

Fisher Information when Attack is Gaussian

Suppose that the signal $\mathbf{s}_{|\theta\rangle}$ is subject to additive Gaussian noise $U_{-N:N} + jV_{-N:N}$, where $U_{-N:N}$ and $V_{-N:N}$ are independent; $U_{-N:N} \sim \mathcal{N}(0, \boldsymbol{\Sigma}_u)$ and $V_{-N:N} \sim \mathcal{N}(0, \boldsymbol{\Sigma}_v)$. The corrupted signal $Y_{-N:N}$ can be written as:

$$Y_{-N:N} = \mathbf{s}_{|\theta\rangle} + U_{-N:N} + jV_{-N:N}, \tag{1.36}$$

and

$$Y_{-N:N} \sim f(\mathbf{y};\theta) = C\exp\left[-\frac{1}{2}(\mathbf{u}^t\boldsymbol{\Sigma}_u^{-1}\mathbf{u} + \mathbf{v}^t\boldsymbol{\Sigma}_v^{-1}\mathbf{v})\right], \tag{1.37}$$

where $\mathbf{u} = \mathrm{Re}\{\mathbf{y} - \mathbf{s}_{|\theta\rangle}\}$, $\mathbf{v} = \mathrm{Im}\{\mathbf{y} - \mathbf{s}_{|\theta\rangle}\}$, and $C = \frac{1}{\sqrt{(2\pi)^{4N+2}|\boldsymbol{\Sigma}_u|\cdot|\boldsymbol{\Sigma}_v|}}$ ensures $\int f(\mathbf{y};\theta)d\mathbf{y} = 1$. After detailed algebra,

$$\mathbf{J}(\theta) = (\nabla\mathbf{s})^t \boldsymbol{\Sigma}^{-1} (\nabla\mathbf{s}), \tag{1.38}$$

where $\mathbf{s} = \left[\mathrm{Re}\{\mathbf{s}_{|\theta\rangle}\}, \mathrm{Im}\{\mathbf{s}_{|\theta\rangle}\}\right]^t$, $[\nabla\mathbf{s}]_{(4N+2)\times K} = \left[\frac{\partial\mathbf{s}}{\partial\theta_1} \; \frac{\partial\mathbf{s}}{\partial\theta_2} \; \cdots \; \frac{\partial\mathbf{s}}{\partial\theta_K}\right]$, and $\left[\frac{\partial\mathbf{s}}{\partial\theta_1} \; \frac{\partial\mathbf{s}}{\partial\theta_2} \; \cdots \; \frac{\partial\mathbf{s}}{\partial\theta_K}\right]$.

Geometric Interpretations

To visualize the Cramér-Rao inequality, a study of single parameter estimation under additive white Gaussian noise (AWGN) is presented next. We shall see Fisher information as the inverse of the variance of linearly approximation error. Then, this interpretation is generalized to cover multiple parameter estimation.

Single-Parameter Signal Modeling

Figure 12 shows a geometric interpretation for single-tone signal synthesis and analysis. Let us write the parametric signal model as:

$$\mathbf{s}_{|\omega\rangle}[n] = \sin(\omega n), \quad n = -N : N. \tag{1.39}$$

Then, the synthesis is a mapping from the unit circle to a closed curve $\Xi(\omega)$ in the signal space R^{2N+1}. If an attack is AWGN $U_n \sim \mathcal{N}(0,\sigma^2)$, the attack can be characterized by an $(2N+1)$ dimensional ball of radius σ around $\mathbf{s}_{|\omega\rangle}$ at each frequency ω. Thus, points on the closed curve $\Xi(\omega)$ are not infinitely distinguishable. More specifically, along a tangential direction $\partial\mathbf{s}/\partial\omega$ the frequency estimation "uncertainty" can be approximated linearly; define a projection vector $\mathbf{v} = \delta\omega \frac{\partial\mathbf{s}}{\partial\omega}$ such that $\mathbf{v}\bullet(\mathbf{u} - \mathbf{v}) = 0$. Then, solving the pseudo-inverse problem, we have:

$$E(\delta\omega)^2 = \frac{\sigma^2}{|\partial\mathbf{s}/\partial\omega|^2}. \tag{1.40}$$

Interestingly, by comparing equations (1.38) and (1.40), we obtain that:

$$J(\omega) = [E(\delta\omega)^2]^{-1}. \tag{1.41}$$

Equation (1.41) shows that the Fisher information is the inverse of the variance of linear approximation error.

Generalization to Multiple Parameter Estimation

Now, consider a K-dimensional signal model $\mathbf{s}_{|\theta\rangle}[n]$, $|\theta\rangle = |\theta_1, \theta_2, \ldots, \theta_K\rangle$. Let $\mathbf{u}$ in R^{2N+1} be an instance of additive Gaussian attack $U_{-N:N} \sim \mathcal{N}(0, \boldsymbol{\Sigma})$. Illustrated in Figure 13, the parallelogram depicts a tangential space to the signal manifold $\Xi(\Theta)$, and the shaded area represents the noise $\mathcal{N}(0, \Sigma)$. With this visualization in mind, the "uncertainty" of multiple parameter estimation can be derived. First, denote the projection of $\mathbf{u}$ on the tangential space as $\mathbf{v} = \sum_i \delta\theta_i \frac{\partial \mathbf{s}}{\partial \theta_i} z$. Each component $\delta\theta_i$ is interpreted as an estimation error $\tilde{\theta}$ due to the presence of $\mathbf{u}$. Then, denote $|\delta\theta\rangle = |\delta\theta_1, \delta\theta_2, \ldots \delta\theta_K\rangle$. Solve the pseudo inverse problem whose solution has a form $|\delta\theta\rangle = (\mathbf{A}^t\mathbf{A})^{-1}\mathbf{A}^t\mathbf{u}$ where $\mathbf{A} = [\nabla\mathbf{s}]_{(2N+1)\times K}$. Then, $E|\delta\theta\rangle\langle\delta\theta| = (\nabla\mathbf{s}^t\nabla\mathbf{s})^{-1}(\nabla\mathbf{s})^t E(\mathbf{u}\mathbf{u}^t)(\nabla\mathbf{s})(\nabla\mathbf{s}^t\nabla\mathbf{s})^{-t} = (\nabla\mathbf{s}^t\nabla\mathbf{s})^{-1}\nabla\mathbf{s}^t\Sigma\,\nabla\mathbf{s}(\nabla\mathbf{s}^t\nabla\mathbf{s})^{-t}$.

If $\Sigma = \sigma^2\mathbf{I}$ in particular (i.e., if the attack is whitened), then:

$$E|\delta\theta\rangle\langle\delta\theta| = \sigma^2(\nabla\mathbf{s}^t\nabla\mathbf{s})^{-1}. \tag{1.42}$$

Interestingly, the Fisher information matrix in equation (1.38) becomes:

$$\mathbf{J} = \frac{1}{\sigma^2}(\nabla\mathbf{s}^t\nabla\mathbf{s}). \tag{1.43}$$

Thus, we have derived that, under AWGN, the Fisher information matrix is the inverse of the covariance matrix of a linear approximation error:

$$\mathbf{J} = (E|\delta\theta\rangle\langle\delta\theta|)^{-1}. \tag{1.44}$$

Intuitively, the following conjecture can be made—in the limit when the signal manifold $\Xi(\Theta)$ is flat, the CRB of parameter estimation is achieved by linear projection.

Figure 12. Audio synthesis maps the frequency domain, a unit circle, to a closed curve in a high dimensional signal space. Gaussian white noise causes an uncertainty in frequency estimation.

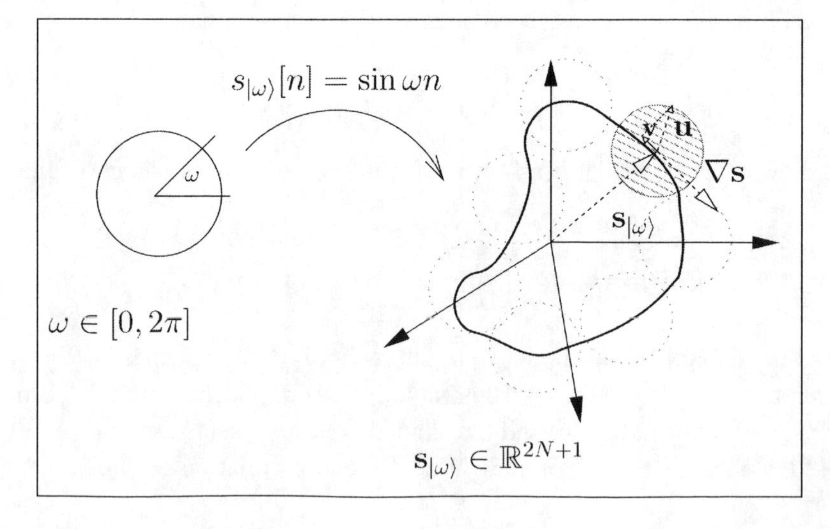

Figure 13. Projection of additive Gaussian noise onto a tangential space of a smooth manifold

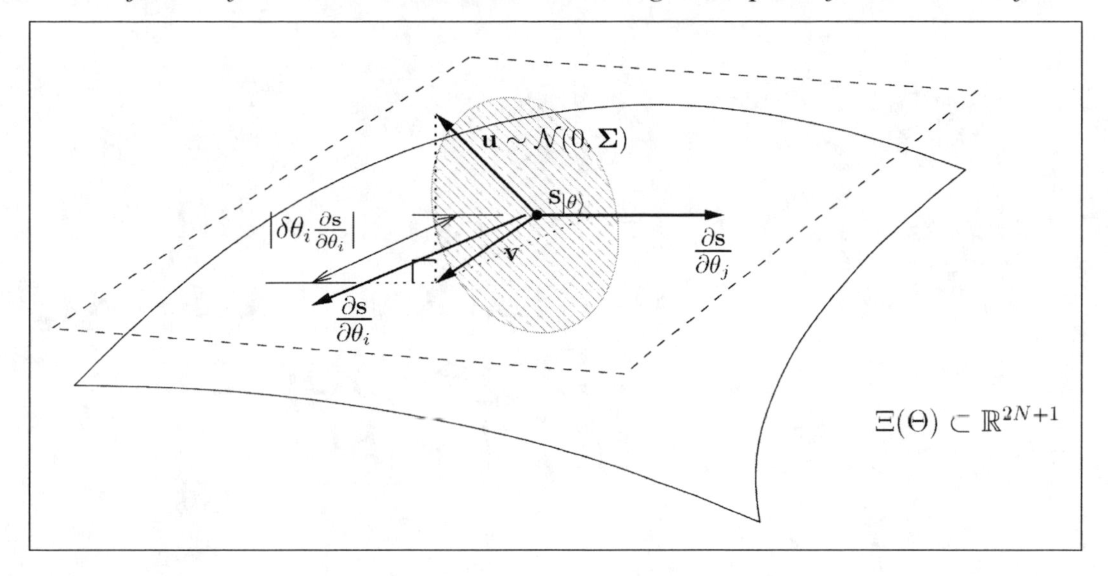

Chapter IV
Robust Zero-Bit and Multi-Bit Audio Watermarking Using Correlation Detection and Chaotic Signals

Nikos Nikolaidis
Aristotle University of Thessaloniki, Greece

Alexia Giannoula
University of Toronto, Canada

ABSTRACT

Digital rights management of audio signals through robust watermarking has received significant attention during the last years. Two approaches for blind robust watermarking of audio signals are presented in this chapter. Both approaches use chaotic maps for the generation of the watermark signals and a correlation-based watermark detection procedure. The first method is a zero-bit method that embeds high-frequency chaotic watermarks in the low frequencies of the discrete Fourier transform (DFT) domain, thus achieving reliable detection and robustness to a number of attacks. The second method operates on the temporal domain and robustly embeds multiple bits of information in the host signal. Experimental results are provided to demonstrate the performance of the two methods.

INTRODUCTION

The full transition from analog to digital audio along with the growing popularity of MPEG-1 Audio Layer 3 (MP3) audio compression format has enabled the fast and easy manipulation and distribution of audio data. Despite the obvious benefits for the involved entities (artists, recording companies, consumers), this development resulted in a tremendous increase in the illegal distribution of such data with a major impact in the music industry. As a consequence, an effort towards the protection of the digital content and the intellectual property rights (IPR) of its creators, owners, and distributors has been launched. This effort has led the scientific community towards digital watermarking as an efficient mean of IPR protection. The two major requirements for audio watermarking techniques intended for IPR protection are robustness to audio signal manipulations (usually referred to as watermarking attacks) and imperceptibility. Since these requirements are competing (a "weaker" watermark might be inaudible but also vulnerable to attacks), devising an efficient audio watermarking method is not trivial at all.

Among the numerous audio watermarking approaches that have been presented within the last decade, techniques based on variants of correlation detectors, that is, techniques where the presence of the watermark is verified by estimating the correlation between the watermark and the watermarked signal (Cox, Miller, & Bloom, 2002), have been widely popular (Bassia, Pitas, & Nikolaidis, 2001; Malvar & Florencio, 2003; Sener & Gunsel, 2004). In this chapter, two approaches towards blind robust watermarking of audio signals that are based on correlation detectors will be presented. The common denominator of these methods is the use of chaotic maps for the generation of the watermark signals. The first method is a zero-bit method, that is, it is capable of conveying information on whether an audio signal is watermarked or not. The method embeds in a multiplicative way high-frequency chaotic watermarks in the low frequencies of the discrete Fourier transform (DFT) domain, thus achieving robustness to a number of attacks and reliable detection. The second method operates on the temporal domain and is capable of robustly embedding multiple bits of information in the host signal. The method is robust to attacks of lowpass nature as well as to cropping of the host signal. Additional details for the presented approaches can be found in Giannoula, Tefas, Nikolaidis, and Pitas (2003); Laftsidis, Tefas, Nikolaidis, and Pitas (2003); and Tefas, Giannoula, Nikolaidis, and Pitas (2005).

Potential applications of the proposed methods include broadcast monitoring (i.e., devising automated royalty collection schemes for copyrighted songs that are aired by broadcasting operators or audience metering), owner identification and proof of ownership (i.e., for notifying or warning a user that audio data are copyrighted, for tracking illegal copies of audio data, or for proving the ownership of an audio file in the case of a legal dispute), transaction tracking (i.e., identifying the entities that received an audio file through a legal transaction and subsequently using this information to track those entities that illegally distributed the audio file), and usage control (i.e., for controlling the terms of use of audio data in conjunction with appropriate compliant devices).

TRANSFORM-DOMAIN CORRELATION-BASED ZERO-BIT AUDIO WATERMARKING USING MARKOV CHAOTIC SEQUENCES

Within the various watermarking techniques that have been proposed in the literature, aiming at the copyright protection of audio signals, little effort has been spent in exploiting the frequency properties of the watermark signal itself. In this section, we describe an enhanced zero-bit audio watermarking technique based on correlation detection. Previous research (Tefas, Nikolaidis, Nikolaidis, Solachidis, Tsekeridou, & Pitas, 2003) demonstrated that high-frequency watermarks exhibit superior detection performance to low-frequency ones. However, such watermarks are vulnerable to lowpass attacks, such as compression or filtering. In order to improve both detection reliability and robustness against lowpass signal manipulations, a method that embeds Markov chaotic watermarks with high-frequency characteristics into the low frequencies of the DFT domain will be presented.

A theoretical performance analysis will be also undertaken. The proposed watermarking methodology will be shown to perform efficiently with respect to both robustness and perceptual quality (inaudibility).

Motivation for Using High-Frequency Chaotic Sequences as Watermark Signals

Research conducted so far (Depovere, Kalker, & Linnartz, 1998; Kalker, Linnartz, & Depovere, 1998; Tefas et al., 2003) highlights the significance of the spectral characteristics of the watermark signal in the detection performance of watermarking techniques that involve correlation detection. In general, high-frequency watermarks demonstrate superior performance if no attacks have been imposed on the original data. A theoretical performance analysis of correlation-based watermarking schemes using additively embedded chaotic sequences was provided in Tefas et al. (2003), where the superiority of high-frequency skew-tent chaotic watermarks against both white and low-frequency ones was demonstrated in cases where the host signals had not been subjected to distortions. High-frequency watermarks, however, tend to be severely affected by lowpass attacks such as compression and lowpass filtering that destroy high-frequency watermark information and lead to deterioration of the overall system performance.

The technique described in this section attempts to surmount the vulnerability of high-frequency watermarks to lowpass attacks and improve the audio watermarking system detection reliability. The basic underlying concept is the development of a watermarking scheme that will exploit the superior correlation properties of high-frequency watermarks and at the same time, will tackle their inefficiency in withstanding lowpass operations (e.g., compression, filtering, etc.). The latter objective will be achieved by embedding the watermark in an appropriate transform domain and not in the time domain. Specifically, a high-frequency watermark will be embedded into the low frequencies of the DFT domain. In this way, the robustness of the technique against lowpass attacks is ensured and the desirable correlation properties of the high-frequency watermarks are preserved, thereby leading to the enhancement of the detection reliability.

Watermarks generated by chaotic sequences have been proposed in Giannoula et al. (2003); Tefas et al. (2003); Tefas et al. (2005); Tsekeridou, Nikolaidis, Sidiropoulos, and Pitas (2000); and Tsekeridou, Solachidis, Nikolaidis, Nikolaidis, and Pitas (2001) as an alternative option to the widely used pseudorandom signals. The key advantage of their utilization is their easily controllable spectral/correlation properties, a fact that renders them appropriate for a wide range of applications. In Tsekeridou et al. (2001), the authors introduced the n-way Bernoulli shift chaotic watermarks, which have been shown to exhibit similar or better performance than the pseudorandom white ones. Moreover, watermarks created by skew tent maps (a member of the piecewise-linear Markov maps family, which can be generated with any desirable spectral characteristics by modifying the chaotic map parameter) have demonstrated superior performance compared to other watermarks having exactly the same spectral characteristics (Tefas et al., 2003). More specifically, low frequency skew tent watermarks have been shown to perform better than low-frequency Bernoulli watermarks, whereas white skew tent watermarks have been found to surpass white pseudorandom watermarks. In this study, the class of piecewise-linear Markov chaotic sequences will be used as watermarks of the audio signals. Special attention will be paid to the skew tent watermarks.

Watermark Generation and Embedding

Several articles of the watermarking literature (Cox, Kilian, Leighton, & Shamoon, 1997; Liang, Xu, & Tran, 2000) propose the embedding of a watermark in the perceptually most significant components of the signal, claiming that the majority of signal processing operations do not affect these components and that a potential malicious attempt to destroy a watermark embedded in these components would significantly degrade the quality of the audio signal itself. However, in order to achieve perceptual transparency, the watermark energy should be concentrated into the perceptually insignificant signal components. In terms of the watermark strength, this should be kept as low as possible for the sake of imperceptibility but, at the same time, it should be sufficiently large so as to guarantee robustness to attacks. Based on the above observations, the watermark in this study is embedded into the low-frequency band of the audio signal, by simultaneously taking care to minimize its effects on the original data. Moreover, the proposed method exploits the convenient correlation properties of the high-frequency chaotic watermarks.

For an audio signal $s(n)$ of length N_s the watermark is embedded in the DFT domain, by applying a multiplicative embedding rule on the magnitude of the DFT coefficients, as follows:

$$F'(n) = F(n) + pW(n)F(n) \qquad (1)$$

where $F'(n)$ and $F(n)$ (n=0... N_s−1) denote the magnitude of the DFT coefficients of the watermarked and the original audio signal respectively. Therefore, the DFT magnitude $F(n)$ will be considered hereafter as the host signal. In the previous equation, $W(n)$ denotes the watermark signal and p is the watermark embedding factor that controls the watermark strength and subsequently the watermark perceptibility. The multiplicative embedding rule of the above equation has been selected due to the simple perceptual masking that it performs, since it alters the DFT coefficients proportionally to their magnitude. The construction of the watermark $W(n)$ of length N_s that is, to be embedded in the signal, is realized by using a principal watermarking sequence $W_0(n)$ which is generated using an appropriate function and a secret watermark key K. More specifically, the sequence $W(n)$ is equal to $W_0(n - aN_s)$ for values of the index n in the range $[aN_s...bN_s]$ and equals $W_0(N_s - n - aN_s)$ for indices in the range $[(1-b)N_s...(1-a)N_s]$. For all other values of n, $W(n)$ equals zero. In other words the watermark is inserted in the low frequency subbands whereas the DFT coefficients outside these subbands are not altered. The limits of the aforementioned DFT subbands, that is, the frequency terms that will be affected by the watermark $W(n)$, are controlled by the parameters a and b ($0 < a < b \leq 0.5$).The selection of these subbands will be detailed in the simulation subsection. The fact that a reflected version of $W_0(n)$ is inserted in the subband $[(1-b)N_s...(1-a)N_s]$ is due to the symmetry of the DFT magnitude with respect to the middle sample.

Regarding the signals that will be used as watermarks, it will be shown in the subsequent subsections that the performance of the correlation detector depends on the first or higher-order correlation statistics of these signals. Therefore, functions that generate watermarks with desirable spectral/correlation properties are required. A variety of 1-D chaotic maps can be shown to have this ability and thus can be used efficiently as watermark generation functions. The class of eventually expanding piecewise-linear Markov maps M:[0,1]→[0,1] has been selected in this study, since they exhibit desirable statistical properties, for example, invariant probability densities and ergodicity under readily verifiable conditions (Boyarsky & Scarowsky, 1979), and their statistics can be derived in a closed form (Isabelle & Wornell, 1997).

A chaotic watermark sequence $x(n)$ is generated by recursively applying such a chaotic 1-D

Equation 2.

$$x(n) = M(x(n-1)) = M^n(x(0)) = \underbrace{M(M(\ldots(M(x(0)))\ldots))}_{n\ times}$$

map M on an initial value $x(0)$, generated by the watermark key, (see equation (2)).

The principal zero-mean chaotic watermark $W_0(n)$ is subsequently obtained by subtracting from $x(n)$ its mean value μ_x:

$$W_0(n) = x(n) - \mu_x \quad (3)$$

Mean value subtraction is performed because zero-mean watermarks are known to exhibit superior detection performance (Linnartz, Kalker & Depovere, 1998). Finally, the watermark $W(n)$ of length N_s, that will be embedded into the host audio signal, is constructed as described in the previous paragraphs.

Watermark Detection

Given an audio signal under test, the signal of the magnitude of its DFT coefficients $F_t(n)$ is constructed and watermark detection is performed in order to verify whether a specific watermark $W_d(n)$ has been embedded in it. $F_t(n)$ can be written as follows:

$$F_t(n) = F_o(n) + pW_e(n)F_o(n) \quad (4)$$

where $F_0(n)$ denotes the DFT magnitude of the original host signal. In the above general expression, three cases (events) can be distinguished: (a) the test signal is watermarked with the investigated watermark, that is, $W_d(n) = W_e(n)$ and $p \neq 0$ (event H_0), (b) the test signal is not watermarked, that is, $p = 0$ (event H_{1a}), and (c) the test signal is watermarked with a different watermark, that is, $W_e(n) \neq W_d(n)$ and $p \neq 0$ (event H_{1b}). The last two events H_{1a} and H_{1b}, can be combined to form event H_1, namely the event that the investigated watermark is not embedded into the test signal. Therefore, a binary hypothesis test (watermark present or not) can be performed at the detection side, using as test statistic the output of a correlation detector:

$$c = \frac{1}{N}\sum_{n=0}^{N-1} F_t(n)W_d(n) = \frac{1}{N}\sum_{n=0}^{N-1}\left(F_o(n)W_d(n) + pW_e(n)F_o(n)W_d(n)\right) \quad (5)$$

To reach a binary decision as of the presence of the watermark, c is compared against a suitably selected threshold T.

The performance of such a correlation-based detection scheme can be assessed with respect to the probability of false alarm or false acceptance $P_{fa}(T)$ (probability of erroneously detecting a specific watermark into an audio signal that has not been watermarked or that has been watermarked with a different watermark) and the probability of false rejection $P_{fr}(T)$ (probability of erroneously rejecting the existence of a specific watermark into an audio signal that has been indeed watermarked). If $f_{c|H_0}$, $f_{c|H_1}$ are the conditional probability density functions (pdf) of the correlation detector output c under the hypotheses H_0, H_1 respectively, P_{fa} and P_{fr} can be expressed as follows:

$$P_{fa}(T) = \int_T^{\infty} f_{c|H_1}(t)dt \quad (6)$$

$$P_{fr}(T) = \int_{-\infty}^{T} f_{c|H_0}(t)dt \quad (7)$$

The plot of P_{fa} vs. P_{fr} for various values of the independent variable T is known as the *receiver operating characteristic* (ROC) curve and is commonly used to characterize the system's detection performance.

For the watermark signals utilized with this method, that is, those generated by piecewise linear Markov chaotic maps, the pdfs $f_{c|H_0}, f_{c|H_1}$ of the correlation detector output can be assumed to be Gaussian (see the following subsection for a justification of this fact). In this case, it can be shown that P_{fa} can be expressed in terms of P_{fr} as follows:

$$P_{fa} = \frac{1}{2}\left[1 - erf\left[\frac{\sqrt{2}\sigma_{c|H_0} erf^{-1}(2P_{fr} - 1) + \mu_{c|H_0} - \mu_{c|H_1}}{\sqrt{2}\sigma_{c|H_1}}\right]\right] \quad (8)$$

This expression defines the ROC curve for the situation at hand. From the above formula one can see that only the mean values $\mu_{c|H_0}$, $\mu_{c|H_1}$ and the variances $\sigma^2_{c|H_0}$, $\sigma^2_{c|H_1}$ are needed in order to define the ROC curve in this case.

Using equation (5), the mean value μ_c of the correlation detector output can be found to be:

$$\mu_c = E[c] = \frac{1}{N}\left(\sum_{n=0}^{N-1} E[F_o(n)]E[W_d(n)] + \sum_{n=0}^{N-1} pE[F_o(n)]E[W_e(n)W_d(n)]\right) \quad (9)$$

Similarly, the variance $\sigma_c^2 = E[c^2] - E^2[c]$ of the correlation detector is given by, (see equation (10)).

It can be readily observed that the above formulas provide the mean and variance values for all three events, H_0 ($W_e(n) = W_d(n)$), H_{1a} ($p = 0$), and H_{1b} ($W_e(n) \neq W_d(n)$). In order to derive the above formulas, the host signal $F_0(n)$ and the watermark signals $W_e(n)$ or $W_d(n)$, have been assumed to be statistically independent.

Analytical Detection Performance Evaluation

In this subsection, closed-form expressions of the mean and variance values μ_c and σ_c^2 in equation (9) and equation (10) will be derived, thereby, enabling the derivation of an analytical description of the ROC curves and a theoretical evaluation of the method performance. Readers interested in the derivation of the closed-form expressions can find more details in Giannoula et al. (2003).

In order to proceed with the derivations, a number of assumptions should be adopted. First of all, stationarity in the wide sense is assumed for the DFT magnitude of the audio signal:

$$E[F_o[n]] = \mu_{F_o} \quad \forall n, \quad n = 0 \ldots N-1 \quad (11)$$

$$E[F_o[n]F_o[n+k]] = R_{F_o}[k] \quad \forall n, \quad n = 0 \ldots N-1 \quad (12)$$

where R_{F_o} denotes the autocorrelation function of the magnitude of the DFT coefficients that is assumed to be exponential (Linnartz, 1998):

Equation 10.

$$\begin{aligned}\sigma_c^2 = & \frac{1}{N^2}\Bigg[\sum_{n=0}^{N-1}\Big(E[F_o^2(n)]E[W_d^2(n)] + p^2E[F_o^2(n)]E[W_d^2(n)W_e^2(n)] + 2pE[F_o^2(n)]E[W_e(n)W_d^2(n)]\Big) \\ & + \sum_{n=0}^{N-1}\sum_{m=0, m\neq n}^{N-1}\Big(E[F_o(n)F_o(m)]E[W_d(n)W_d(m)] + pE[F_o(n)F_o(m)]E[W_e(m)W_d(n)W_d(m)] \\ & + pE[F_o(n)F_o(m)]E[W_e(n)W_d(n)W_d(m)] + p^2E[F_o(n)F_o(m)]E[W_e(n)W_e(m)W_d(n)W_d(m)]\Big)\Bigg] \\ & - \mu_c^2\end{aligned}$$

$$R_{F_o}[k] = \mu^2_{F_o} + \sigma^2_{F_o}\beta^k, \quad k \geq 0, \ |\beta| \leq 1 \tag{13}$$

where β is the parameter of the autocorrelation function and $\sigma^2_{F_o}$ is the variance of the host signal:

$$\sigma^2_{F_o} = E[F_o^2[n]] - E[F_o[n]]^2 \tag{14}$$

Although the autocorrelation model of equation (13) appears to be rather simplistic, its use towards the theoretical evaluation of the method resulted in ROC curves sufficiently close to the experimentally derived.

Before proceeding with the derivations, we will consider that the mean value of the test signal $E[F_t(n)]$ is subtracted from $F_t(n)$ before the evaluation of the correlation detector output in equation (5), since this can be shown to improve the performance of the method (Giannoula et al., 2003; Tefaset et al., 2003).

With respect to the watermark signals whose statistics are involved in equations (9) and (10), it can be shown that if Markov maps are employed for their generation, these signals attain an exponential autocorrelation function of the form shown in equation (13), where β is an eigenvalue of the corresponding Frobenius-Perron (FP) matrix (Kohda, Fukisaki, & Ideue, 2000). Furthermore, it can be shown that, under certain conditions (Tefas et al., 2003), if the watermarks $W_d(n)$, $W_e(n)$ have been generated by the same chaotic map using equation (2) with two different initial conditions, then $W_d(n) = W_e(n + k)$, where $k > 0$ is an integer known as the sequence *shift*. Thus, a watermark sequence can be considered as a shifted version of another sequence. This fact, along with the exponential autocorrelation function of chaotic watermarks generated by piecewise-linear Markov maps implies that such signals are correlated for a small sequence shift $k > 0$ but this correlation decreases rapidly as k increases. Thus, the Central Limit Theorem for random variables with small dependency (Billingsley, 1995) can be used in order to establish that the correlation detector of equation (5) attains a Gaussian probability distribution, even in the worst-case scenario (event H_{1b}).

By taking into account the above, the mean and variance of the correlation detector given by equations (9) and (10) can be expressed as follows:

Equation 16.

$$\begin{aligned}
\sigma_c^2 \;=\;& \frac{1}{N}\left(R_x[0,k,k] - 2\mu_x R_x[0,k] + 2\mu_x^2 R_x[0] - 2\mu_x R_x[k,k] + 4\mu_x^2 R_x[k] - 3\mu_x^4\right)p^2 R_{F_o}[0] \\
+\;& \frac{1}{N}\left(2pR_x[k,k] - 4p\mu_x R_x[k] + (1-2p\mu_x)R_x[0] + 4p\mu_x^3 - \mu_x^2\right)\sigma^2_{F_o} \\
+\;& \frac{2}{N^2}\Bigg[\sum_{m=1}^{N-1}(N-m)(R_{F_o}[m] - \mu^2_{F_o})R_x[m] + (4p\mu_x^3 - \mu_x^2)\sum_{m=1}^{N-1}(N-m)(R_{F_o}[m] - \mu^2_{F_o}) \\
-\;& p\mu_x \sum_{m=1}^{N-1}(N-m)(R_{F_o}[m] - \mu^2_{F_o})(2R_x[m] + 2R_x[k] + R_x[m+k] + R_x[k-m]) \\
+\;& p\sum_{m=1}^{N-1}(N-m)(R_{F_o}[m] - \mu^2_{F_o})(R_x[k,k-m] + R_x[k,m+k]) + \sum_{m=1}^{N-1}(N-m)R_{F_o}[m] \times \\
\times\;& \Big\{p^2 R_x[m,k,m+k] - p^2\mu_x(R_x[m,k] + R_x[m,m+k] + R_x[k,k-m] + R_x[k,m+k]) \\
+\;& p^2\mu_x^2(2R_x[m] + 2R_x[k] + R_x[k-m] + R_x[m+k])\Big\} - 3p^2\mu_x^4\sum_{m=1}^{N-1}(N-m)R_{F_o}[m]\Bigg] - \mu_c^2
\end{aligned}$$

$$\mu_c = p\mu_{F_o}(R_x[k] - \mu_x^2) \quad (15)$$

(see equation (16)).

In the previous equations $R_q[k_1, k_2, ..., k_r]$ is used to denote the r-th order correlation statistic of a wide-sense stationary signal q, in our case the watermark signal x:

$$R_q[k_1, k_2, \ldots, k_r] = E[q[n]q[n+k_1]q[n+k_2]\ldots q[n+k_r]] \quad (17)$$

The statistical moments of watermark signals generated by Markov maps that are required to evaluate expressions (15) and (16), have been derived in closed forms in Isabelle and Wornell (1997), Schimming, Gotz, and Schwarz (1998). The above equations can describe all events H_0, H_{1a}, H_{1b} involved in the hypothesis test at the detection side. Specifically, event H_0 can be described by setting the sequence shift k equal to zero and the embedding factor $p > 0$. Event H_{1a} (absence of the watermark) can be obtained by using an embedding factor $p = 0$, and finally, event H_{1b} (presence of a different watermark) is represented by setting $p \neq 0$ and a sequence shift $k > 0$.

By observing the above formulas, it can be readily shown that the mean value μ_c of the correlation detector is zero for the event H_{1a} and converges quickly to zero with increasing k for the event H_{1b}. For the event H_0, μ_c depends on the power and variance of the watermark and the mean value μ_{F_0} of the host signal and is independent of the correlation statistics of the watermark. Conversely, the variance σ_c^2 of the correlation detector is related to the correlation statistics of the watermark itself. However, the correlation statistics of the watermark and more specifically its autocorrelation function $R_x[m]$ are directly related to its spectral properties, namely its power *spectral density* (PSD) $S_x(\omega)$ through the well-known formula:

$$S_x(\omega) = \sum_{m=-\infty}^{\infty} R_x[m]e^{-j\omega m} \quad (18)$$

Therefore, the variance of the correlator output and subsequently its detection performance as defined by its ROC curve (see equation (8)) depend on the spectral characteristics of the watermark signal, thus justifying the use of chaotic watermark whose spectral characteristics can be easily controlled. Indeed, in the next subsection it will be shown that in the proposed multiplicative watermarking method, high-frequency chaotic watermarks outperform white-spectrum pseudorandom ones.

A class of piecewise-linear Markov maps with interesting properties is the *skew tent* maps, described by the following expression, (see equation (19)).

Analytical expressions for the first, second, and third-order correlation statistics of the skew tent map have been derived in Tefas et al. (2003). The power spectral density of such chaotic sequences has been found to be:

$$S_t(\omega) = \frac{1 - e_2^2}{12(1 + e_2^2 - 2e_2 \cos\omega)} \quad (20)$$

Equation 19.

$$T : [0,1] \to [0,1]$$

$$\text{where } T(x) = \begin{cases} \frac{1}{\lambda}x & , 0 \le x \le \lambda \\ \frac{1}{\lambda - 1}x + \frac{1}{1-\lambda} & , \lambda < x \le 1 \end{cases}, \lambda \in (0,1)$$

where $e_2 = 2\lambda - 1$ is an eigenvalue of the Frobenius-Perron (FP) matrix (Kohda, Fujisaki & Ideue, 2000). Furthermore, analytical expressions for the mean and variance of the correlation detector when multiplicatively embedded skew tent chaotic watermarks are used have been evaluated. The interested reader can consult Giannoula et al. (2003). Equation (20) shows that the spectral characteristics of this map are controlled by the parameter λ thereby enabling the construction of watermark signals with desirable spectral properties. For the proposed watermarking method, an appropriate high-frequency sequence ($\lambda < 0.5$) is selected.

Simulation Results

Various simulations were conducted on a number of audio signals to demonstrate the efficiency of the proposed audio watermarking scheme. Results presented below correspond to a music audio signal of approximately 8 seconds duration, sampled at 44.1 kHz with 16 bits per sample. In all sets of simulations, high-frequency chaotic watermark signals generated by the skew tent map with a value λ equal to 0.3 were used. The detection performance of the proposed watermarking method was assessed in terms of the ROC curves.

First, the just audible distortion (JAD) level was estimated for five different frequency subbands in order to select the one that is more suitable for embedding. The length of each subband was equal to $0.1N_s$, that is, 10% of the entire host signal. For each frequency band, a symmetric band was also used, leading to a total number of watermarked samples equal to $0.2N_s$. The embedding power p was gradually increased in small steps and the watermarked signal was presented to a panel of listeners which were asked to select the largest p value for which the distortions with respect to the original signal could not be perceived.

This value was considered as the maximum tolerable embedding power for the specific frequency band, that is, the JAD level. The corresponding SNR values (*in* dB) and embedding factors for the five subbands of the signal spectrum are shown in Table 1.

It can be easily observed that embedding the watermark into the lowest frequency band (1%-11%), provides significant perceptual capacity and allows the use of a large embedding power p without affecting the perceptual quality of the signal. As the embedding frequency band moves towards higher frequencies, one should use watermarks of smaller strength to prevent audible distortions.

However, since one of the watermark requirements is robustness against lowpass attacks, one should also verify that the lowest frequency band is sufficiently immune to such attacks before finally selecting it as the band of choice for watermark embedding. For this reason, the proposed method was tested against 64kbps MPEG-1 Audio layer III (MP3) compression, for the five frequency bands defined in Table 1.

Table 1. SNR and embedding power values p that result in inaudible distortions for five frequency bands

Frequency Band	SNR (dB)	Embedding Power p
a = 1%, b = 11%	23.08	0.27
a = 5%, b =15%	41.39	0.15
a = 10%, b = 20%	41.66	0.18
a = 15%, b = 25%	45.02	0.15
a = 20%, b = 30%	45.17	0.25

Inspection of the corresponding ROC curves revealed, as expected, that the lowest-frequency subband exhibits superior robustness with respect to the other subbands in case of MPEG compression. Based on the above observations, the lowest frequency subband (a=1%, b=11%) was selected as the most suitable for the watermark embedding and was used in all subsequent simulations.

The robustness of the proposed watermarking scheme with the embedding parameters selected through the previous experiments was next evaluated against lowpass attacks. In order to compare the proposed audio watermarking method with other techniques, simulations were conducted for two alternative audio watermarking methods. The first alternative embedding scheme utilizes white pseudorandom watermark sequences multiplicatively embedded in the same low frequency subband (1%-11%) of the DFT domain, using an embedding power that leads to inaudible watermarked signals of SNR = 23 dB. Correlation detection is subsequently used to detect the watermark. The second scheme is based on the time-domain audio watermarking technique presented in Bassia et al. (2001): a bipolar white pseudorandom watermark $w(n)\in\{-1,1\}$ is first modulated by the magnitude of the original audio samples $m(n)$:

$$w'(n) = p\,|\,m(n)\,|\,w(n) \tag{21}$$

where p denotes the watermark embedding strength. Subsequently, $w'(n)$ is filtered with a lowpass 25-th order Hamming filter. The resulting filtered watermark signal $w''(n)$ is additively embedded in the time domain of the original signal to produce an inaudible watermarked signal of SNR = 22 dB:

Figure 1. ROC curves for the three watermarking methods (skew tent and pseudorandom white watermarks embedded in the DFT domain and lowpass-filtered watermarks embedded in the time domain) after 64kbps MPEG compression

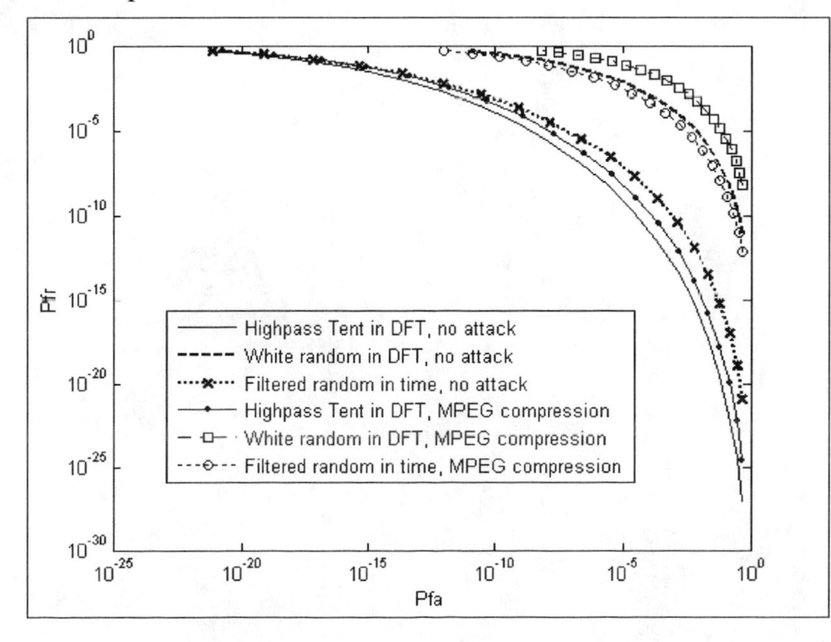

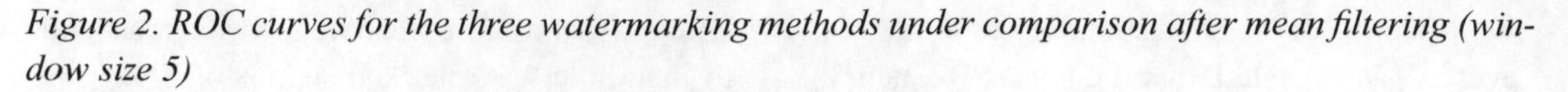

Figure 2. ROC curves for the three watermarking methods under comparison after mean filtering (window size 5)

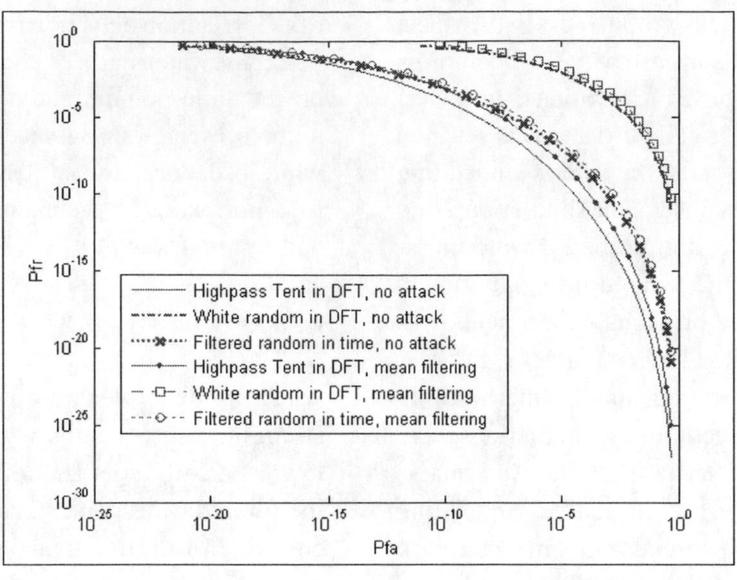

Figure 3. ROC curves for the three watermarking methods under comparison after median filtering (window size 5)

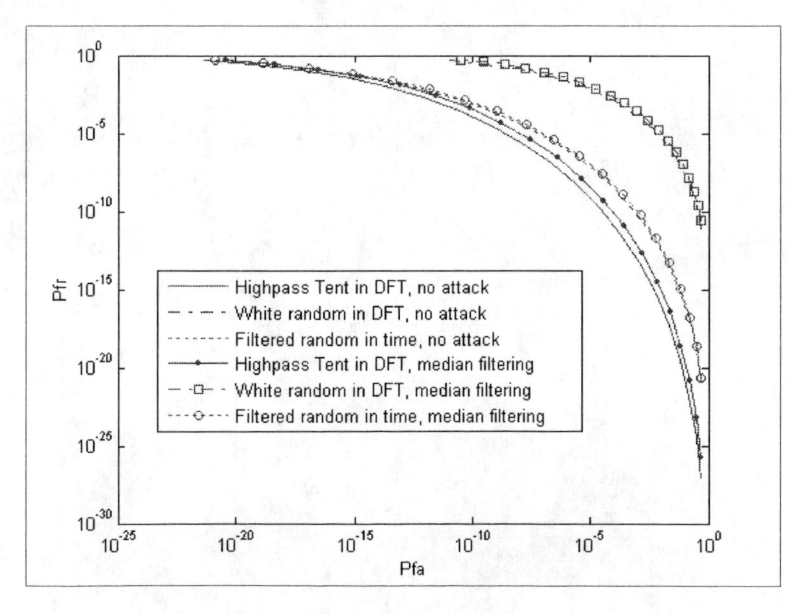

$$m_w(n) = m(n) + w''(n) \qquad (22)$$

Detection was performed through correlation.

The superior performance of high-frequency chaotic tent watermarks embedded into the low DFT frequencies, against the two alternative techniques, is illustrated in Figure 1 for the case of MPEG-I layer III encoding at 64 kbps.

Further simulations attempted to evaluate the detection performance of the proposed method in case of filtering using mean and median filters with a window of size equal to 5. All three techniques proved to be robust against filtering, as shown in Figure 2 and Figure 3, but the proposed watermarking method still outperforms the other two methods. The watermarks of the two alternative watermarking techniques withstand these distortions, since they are also of lowpass nature as they are either embedded in the low frequencies of the host signal (DFT-domain method) or filtered with a lowpass Hamming filter (time-domain method).

The watermarked audio signals generated by the three methods were next sub-sampled down to 11.025 KHz (i.e., sub-sampled with a ratio of 1:4), and then restored through linear interpolation to their initial 44.1 KHz sampling rate. The corresponding ROC curves are depicted in Figure 4. Observation of these curves leads to the conclusion that the proposed technique is very robust to resampling and outperforms the other two techniques.

Then quantization of the watermarked 16-bit audio signals generated by the three methods down to 8 bits per sample and backwards was performed. This operation had no effect on the watermarks for all three methods, as shown in Figure 5.

In a last set of simulations, the watermarking techniques under comparison were tested against cropping. In these simulations, a number of samples were removed from the beginning of the audio signal and an equal number of samples were inserted at its end in order to preserve the total number of samples of the signal. Such an

Figure 4. ROC curves for the three watermarking methods under comparison after sub-sampling down to 11025 Hz and interpolation back to 44100 Hz

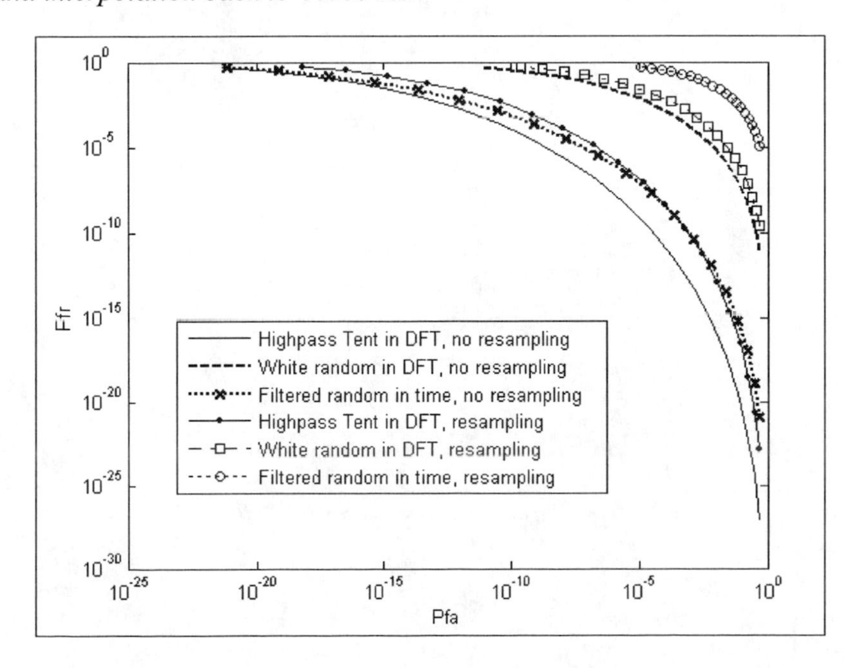

Figure 5. ROC curves for the three watermarking methods under comparison after quantization down to 8 bits per sample and backwards

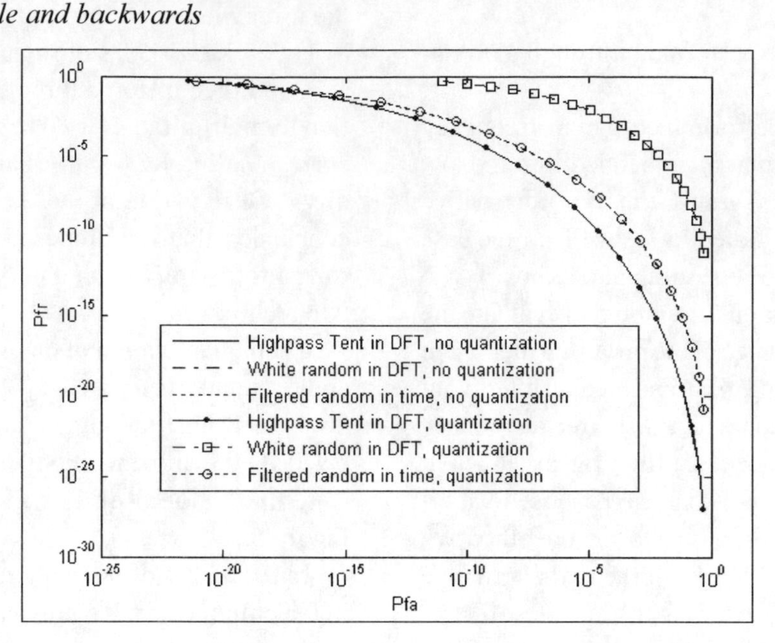

Figure 6. ROC curves for the proposed method after cropping 5%, 10%, 15% and 20% of the samples at the beginning of the signal

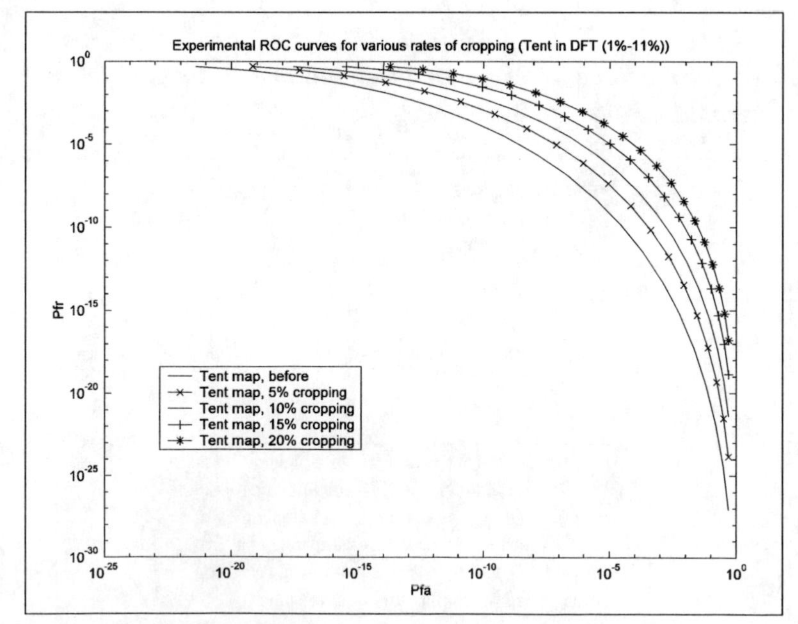

Note. Reproduced from: Tefas, A., Giannoula, A., Nikolaidis, N., & Pitas, I. ((c) 2005 IEEE)

experimental setup was used because watermark embedding is usually performed by repeatedly applying the watermark in consecutive, fixed-length audio segments and detection follows the same pattern. Thus this setup simulates sufficiently well the case where cropping occurs on the signal and therefore the detection is applied on a segment of the signal which is not aligned with the segments used during detection. Simulations with cropping rates equal to 5%, 10%, 15%, and 20% (indicating the percentage of samples that were removed from the signal) were performed. The time-domain watermark was seriously degraded by cropping of even the smallest ratio due to synchronization loss. However, the two frequency domain watermarking schemes appeared to withstand cropping efficiently with the proposed method being superior. More specifically, the proposed audio watermarking method proved to be significantly robust even at the highest cropping rate of 20% (Figure 6), whereas the performance of the other frequency domain method deteriorated significantly at this rate.

TEMPORAL DOMAIN MULTIPLE-BIT AUDIO WATERMARKING

The audio watermarking method that is presented in this section is a multiple-bit extension of the watermarking technique presented in Bassia et al. (2001). The method is a blind one and is capable of withstanding attacks of lowpass nature, such as compression and cropping. Chaotic number generators are used in the watermark generation procedure since they attain controllable spectral characteristics. The following subsections will present in detail the various modules of the method that is watermark generation, embedding, and detection.

Watermark Generation

The multiple-bit watermark is generated through a chaotic random number generator. More specifically, a chaotic function G, that is, a nonlinear transformation, that maps scalars to scalars is applied recursively using as a starting point the watermark key $z[0]$ thus producing a sequence $z[i]$ of length N:

$$z[i+1] = G(z[i]) \quad i = 0,..,N-2 \qquad =(23)$$

The Renyi map $G: U \rightarrow U, U \subset \mathcal{R}$ was used in our case:

$$z[i+1] = (\lambda \cdot z[i]) mod\ \pi, \quad \lambda \in \mathcal{R} \qquad (24)$$

The parameter λ in the previous equation controls the spectral characteristics of the sequence. For $\lambda > 1$ and close to 1 low-pass behavior can be obtained, whereas for $\lambda \approx 2$ an almost white signal results. Subsequently $z[i]$ is thresholded in order to generate a bipolar signal $w(i)$ of values +1 and –1. The threshold applied on the sequence is selected so that symbols +1 and –1 are generated with equal probabilities. The outcome of this procedure is the basis watermark that is used to generate the multiple-bit watermark.

The fact that a chaotic watermark is used ensures that the correlation between different watermarks would be low. In the specific implementation, a value of λ that is close to 1 and also such that $\lambda > 1$, was used in order to generate a watermark that is inaudible and also robust to lowpass processing.

Embedding Procedure

During embedding, the basis watermark $w(i)$ described in the previous subsection is modified and repeated multiple times, so that an M-bit long message m in binary format can be encoded.

This new watermark is subsequently embedded in the host signal f and the watermarked signal f_w is obtained. In order to embed the M bits once, the host signal f is divided into $M+1$ consecutive non-overlapping segments each containing N samples, N being a user-selected parameter:

$$f_k(i) = f(k \cdot N + i),\ i = 0,\ldots,N-1,\ k = 0,\ldots,M \quad (25)$$

The M segments are used to host the message bits whereas the additional segment is needed in order to achieve synchronization when the host signal is cropped as will be explained below. Thus, an audio signal consisting of at least $(M+1)\cdot N$ samples is required for embedding a message of M bits and consequently the payload of the method is $M/((M+1)\cdot N)$ bits/sample. If the host signal has more than $(M+1)N$ samples (which is usually the case), then the watermark is repeatedly embedded in the signal. This redundancy ensures that a copy of the watermark will be available after cropping.

In the proposed method, the embedding of the basis watermark does not start from the first sample of the audio signal. Instead, the audio sample, among the first N ones, that is, the most appropriate (in terms of the correlation-based detection that will follow) for being the starting point of the embedding is sought. This ensures minimal interference from the host signal and increases the detection performance. In order to find the most appropriate sample for starting the embedding, the following procedure is used.

The correlation $c_k(i)$ between the first $M+2$ segments $f_k(i)$ and the basis watermark $w(j)$ is calculated for different starting positions defined by index i, and the sum $C(i)$ with respect to k of the absolute correlation values is evaluated:

$$c_k(i) = \sum_{j=0}^{N-1} f(k \cdot N + i + j) \cdot w(j) \quad i = 0,\ldots,N-1 \quad (26)$$

$$C(i) = \sum_{k=0}^{M+1} |c_k(i)| \quad (27)$$

Consequently, the sample p, among the first N samples of the signal that is related to the smallest value of $C(i)$ is chosen as the starting sample for the embedding procedure:

$$p = \arg\min_i(C(i)) \quad (28)$$

In order to be able to locate the beginning of the embedded message in case of a cropped signal, the bits of the message should be encoded in a way that facilitates their recovery even if only a part of the audio signal is available during detection, provided that it is long enough to contain the full message. The following paragraphs describe the method that has been used to cope with this issue. It should be noted that this method is used only in cases where the audio signal is long enough to accommodate multiple instances of the watermark.

According to this method, a segment of the signal that should have hosted a bit of the message is left unwatermarked. This segment is called the synchronization-bit segment and is used to mark the fact that the next segment of the signal contains the first bit of the sequence. Obviously, it is important that this synchronization-bit segment is indeed detected during the correlation-based detection as an unwatermarked "gap" among the watermarked segments of the signal. In order to achieve this, this segment is selected so that the correlation of the basis watermark and the audio signal, at that segment, is significantly smaller than the correlation values obtained from the other watermarked segments. Thus the first $M+1$ segments of the signal are examined and the one that gives the minimum absolute correlation value with the watermark is selected as the position g of the synchronization bit:

$$g = \arg\min_k(|c_k(p)|) \quad (29)$$

The procedure continues by repeatedly embedding the watermark until the end of the signal is reached. However, the next synchronization-bit segment, which will have to appear in the $g + M + 1$ segment, will probably not be an appropriate one, that is, it might not be such that its correlation with the watermark is sufficiently small. Therefore, small modifications are applied to its samples so that this desired condition is enforced, thus ensuring that this segment will be detected as a synchronization-bit in case the first synchronization-bit is cropped. The same procedure is applied to all subsequent signal segments that are to host a synchronization-bit.

When finished with the above preparatory steps, the actual watermark embedding can be performed. Starting from the p-th sample, the host signal is again split in segments $x_k(i)$ of length N:

$$x_k(i) = f(p + k \cdot N + i), \quad i = 0,\dots,N-1 \qquad (30)$$

In order to embed the message, the basis watermark is embedded multiple times in the audio signal. Each embedded basis watermark (i.e., each audio segment) encodes one bit of the message. If the bit to be embedded is equal to one, then $w(i)$ is embedded, otherwise $-w(i)$ is used. Therefore, the watermarked signal is generated using the formula, in equation (31), where constant a controls the watermark's strength and the function $t_k(i)$ is defined as follows:

$$t_k(i) = \begin{cases} 1 & \text{if bit 1 is encoded in segment } k \\ -1 & \text{if bit -1 is encoded in segment } k \\ 0 & \text{if } k = g \end{cases} \qquad (32)$$

The watermark is repeatedly embedded in the signal until all signal samples are utilized.

Detection Procedure

The goal of the detection is twofold, that is, to test whether the audio signal y under investigation is watermarked, and, provided that this is indeed the case, proceed in the extraction of the message $\hat{m}$. The detection procedure for the first instance of the watermark proceeds as follows. The correlation of the test signal with the basis watermark is calculated for every segment $k = 0, \dots, M$ and for different starting positions defined by the index i. This correlation is normalized to the range [0, 1] by dividing its value with an approximation of the maximum correlation value that is expected in case y is watermarked:

$$c'_k(i) = \frac{\sum_{j=0}^{N-1} y(k \cdot N + i + j) \cdot w(j)}{\alpha \sum_{j=0}^{N-1} | y(k \cdot N + i + j) |} \qquad (33)$$

The normalized correlation $c'_k(i)$ is then used to derive the mean absolute correlation $C'(i)$ for all segments:

$$C'(i) = \frac{\sum_{k=0}^{M} | c'_k(i) |}{M+1} \qquad (34)$$

If a significantly high peak exists in $C'(i)$ a watermark is indeed embedded in y. Thus, the maximum value of $C'(i)$ for all i is evaluated and compared to an appropriate threshold. If this value is above the threshold, the test signal is declared as watermarked and the method proceeds to extract the message that is encoded in it. In order to do so, the value $\hat{p}$ of the index i where the maximum value of C' occurs is found:

$$\hat{p} = \arg\max_i (C'(i)) \qquad (35)$$

Equation 31.

$$f_w(p + k \cdot N + i) = x_k(i) + t_k(i) \cdot \alpha \, | x_k(i) | \, w(i), \quad i = 0,\dots,N-1$$

Obviously, in order to have a successful detection, $\hat{p}$ should be the same as the index p, which was calculated while embedding the watermark. The embedded bits are then derived judging on the sign of samples $c_k'(\hat{p})$ for every segment $k \in [0,...,M]$ and for the offset $\hat{p}$ found in the previous step. In more detail, the sequence $b(k)$ of $c_k'(i)$ values at index $\hat{p}$ is evaluated:

$$b(k) = c_k'(\hat{p}),\ k = 0,...,M \qquad (36)$$

Subsequently, the index of the synchronization-bit (i.e., the segment where the synchronization-bit is encoded) is evaluated as the index k of the minimum absolute value of $b(k)$:

$$\hat{g} = \arg\min_k(|b(k)|) \qquad (37)$$

Finally, the message bits are extracted in the correct order by shifting sequence $b(k)$ according to the index of the synchronization-bit and detecting the sign of its elements, (see equation (38)).

The above procedure is repeatedly applied if the audio signal is of sufficient length to host multiple instances of the watermark.

Experimental Performance Evaluation

A number of experiments have been carried out in order to calculate optimal values for the algorithm's parameters and evaluate its performance in terms of watermark detection (verifying whether the signal is watermarked or not), and message decoding (retrieving the bits of the message that has been embedded in the signal). The audio signals which were used in the experiments were classical and ethnic music excerpts sampled at *44.1* KHz. The length N of the basis watermark, and therefore of the signal segment that hosts it, has been chosen to be 2^{15} samples. The selection of this value was experimentally proven to be a good compromise between the detection performance (which increases as N increases) and the signal length required to encode one bit of the message, which obviously affects the payload of the method. With the chosen segment length and for moderately large messages (so that $M/(M+1) \approx 1$), the payload of the method according to the formula provided in the subsection "Embedding Procedure" is approximately 2^{-15} message bits per audio sample. Obviously this payload figure does not take into account repetitive embedding of the watermark in case of signals that are long enough to hold more than one instances of the watermark. The parameter α that controls the watermark strength and thus the alterations induced on the host signal has been assigned values which resulted in watermarked audio signals with an SNR of approximately 25.5dB. This SNR value was chosen by conducting listening tests during which the watermarked signals and the original signals were presented to several people of medium and higher music education. These experiments verified that indeed watermarked signals with such an SNR value cannot be distinguished form the original signals. Figure 7 presents the ROC curves obtained for watermark detection when messages of different sizes (in bits) were used. The length of the host audio signal in this experiment was equal to the minimum length required to host the message, for example, 1,638,400 audio samples (or equivalently 37.15 seconds in a 44.1 KHz signal) for a message comprising of *48* bits. From this figure it is obvious that the detection results improve as the length of the encoded message increases. This is due to the fact that longer messages require more

Equation 38.

$$\hat{m}(k) = (1 + sgn(b((\hat{g} + k) mod(M+1))))/2,\quad k = 0,...,M-1$$

signal samples in order to be embedded, which in turn results in improved performance of the correlation detector. Messages longer than 16 bits provide acceptable detection results since in such cases the equal error rate (EER), that is, the point of the ROC curve where the two types of error (P_{fa} and P_{fr}) are equal, is smaller that 10^{-16}. In terms of decoding performance, the experiments showed that the bit error rate (BER) is 0.4% in the case of encoding 8 bits, whereas when longer messages are encoded the BER is even smaller.

As expected, the technique has been found to be robust to cropping. Furthermore, experiments proved the method's robustness to sampling to a different sampling rate and back as well as quantization to different bits per sample and back. The robustness of the method to MPEG Audio-I layer III (MP3) compression has also been investigated. The detection results after MP3 compression for a signal hosting a 64-bit message and for various compression ratios are depicted in Figure 8. Inspection of this figure reveals that the detection results after MP3 compression are in general improved with respect to those achieved when no compression occurs. This can be attributed to the lowpass nature of the watermark. Being essentially a lowpass operation, the compression does not affect significantly the lowpass watermark, whereas some high-frequency information of the host signal is removed. As a result, the power of the watermark is increased with respect to the power of the host signal. Thus, the performance of the proposed watermarking method is very good even for large compression ratios (e.g., 40kbps) that degrade significantly the quality of the audio signal.

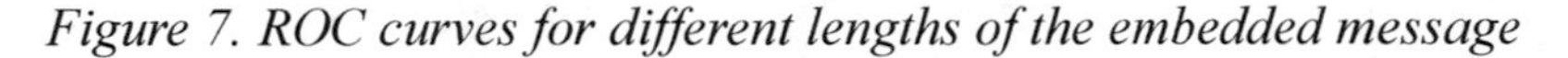

Figure 7. ROC curves for different lengths of the embedded message

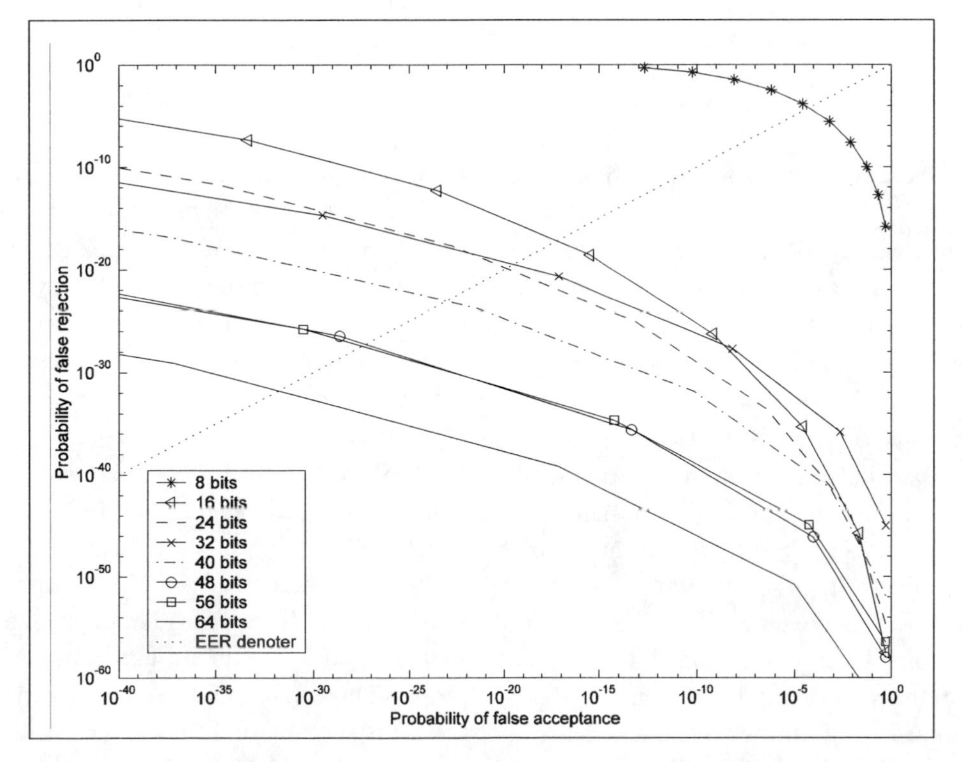

Note. The point where the dotted straight line intersects a ROC curve is the EER point for this curve (Reproduced from: Laftsidis, C., Tefas, A., Nikolaidis, N., & Pitas, I. ((c) 2003 IEEE).

Figure 8. ROC curves for a signal hosting a message of 64 bits, compressed using the MP3 algorithm at various compression ratios.

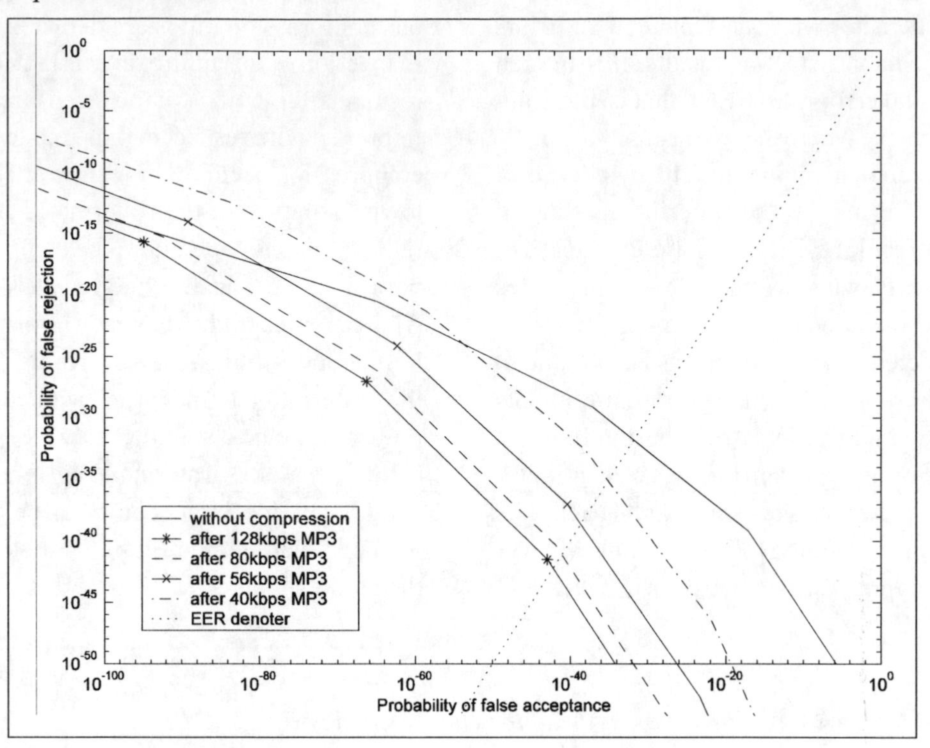

Note. Reproduced from: Laftsidis, C., Tefas, A., Nikolaidis, N., & Pitas, I. ((c) 2003 IEEE)

FUTURE RESEARCH DIRECTIONS

The field of audio watermarking is a relatively new research area that poses significant challenges. Despite the progress witnessed during the last years, there are still many interesting issues to be dealt with. Two approaches towards blind robust watermarking of audio signals that are based on a correlation detector and utilize chaotic watermarks were presented in this chapter. The first method embeds high-frequency chaotic watermarks in the low frequencies of the DFT domain whereas the second operates on the temporal domain and is capable of robustly embedding multiple bits of information in the host signal. Both methods are sufficiently robust to a number of attacks such as compression, resampling, requantization, and cropping and can thus be used for digital rights management of audio signals. Currently, both methods are not robust to random removal of samples from the host audio signal, since such an operation would result in a loss of synchronization during detection. Future research will aim at improving the methods robustness to such a manipulation. Furthermore, multiple-bit variants of the first method will be pursued.

As shown in this chapter as well as in other publications (Tefas et al., 2003; Tefas et al., 2005), chaotic watermarks posses certain interesting properties when used within a watermarking system that utilizes a correlation detector. Thus, such watermarks can be used, along with appropriately devised attack countermeasures, for the construction of efficient correlation-based watermarking schemes. Moreover, the theoretical analysis presented in section "Transform-Domain

Correlation-Based Zero-Bit Audio Watermarking Using Markov Chaotic Sequences" can be applied, after appropriate modifications, for the study of the performance of similar correlation-based schemes that utilize other types of watermark signals.

ACKNOWLEDGMENT

The authoring of this chapter has been supported in part by the European Commission through the IST Programme under Contract IST-2002-507932 ECRYPT.

REFERENCES

Bassia, P., Pitas, I., & Nikolaidis, N. (2001). Robust audio watermarking in the time domain. *IEEE Transactions on Multimedia, 3*(2), 232-241.

Billingsley, P. (1995). *Probability and measure.* New York: Wiley.

Boyarsky, A., & Scarowsky, M. (1979). On a class of transformations which have unique absolutely continues invariant measures. *Transactions of American Mathematical Society, 255*, 243-262.

Cox, I., Kilian, J., Leighton, F., & Shamoon, T. (1997). Secure spread spectrum watermarking for multimedia. *IEEE Transactions on Image Processing, 6*(12), 1673-1687.

Cox, I., Miller, M., & Bloom, J. (2002). *Digital watermarking.* San Francisco: Morgan Kaufmann Publishers.

Depovere, G., Kalker, T., & Linnartz, J. P. (1998). Improved watermark detection reliability using filtering before correlation. In *Proceedings of the IEEE International Conference on Image Processing* (Vol. 1, pp. 430-434).

Giannoula, A., Tefas, A., Nikolaidis, N., & Pitas, I. (2003). Improving the detection reliability of correlation-based watermarking techniques. In *Proceedings of the IEEE International Conference on Multimedia and Expo* (Vol. 1, pp. 209-212).

Isabelle, S., & Wornell, G. (1997). Statistical analysis and spectral estimation techniques for one-dimensional chaotic signals. *IEEE Transactions on Signal Processing, 45*(6), 1495-1506.

Kalker, T., Linnartz, J.P., & Depovere, G. (1998). On the reliability of detecting electronic watermarks in digital images. In *Proceedings of the European Signal Processing Conference.*

Kohda, T., Fujisaki, H., & Ideue, S. (2000). On distributions of correlation values of spreading sequences based on Markov information sources. In *Proceeding of the IEEE International Symposium on Circuits and Systems* (Vol. 5, pp. 225-228).

Laftsidis, C., Tefas, A., Nikolaidis, N., & Pitas, I. (2003). Robust multibit audio watermarking in the temporal domain. In *Proceedings of the IEEE International Symposium on Circuits and Systems* (pp. 944-947).

Liang, J., Xu, P., & Tran, T.D. (2000). A Robust DCT-based low frequency watermarking scheme. In *Proceedings of the 34th Annual Conference in Information Systems and Science* (Vol. 1, pp. 1-6).

Linnartz, J.P., Kalker, T., & Depovere, G. (1998). Modeling the false alarm and missed detection rate for electronic watermarks. In *Proceedings of the Second Information Hiding Workshop* (pp. 329-343).

Malvar, H.S., & Florencio, D.F. (2003). Improved spread spectrum: A new modulation technique for robust watermarking. *IEEE Transactions on Signal Processing, 51*(4), 898-905.

Schimming, T., Gotz, M., & Schwarz, W. (1998). Signal modeling using piecewise linear chaotic generators. In *Proceedings of the European Signal Processing Conference* (pp. 1377-1380).

Sener, S., & Gunsel, B. (2004). Blind audio watermark decoding using independent component analysis. In *Proceedings of the IEEE International Conference on Pattern Recognition* (Vol. 2, pp. 875-878).

Tefas, A., Giannoula, A., Nikolaidis, N., & Pitas, I. (2005). Enhanced transform-domain correlation-based audio watermarking. In *Proceedings of the IEEE International Conference on Acoustics, Speech, and Signal Processing* (Vol. 2, pp. 1049-1052).

Tefas, A., Nikolaidis, A., Nikolaidis, N., Solachidis, V., Tsekeridou, S., & Pitas, I. (2003). Performance analysis of correlation-based watermarking schemes employing Markov chaotic sequences. *IEEE Transactions on Signal Processing, 51*(7), 1979-1994.

Tsekeridou, S., Nikolaidis, N., Sidiropoulos, N., & Pitas, I. (2000). Copyright protection of still images using self-similar chaotic watermarks. In *Proceedings of the IEEE International Conference on Image Processing* (Vol. 1, pp. 411-414).

Tsekeridou, S., Solachidis, V., Nikolaidis, N., Nikolaidis, A., & Pitas, I. (2001). Statistical analysis of a watermarking system based on Bernoulli chaotic sequences. *Signal Processing: Special Issue on Information Theoretic Issues in Digital Watermarking, 81*(6), 1273-1293.

ADDITIONAL READING

A Detailed Treatment of the Field of Nonlinear Dynamics and Chaos can be Found in:

Hilborn, R.C. (1994). *Chaos and nonlinear dynamics.* Oxford: Oxford University Press.

Ott, E. (1993). *Chaos in dynamical systems.* Cambridge: Cambridge University Press.

Audio and Image Watermarking Methods that involve Chaotic Watermark Signals are Detailed in:

Dawei, Z., Guanrong, C., & Wenbo, L. (2004). A chaos-based robust wavelet-domain watermarking algorithm. *Chaos, Solutions and Fractals, 22*(1), 47-54.

Nikolaidis, A., & Pitas, I. (2000). Comparison of different chaotic maps with application to image watermarking. In *Proceedings of the IEEE International Symposium on Circuits and Systems* (Vol. 5, pp. 509-512).

Nikoladis, N., Tefas, A., & Pitas, I. (2006). Chaotic sequences for digital watermarking. In S. Marshall and G. Sicuranza (Eds.), *Advances in nonlinear signal and image processing.* New York: Hindawi Publishing Co.

Silvestre, G., Hurley, N., Hanau, G., & Dowling, W. (2001). Informed audio watermarking scheme using digital chaotic signals. In *Proceedings of the IEEE International Conference on Acoustics, Speech, and Signal Processing* (Vol. 3, pp. 1361-1364).

Voyatzis, G., & Pitas, I. (1998). Chaotic watermarks for embedding in the spatial digital image domain. In *Proceedings of the IEEE International Conference on Image Processing* (Vol. 2, pp. 432-436).

Wang, D.J., Jiang, L.G., & Feng, G.R. (2004). Novel blind non-additive robust watermarking using 1-D chaotic map. In *Proceedings of the IEEE International Conference on Acoustics, Speech, and Signal Processing* (Vol. 3, pp. 417-420).

Additional Audio Watermarking Methods that are Based on Correlation Detection are Described in:

Cvejic, N., Keskinarkaus, A., & Seppanen, T. (2001). Audio watermarking using m-sequences and temporal masking. In *Proceedings of the IEEE Workshop on Applications of Signal Processing to Audio and Acoustics* (pp. 227-230).

Erkucuk, S., Krishnan, S., & Zeytinoglu, M. (2006). Robust audio watermark representation based on linear vhirps. *IEEE Transactions on Multimedia, 8*(5), 925-936.

Kim, H. (2000). Stochastic model based audio watermark and whitening filter for improved detection. In *Proceedings of the IEEE International Conference on Acoustics, Speech, and Signal Processing* (Vol. 4, pp. 1971-1974).

Liu, Z., & Inoue, A. (2003). Audio watermarking techniques using sinusoidal patterns based on pseudorandom sequences. *IEEE Transactions on Circuits and Systems for Video Technology, 13*(8), 801-812

Swanson, M.D., Zhu, B., Tewfik, A.H., & Boney, L. (1998). Robust audio watermarking using perceptual masking. *Signal Processing, 66*, 337-355.

Tachibana, R., Shimizu, S., Nakamura, T., & Kobayashi, S. (2001). An audio watermarking method robust against time- and frequency-fluctuation. In *Proceedings of the SPIE Conference on Security and Watermarking of Multimedia Contents III* (Vol. 4314, pp. 104-115).

Wu, C.P., Su, P.C., & Kuo, C.C.J. (2000). Robust and efficient digital audio watermarking using audio content analysis. In *Proceedings of the SPIE Security and Watermarking of Multimedia Contents II* (Vol. 3971, pp. 382-392).

Yaslan, Y., & Gunsel, B. (2004). An integrated decoding framework for audio watermark extraction. In *Proceedings of the IEEE International Conference on Pattern Recognition* (Vol. 2, pp. 879-882).

Chapter V
Three Techniques of Digital Audio Watermarking

Say Wei Foo
Nanyang Technological University, Singapore

ABSTRACT

Based on the requirement of watermark recovery, watermarking techniques may be classified under one of three schemes: nonblind watermarking scheme, blind watermarking schemes with and without synchronization information. For the nonblind watermarking scheme, the original signal is required for extracting the watermark and hence only the owner of the original signal will be able to perform the task. For the blind watermarking schemes, the embedded watermark can be extracted even if the original signal is not readily available. Thus, the owner does not have to keep a copy of the original signal. In this chapter, three audio watermarking techniques are described to illustrate the three different schemes. The time-frequency technique belongs to the nonblind watermarking scheme; the multiple-echo hiding technique and the peak-point extraction technique fall under the blind watermarking schemes with and without synchronization information respectively.

INTRODUCTION

The ease with which digital audio data such as songs and music can be copied and distributed has created a demand for methods for copyright protection and identification of ownership. Audio watermarking techniques are possible solutions to these problems. Watermark can be considered as special information hidden within the audio data. This information should be readily recovered by the owner, transparent to the user, resistant to unauthorized removal, and robust against attacks. In other words, the audio watermark should not be perceptible even to golden ears. In addition, the watermark should be robust to manipulation and common signal processing operations such as filtering, re-sampling, addition of noise, cropping, digital-to-analog and analog-to-digital conversions, and lossless / lossy compression. It should be tamperproof—resistant to active, passive, collusion and forgery attacks. On the other hand, watermark detection should unambiguously identify the owner.

Several methods have been proposed for audio watermarking. The least significant bit (LSB) method (Bassia & Pitas, 1998) is based on the substitution of the LSB of the carrier signal with the bit pattern from the watermark. Retrieval of the hidden information is done by reading the LSB value of each sample. This time-domain method offers a high data rate. However, the basic method is not robust against compression, requantization, or random changes to the marked LSB. A method to improve the LSB method has been investigated (Cvejic & Seppänen, 2005). Several researchers have investigated the embedding of watermarks in the transform domain (Hsieh & Tsou, 2002; Vladimir & Rao, 2001). Spread spectrum watermarking (Garcia, 1999; Kirovski & Malvar) is resilient to additive and multiplicative noise and does not require the original signal for recovery. However, the watermarked signal and the watermark have to be perfectly synchronized for recovery.

As one of the challenges of audio watermarking is the perceptibility of the added information, the psychoacoustic auditory model that imitates the human hearing mechanism is often explored to assess how best to hide the watermark (Swanson, Zhu, Tewfik, & Boney; Tewfik & Hamdy, 1998). The most pertinent aspect of the model is simultaneous frequency masking. Simultaneous masking of sound occurs when two sounds are played at the same time and one of them may be masked or made inaudible by the other if the other has sufficiently higher sound pressure level. The sound that is masked is called maskee and the other sound is called masker. The difference in sound pressure level required for masking is called the masking level. By embedding the watermarks below the masking level, the perceptual quality of the original audio signal will not be affected. A set of procedure is available to determine the masking level across the frequency range of interest for a given frame of audio signal. Details of the psychoacoustic auditory model may be found in Zwicker and Fastl (1990), and Zwicker and Zwicker (1991).

Based on the requirement of watermark recovery, watermarking techniques may be classified under one of three schemes: nonblind watermarking scheme, blind watermarking schemes with and without synchronization information. For the nonblind watermarking scheme, the original signal is required for comparison with the watermarked signal to extract the watermark. For the blind watermarking schemes, the embedded watermark can be extracted even if the original signal is not readily available. In the case of blind watermarking, synchronization information is crucial to identify where the watermarks are embedded. There are two approaches to provide synchronization information. One extracts special points and regions from the audio signals using their own special features as synchronization information while the other depends on additional synchronization information. The latter is suitable for all types of audio signals, but it may distort the original audio signal and draw the attention of attackers. The former, in contrast, does not have such problems. However, the special features of different types of audio signals vary and hence it is very difficult to have a feature-based synchronization scheme suitable for all types of audio signals.

The different schemes have their own merits and applications. For the nonblind watermarking scheme, the original signal is required for extracting the watermark and hence only the owner of the original signal will be able to perform the task. For the blind watermarking schemes, with or without synchronization, the embedded watermark can be extracted even if the original signal is not available. Thus, the owner does not have to keep an inventory of the original signal.

In this chapter, three techniques of audio watermarking investigated by the author are described. These include one nonblind audio watermarking technique (Foo, Xue, & Li, 2005) and two blind audio watermarking techniques (Foo, Yeo, & Huang, 2001; Foo, Ho, & Ng, 2004) using the two different synchronization approaches mentioned above.

NONBLIND WATERMARKING TECHNIQUE

Overview of the Technique

A nonblind audio watermarking technique (Foo et al., 2005) that makes use of both time and frequency based modulation is presented in this section. For this technique, segments of an audio signal are expanded or compressed by adding or removing samples in the frequency domain. The addition and removal of samples in the frequency domain is manifested in the time domain as the lengthening and shortening of segments of the audio signal.

Watermark information is represented by the pattern of expansion and compression embedded in the audio signal. An expanded (lengthened) frame followed by a compressed (shortened) frame of samples may be used to indicate the binary code '1' and a normal frame in between may be used to indicate the binary code '0'. A sequence of letters of the alphabet, such as FSW may be used as the watermark. To convert the letters to binary format, the ASCII code is adopted. Thus, FSW is represented in binary form as 01000110 0101001101010111. This binary signature is then embedded in the signal as a watermark.

To minimise the perceptual distortion of the coded audio signal, a psychoacoustic model (Zwicker & Fastl, 1990) is used to guide the processes of expansion and compression. Samples removed are those with amplitude below the threshold limits calculated with a motion picture expert group (MPEG) psychoacoustic model. These samples are selected as their removal does not significantly affect the perceptual quality of the audio signal.

For the recovery of the signature from the watermarked signal, the watermarked signal is subtracted from the original signal and the difference signal is obtained. From the difference signal, the frames that are subjected to expansion and compression are identified. The binary codes are then extracted and hence the letters and the signature. The technique as proposed is a nonblind method as the original signal is required as a reference in order to determine the difference signal.

Watermark Embedding

The original audio signal is sampled at a sampling frequency of 44.1 k samples per second quantised with 16 bits per sample. The digitised samples are segmented into frames of 1024 samples per frame. Successive frames have 512 overlapping samples. Each frame of audio samples is windowed with a Hanning window, as shown in Figure 1.

The energy level of the signal is determined. The signature is embedded only if the frame has an energy level above a given threshold. For frames that are to be expanded or compressed, fast Fourier transform is performed. The psychoacoustic model is then used to determine the masking threshold in each subband. The length of the frame of samples may be expanded by inserting samples with amplitude not exceeding the threshold of their respective subbands. Similarly, the length of the frame of samples may be shortened by removing samples with amplitude not exceeding the threshold of their respective subbands.

Inverse fast Fourier transform is subsequently performed to recover the signal in the time domain. If additional samples are added in the frequency domain, there will be additional samples recovered in the time domain. Similarly, if samples are removed in the frequency domain, the number of samples in the time domain is decreased. In short, the number of samples added or removed in the frequency domain will translate into extra or reduced number of discrete samples in the time domain.

The recovered frame of time domain samples is then added to the time domain samples of the preceding frame. The choice of the number of samples to be slotted in or removed is a compromise between the quality of the watermarked signal and the ease of recovery of the watermark. After trial and error, it is assessed that the addition or removal of four samples per frame of 1024 samples is sufficient for coding purposes while at the same time the expanded or compressed frames do not add to perceptible distortion.

In order to not alter the total length of the watermarked audio signal, the number of expanded frames is made equal to the number of compressed frames. A simple technique is to have an expanded frame followed immediately by a compressed frame, as shown in Figure 1. After a set of expansion and compression, the altered signal is in synchronisation with the original signal again. The pair of frames, which shall be referred to as diamond-frames, may be used to represent a binary '1' while two consecutive normal frames may be used to represent a binary '0'.

For audio signal of sufficiently long duration, the watermark can be repeatedly embedded in the signal. Such redundancy is practised in some digital watermarking techniques (Kim, Choi, Seok, & Hong, 2004; Wu, Wu, Su, & Kuo, 1999) as it increases the accuracy of recovery.

Watermark Recovery

For watermark recovery, the original digital audio samples are matched against the digital watermarked samples. The difference samples are then obtained by subtracting the amplitudes of the watermarked samples from the amplitudes of the corresponding samples of the original signal.

A typical difference signal is shown in Figure 2. The difference between the original samples and each pair of expanded and compressed frames or diamond-frames shall normally take the shape of a diamond, while unmarked frames (normal frames) will return a difference that is practically zero. This is because there is a gradual increase in difference between the original samples and samples of the expanded frame with time and decrease in difference between the original samples and the samples of the compressed frame with time. For the recovery process, the first task is to identify the diamond-frames. The diamonds are differentiated from the unmarked frames by the

Figure 1. Concatenation of frames

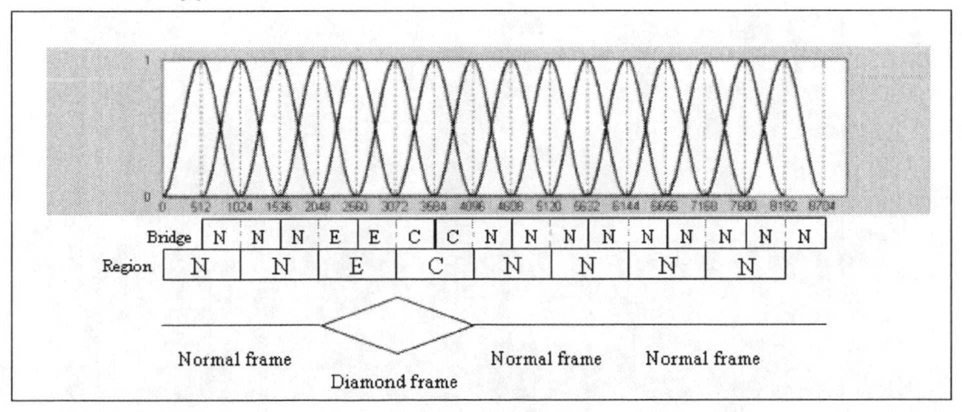

sloping sides. Theoretically, a diamond should have a slope greater than zero while an unmarked frame should have a slope close to zero. Once the binary symbols are identified, the next task is to determine the letters of alphabet and hence the signature from the binary sequence.

Performance of the Technique

The method is tested using excerpts from the following five audio pieces: (1) classical music, (2) instrumental music, (3) rock music, (4) pop music, and (5) male speech. Rock music and pop music have very high signal energy while classical music and instrumental music have moderate signal energy. The speech clip has a relatively high percentage of silent intervals.

The audio quality of the watermarked signals is evaluated through listening tests. The listeners were presented with the original signals and the watermarked signals in random, and asked to indicate their preference. It is found that the embedded watermark does not affect the audio quality to any significant degree.

Recovery is perfect for all audio signals tested when they are not attacked by further signal processing.

The following five types of attack (further signal processing) are applied on the watermarked signals.

1. **Addition of white Gaussian noise:** White Gaussian noise is added so that the resulting signal has a signal-to-noise ratio of 30dB.
2. **Resampling:** The test signal originally sampled at 44.1 ksamples/s is re-sampled at 22.05 ksamples/s, and then restored by sampling again at 44.1 ksamples/s.
3. **Cropping:** Some segments of the watermarked signal are replaced with segments of the signal attacked with filtering and additive noise.
4. **Low-pass filtering:** A finite impulse response (FIR) low pass filter with cutoff frequency at 10 kHz, passband ripple 0.173dB and stopband attenuation 40dB is used.
5. **MP3 conversion:** The watermarked test signal is encoded into a 128kbits/s MP3 format and then restored.

The watermark or signature embedded was 'FSW.' The rate of recovery of a watermark, defined as the ratio of the number of watermark

Figure 2. The difference signal

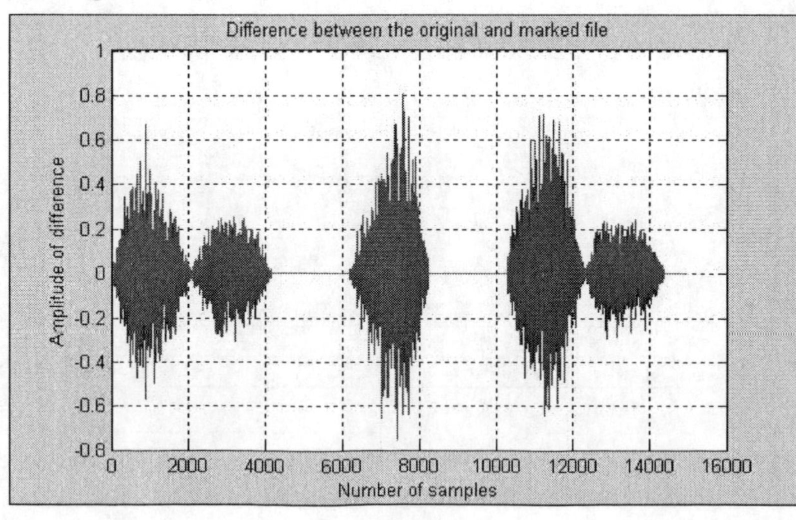

Table 1. Rate of recovery of watermark

Types of Attack	Types of Signal	Signature	Letter
Addition of noise	*Classical*	100%	100%
	Instrumental	89%	93%
	Rock	100%	100%
	Pop	90%	97%
	Speech	100%	100%
Re-sampling	*Classical*	90%	97%
	Instrumental	100%	100%
	Rock	78%	93%
	Pop	22%	57%
	Speech	80%	93%
Cropping	*Classical*	100%	100%
	Instrumental	100%	100%
	Rock	100%	100%
	Pop	100%	100%
	Speech	60%	67%
Low-pass filtering	*Classical*	90%	97%
	Instrumental	89%	97%
	Rock	89%	97%
	Pop	22%	57%
	Speech	80%	93%
MP3 compression	*Classical*	100%	100%
	Instrumental	97%	100%
	Rock	11%	69%
	Pop	11%	57%
	Speech	60%	87%

recovered over the total number of watermark present, is assessed in two ways: recovery of a *signature* and recovery of a *letter*. A signature is a complete string of three letters, for example 'FSW.' A 50% rate of signature recovery indicates that for a complete string of ten 'FSW's, five complete 'FSW's are recovered. Successful recovery of a *letter* means that the letter is recovered at the right position in the signature. If 'WSF' is recovered in place of 'FSW,' only the letter 'S' is deemed correctly recovered and the percentage recovery is taken as 33%.

The rates of watermark recovery are presented in Table 1. It can be seen that when the watermarked signal is under attack, certain audio signals like instrumental music and classical music stand out as good host signals for the watermarking technique, in that the various attacks have not affected the watermark recovery significantly. Other audio signals like rock music and pop song exhibit good percentage recovery of complete signatures under some attacks like the addition of noise and cropping, but are inadequate under others like filtering and MP3 compression. For male speech, cropping and MP3 conversion affect the rate of recovery the most.

In all cases under investigation, although the rate of recovery is affected after attacks, the watermark can still be identified as there are many copies of the watermark embedded throughout the duration of the signal. The method thus achieves a balance between the aural imperceptibility of the introduced distortion and the robustness of the watermark.

BLIND WATERMARKING TECHNIQUE WITHOUT ADDITIONAL SYNCHRONISATION INFORMATION

Overview of the Technique

A blind audio watermarking technique that makes use of the energy feature of the signal itself for synchronisation, the peak-point extraction technique, is described in this section. For this blind digital audio watermarking technique, the watermark is embedded in the discrete cosine transform (DCT) domain. The psychoacoustic model is incorporated to ensure the imperceptibility of the embedded watermark. Error correcting codes are used to improve the rate of correct watermark recovered. Details of the various steps are presented in the following subsections.

Identification of Synchronisation Points

First, a set of synchronisation points are identified in the time domain. An energy-based synchronisation technique called the peak points extraction (PPE) technique is used. For this technique, the power of the original signal is specially shaped by raising the sample value to a high power as exemplified by the following equation:

Figure 3. Audio signal before and after shaping

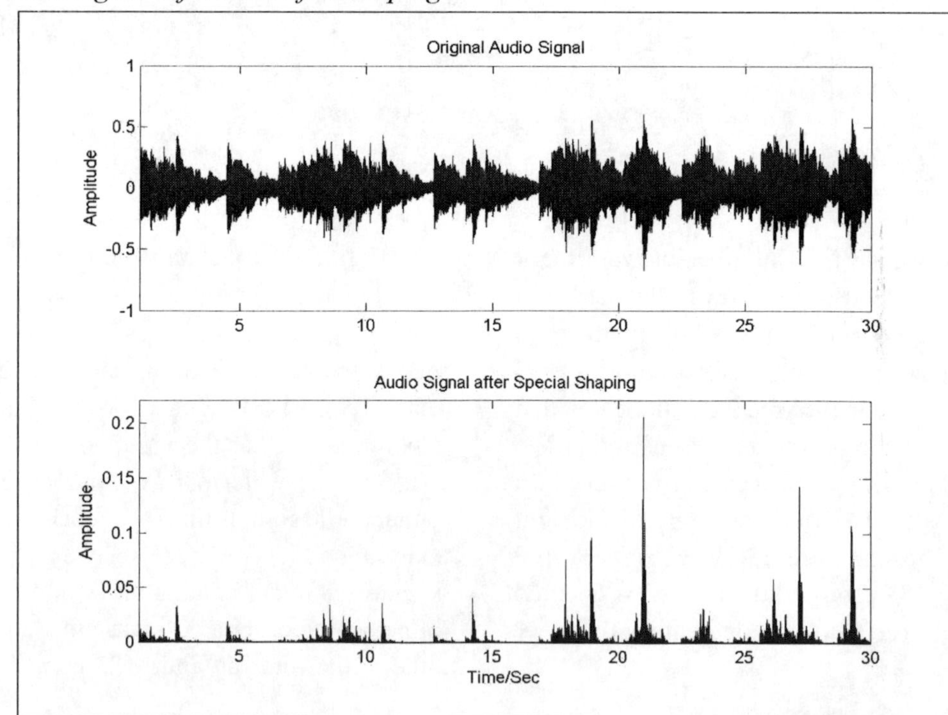

$$x'(n) = x^4(n) \qquad (1)$$

where $x(n)$ is the original audio signal, and $x'(n)$ is the signal after shaping. Various power factors (e.g., power of 2, power of 3, etc.) are tested, and power of 4 is found to give the best compromise between the ability to extract the peak points and the efficiency of computation.

This process intentionally modifies the signal shape to exaggerate the energy differences between the peak regions and low-energy regions (Figure 3). After special signal shaping, the specific regions for watermark insertion are then identified by comparing the resulting signal with a threshold.

The threshold is determined adaptively for different audio signals. The threshold is chosen at 15% of the sample value of the highest peak of every piece of signal. Samples that have value higher than the threshold are extracted as the peak points. A raw set of peak points is thus obtained.

There is usually a group of samples appearing at the peak points. The last point of every group is taken as a synchronisation point. As one frame is required to encode one bit of the watermark, there must be sufficient number of frames between two peak points. If the number of frames between two consecutive peak points does not satisfy the requirement stated above, the last peak point is considered a redundant synchronisation point, and is removed from the raw set of peak points. At the end of the process, a set of synchronisation points is obtained.

WATERMARK EMBEDDING

The watermark is embedded into the audio signal in binary form. If the signature or the copyright information consists of alphanumeric characters, the characters are first encoded using binary codes before embedding. For the experiments carried out, a 6-bit code is used to represent the alphanumeric characters. Error correcting codes (Lathi, 1998) are employed to improve the accuracy of detection. A (6, 10) linear block code is adopted for the experiments. Such a code is able to correct all single bit errors, and up to five double bit errors.

The watermark is embedded in the frequency domain. Embedding the watermark in the frequency domain has advantages over embedding in the time domain in terms of imperceptibility and robustness. Discrete cosine transform is adopted for transforming the samples into frequency domain.

The block diagram of the watermark embedding process is given in Figure 4. The actual watermark embedding consists of three stages:

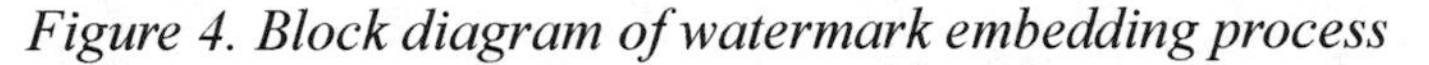

Figure 4. Block diagram of watermark embedding process

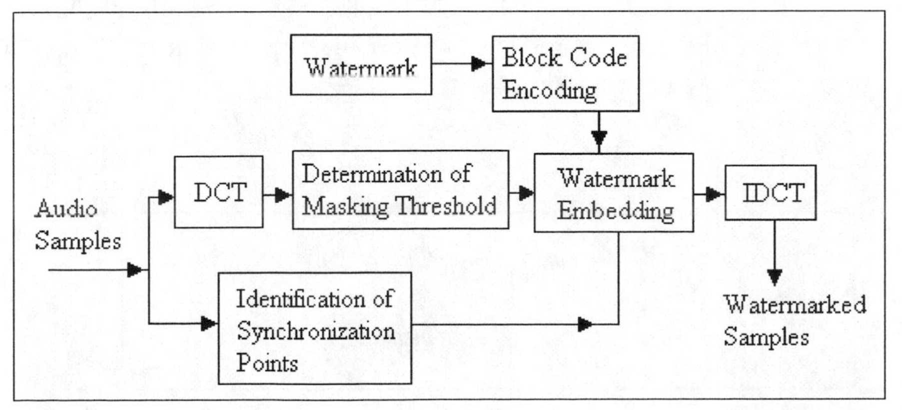

pre-processing, layer one embedding, and layer two embedding.

1. **Preprocessing:** The alphanumeric symbols of the signature are encoded into 6-bit binary data words (without error correction) or 10-bit binary code words using (6, 10) linear block codes.
2. **Layer one embedding (symbol embedding):** The locations for embedding the alphanumeric symbols are identified. Each synchronisation point is the starting point of one embedded alphanumeric symbol. The *n*-bit code word representing the alphanumeric symbol is embedded after the synchronisation point using *n* consecutive non-overlap bit-embedding frames. A bit-embedding frame is a frame of samples used to embed one binary bit. The *n* consecutive bit-embedding frames are together referred as a symbol-embedding frame. In most cases, the number of synchronisation points obtained is greater than the number of alphanumeric symbols in the signature. So the signature is repeatedly embedded throughout the audio signal.
3. **Layer two embedding (bit embedding):** After the symbol-embedding frames are determined, the locations of all bit-embedding frames are also confirmed. The binary bit '1' or '0' is embedded into the bit-embedding frames in the layer two embedding process. Every bit-embedding frame has 1024 samples. The psychoacoustic model is applied to each bit-embedding frame. The mask threshold of each bit-embedding frame is determined. The binary bit is embedded in the DCT domain by manipulating those frequency components below the masking threshold, which are known as the imperceptible components. More robustness can be achieved if hidden data are placed among the low frequency coefficients in the DCT domain (Cox, Kilan, Leighton, & Shamoon, 1997).

For each bit-embedding frame, the imperceptible component with the maximum power is selected from among all the imperceptible components and is used as a reference point. The reference point is always found in the low frequency range. Five succeeding imperceptible components after the reference point are used to code the binary bit. They are referred to as the information carriers. Five information carriers are used to add in redundancy to reduce the probability of error in the recovery process.

If a binary '1' is to be embedded in the bit-embedding frame, the signs of all five information carriers are made positive, regardless of the original signs. On the other hand, if a binary '0' is to be embedded in the bit-embedding frame, the signs of all five information carriers are made negative, regardless of the original signs.

Mathematically, (see equation (2)), where C_i is the value of the information carriers before embedding, C_i' is the value of the information carries after embedding.

This bit-embedding mechanism preserves the original power spectrum. The bit is embedded into

Equation 2.

$$C_i' = \begin{cases} +|C_i| & if \quad '1' \quad to \quad be \quad embedded \\ -|C_i| & if \quad '0' \quad to \quad be \quad embedded \end{cases} \qquad i = 1,2,3,4,5$$

the information carriers by changing their signs (the phases). The embedding process does not alter the amplitudes of those information carriers. The preservation of the power spectrum ensures that the reference point and the five information carriers can be determined without error or displacement in the watermark recovery process.

After the watermark is embedded, inverse discrete cosine transform (IDCT) is carried out to obtain the watermarked signal.

Watermark Recovery

The block diagram of watermark recovery process is given in Figure 5. The watermark recovery process is basically the reverse of the watermark embedding process. It also consists of three stages:

1. **Preprocessing:** The primary task of the preprocessing stage is to recover the synchronisation points. The synchronisation points are extracted from the watermarked audio signal. Subsequently, the locations of all bit-embedding frames are determined. The recovery process starts with recovering the binary bit embedded in every bit-embedding frame.
2. **Layer two recovery (bit recovery):** The psychoacoustic model is applied to every bit-embedding frame. The masking threshold is estimated and the imperceptible components are determined. The reference point and information carriers are then determined in the same manner as in the embedding process.
   The signs of the DCT coefficients of the five information carriers are examined. The decision is made based on the signs of the majority of the five information carriers. A majority of positive signs is taken to indicate binary '1' and a majority of negative signs is taken to indicate a binary '0'.
3. **Layer one recovery (symbol recovery):** Every symbol-embedding frame consists of *n* bit-embedding frames. After the bit-recovery process in the *n* consecutive bit-embedding frames, an *n*-bit binary code word is recovered from each letter-embedding frame. The alphanumeric symbols are then decoded and the watermark recovered.

Performance of the Technique

Audio signals of 30 seconds each derived from five diverse sources are selected to test the per-

Figure 5. Block diagram of watermark recovery process

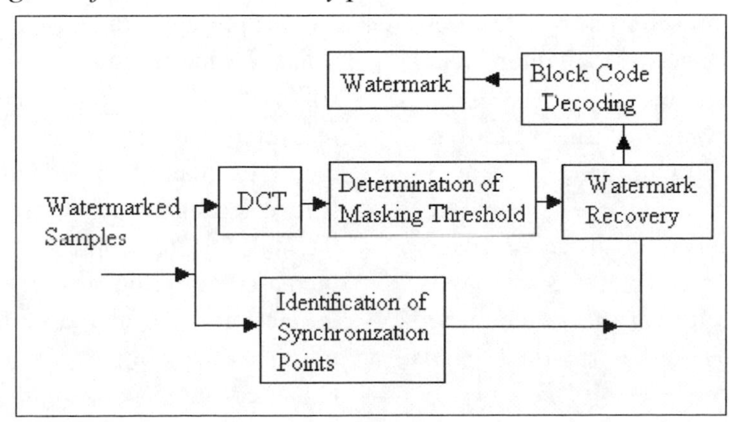

Table 2. Rate of recovery of watermark

Types of Attack	Types of Signal	ROCLR	ROCBR
Addition of noise	*Classical*	67%	92%
	Instrumental	88%	98%
	Piano	22%	68%
	Pop	59%	83%
	Speech	70%	83%
Re-sampling	*Classical*	83%	92%
	Instrumental	75%	81%
	Piano	94%	99%
	Pop	62%	84%
	Speech	81%	90%
Cropping	*Classical*	100%	100%
	Instrumental	100%	100%
	Piano	100%	100%
	Pop	88%	96%
	Speech	93%	99%
Low-pass filtering	*Classical*	33%	69%
	Instrumental	50%	73%
	Piano	39%	79%
	Pop	35%	71%
	Speech	75%	85%
MP3 compression	*Classical*	17%	42%
	Instrumental	38%	71%
	Piano	56%	82%
	Pop	47%	75%
	Speech	70%	82%

formance of the proposed technique. The five sources consist of Classical Music, Instrumental Music, Piano Music, Pop Song, and Male Speech. A (6, 10) linear block codes with bit interleaving is used for error correction.

Listening tests reveal that the watermark does not affect the quality of the original audio signals.

With no added signal processing, the recovery of the watermark is perfect for all five test signals. The effects of the five types of attacks mentioned earlier in the chapter on the watermarked signals are investigated.

The signature 'FSW' is used as the watermark. The rates of recovery under different types of attack are presented in Table 2. In the table, *ROCLR* stands for the ratio of correct letters recovered, while *ROCBR* stands for the ratio of correct bits recovered. The ratios are computed based on the first signature embedded in the signal only. If repeated signatures in the signal are used and the majority rule applied, the accuracy may be increased. As the bit recovery rate is much higher, the performance of the system can further be improved by using error correcting codes with higher correction capability.

It can be seen that the ability to recover the signature is dependent on the type of audio signal and the nature of the attack. Results show that the technique is robust to cropping.

BLIND WATERMARKING TECHNIQUE WITH INFORMATION ON SYNCHRONISATION

Overview of the Technique

By adding a delayed version of the signal to itself, creating a so-called echo, binary information can be embedded using two different values of delay. It is assumed that by limiting the values of the delay, the echoes shall be inaudible. However, the perceptual quality of the audio signal watermarked using simple echo hiding technique (Gruhl, Lu, & Bender, 1996) is poor; the echoes are at time highly audible.

The audibility of the echoes may be reduced using adaptive and content-based audio watermarking techniques based on single and multiple echo hiding. The adaptation is carried out by changing the decay rates of the echoes according to the energy level of segments of the signal.

Watermark Embedding

Segmentation

The original audio is first divided into segments of 0.25s duration. The segment that satisfies a set of given criteria will be coded with one bit of the watermark represented by the echo kernel.

Figure 6. Single echo kernels

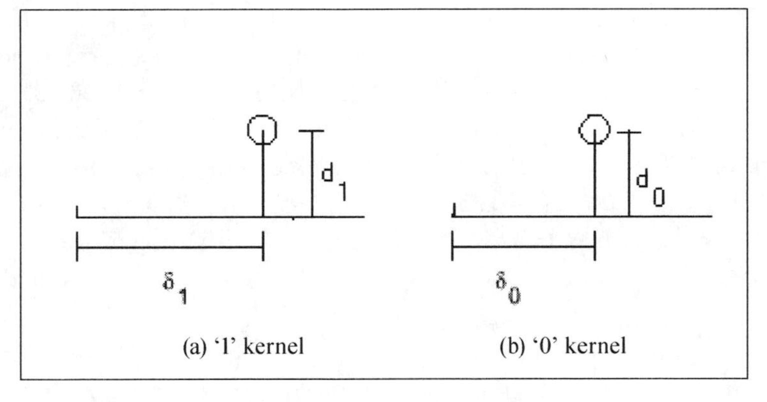

Figure 7. Multiple echo kernels

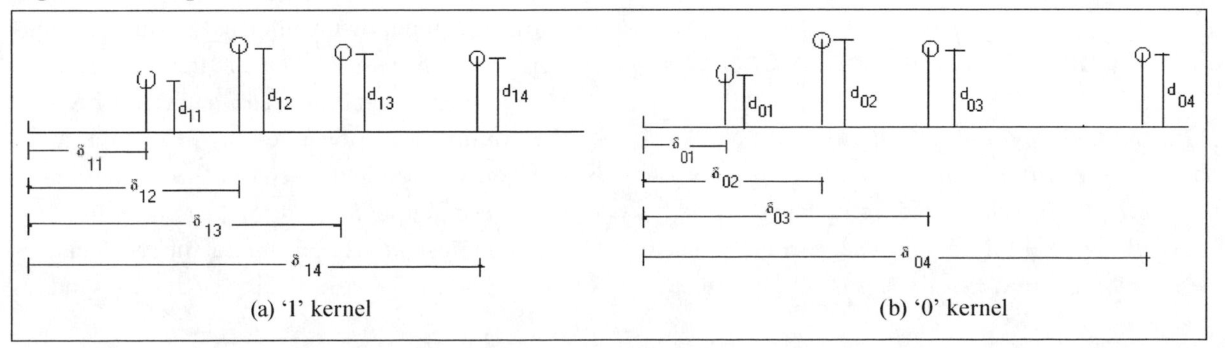

Two forms of echo kernels, the single echo kernels and the multiple echo kernels, are investigated. They are depicted in Figure 6 and Figure 7 respectively.

The parameters used for the single echo kernels are given below:

- $\delta_1 = 0.001$s (Delay for '1' kernel)
- $\delta_0 = 0.0013$s (Delay for '0' kernel)
- $d_1 = 0.5$ (Decay rate for '1' kernel to be adaptively modified by masking)
- $d_0 = 0.5$ (Decay rate for '0' kernel to be adaptively modified by masking)

(For simple echo hiding technique, d_1 and d_0 are fixed).

For multiple echo hiding, four impulses are used to create four echoes as shown in Figure 7. The parameters used for the echo kernels are:

- $\delta_{11}=0.00104$s, $\delta_{12}=0.00183$s, $\delta_{13}=0.00220$s, $\delta_{14} = 0.00262$s (Delays for '1' kernel)
- $\delta_{01}=0.00127$s, $\delta_{02}=0.00133$s, $\delta_{03}=0.00136$s, $\delta_{04} = 0.00238$s (Delays for '0' kernel)
- $d_{11} = 0.15$, $d_{12} = 0.45$, $d_{13} = 0.40$, $d_{14} = 0.50$ (Decay rates for '1' kernel to be adaptively modified by masking)
- $d_{01} = 0.45$, $d_{02} = 0.35$, $d_{03} = 0.20$, $d_{04} = 0.35$ (Decay rates for '0' kernel to be adaptively modified by masking)

The values chosen for the parameters are determined using the multi-objective evolutionary algorithm (MOEA) (Coelle, Veldhuizen, & Lamont, 2002), an efficient search algorithm.

Mask Computation and Echo Encoding

Different segments contain different levels of energy; the 'quiet' segments have extremely low level of energy. It is problematic if a constant decay rate is used for all audio segments. As such, assessment is made of the signal energy of each segment to determine the maximum decay rates of the impulses in the kernel for the echoes. The root-mean-square signal energy for a segment of signal, E_{rms}, is computed using the following expression:

$$E_{rms} = \sqrt{\frac{1}{N}\sum_{i=1}^{N} x_i^2} \tag{3}$$

where x_i, i=1, 2, ..., N are the values of the N samples of the segment. The value so obtained is compared with a threshold. Only segments with signal energy higher than the threshold are coded with the watermark bit. The decay rates for the echoes in each audio segment are adaptively determined based on the masking level determined using MPEG psychoacoustic model.

Watermark Checking

To ensure that the watermark can be decoded, decoding is carried out on the watermarked segment. If the peaks corresponding to the echo kernel being used are detected, the watermarked segment is accepted and the segment number is registered; otherwise, this segment is skipped and the watermark is applied to the next selected segment. Information on the positions of the watermarked segments is recorded for watermark recovery.

Watermark Recovery

Watermark recovery is carried out on the watermarked segments using information on watermark positions. The essential task in the recovery process is peak detection. Peak detection involves the computation of autocepstrum on the audio segments stored in the vector of watermark positions. For single echo hiding, the magnitude of the autocepstrum is examined at the two locations corresponding to the delays of the '1' and '0' kernel respectively. If the autocepstrum is greater at δ_1 than it is at δ_0, it is decoded as '1' else it is decoded as '0'.

For multiple echo hiding, all peaks present in the autocepstrum are detected. The number of peaks corresponding to the delay locations of the '1' and '0' kernels are then counted and compared. If there are more peaks at the delay locations for the '1' echo kernel, it is decoded as '1'. On the other hand, if there are more peaks at the delay locations for the '0' echo kernel, it is decoded as '0'.

Performance of the Technique

The audio quality and robustness of the various encoding options for echo hiding are tested. The following five audio pieces are used: (1) pop song, (2) saxophone music, (3) er hu music, (4) a cappella song, and (5) male speech.

Pop song has very high signal energy while saxophone music and er hu music have moderate signal energy. A cappella song has low signal energy and contains noticeable periods of silence. The speech clip, as in most speech, consists of a high percentage of silent intervals.

Listening tests reveal that in general, the effect of watermarking is less audible in the adaptively modified watermarked signal than in the watermarked signal obtained using the simple echo hiding technique. Of the two simple echo hiding techniques investigated, the quality of watermarked audio with decay = 0.2 is much better than that with decay = 0.8. Echoes can readily be discerned for watermarked audio with decay = 0.8. It is also observed that in terms of inaudibility of watermark, multiple echo hiding yields better result than single echo hiding because the decay rate for each echo per segment is significantly reduced.

The rates of recovery of watermark for the five audio pieces using different echo hiding techniques and after different forms of attack are presented in Tables 3-7.

It can be seen that the detection accuracy is dependent on additional signal processing performed. One hundred percent accuracy can be achieved if there is no additional signal process-

Table 3. Rate of recovery of watermark for pop song

	Addition of Noise	Re-sampling	Cropping	Filtering	MP3 Compression
Simple Echo Hiding (decay=0.8)	100%	97%	97%	100%	91%
Simple Echo Hiding (decay=0.2)	71%	74%	74%	80%	74%
Adaptive (single echo hiding)	94%	91%	100%	89%	80%
Adaptive (multiple echo hiding)	97%	91%	100%	91%	74.%

Table 4. Rate of recovery of watermark for saxophone music

	Addition of Noise	Re-sampling	Cropping	Filtering	MP3 Compression
Simple Echo Hiding (decay=0.8)	97%	91%	94%	91%	89%
Simple Echo Hiding (decay=0.2)	69%	57%	77%	63%	63%
Adaptive (single echo hiding)	94%	91%	100%	86%	91%
Adaptive (multiple echo hiding)	89%	80%	100%	74%	71%

ing. The most damaging additional operation is MP3 compression.

It shall be noted that on the average, the simple form of echo hiding with decay = 0.8 is robust to the various forms of processing mentioned and the average rate of recovery is very high. However, this simple form of echo hiding loses out in audibility. When the decay rate is reduced to 0.2 for the simple echo hiding technique, the rates of recovery are generally below those achieved using the adaptive techniques.

Signal characteristics become important consideration if adverse signal processing is imposed. Pop song has very high signal energy with no gaps of silence in general. For this test clip, an excellent average detection accuracy of over 90% is achieved. The watermark is also easily recoverable even after common signal processing operations. On the other extreme, the average detection accuracy for male speech, a short speech with many silent intervals, is only 70%.

Though multiple echo hiding introduces more echoes per segment, the decay rate for each echo is much lower compared to single echo hiding. The watermarking techniques are more robust to cropping (average detection accuracy = 97.48%), noise addition (average detection accuracy = 84.82%), and resampling (average detection accuracy =

Table 5. Rate of recovery of watermark for er hu music

	Addition of Noise	Re-sampling	Cropping	Filtering	MP3 Compression
Simple Echo Hiding (decay=0.8)	100%	100%	97%	97%	94%
Simple Echo Hiding (decay=0.2)	77%	60%	74%	69%	60%
Adaptive (single echo hiding)	91%	91%	100%	80%	63%
Adaptive (multiple echo hiding)	86%	83%	97%	77%	69.%

Table 6. Rate of recovery of watermark for a cappella song

	Addition of Noise	Re-sampling	Cropping	Filtering	MP3 Compression
Simple Echo Hiding (decay=0.8)	100%	94%	100%	97%	89%
Simple Echo Hiding (decay=0.2)	74%	69%	83%	74%	63%
Adaptive (single echo hiding)	79%	79%	100%	79%	86%
Adaptive (multiple echo hiding)	93%	86%	93%	71%	71%

Table 7. Rate of recovery of watermark for male speech

	Addition of Noise	Re-sampling	Cropping	Filtering	MP3 Compression
Simple Echo Hiding (decay=0.8)	91%	86%	97%	89%	91%
Simple Echo Hiding (decay=0.2)	60%	69%	80%	66%	54%
Adaptive (single echo hiding)	69%	77%	94%	74%	69%
Adaptive (multiple echo hiding)	74%	80%	97%	69%	54%

81.74%) than the other signal processing operations because filtering and MPEG coding/decoding remove much more signal information.

Although multiple echo hiding improves inaudibility in general, the rate of recovery is compromised when additional signal processing is imposed. If inaudibility and accuracy in detection are of paramount importance, psychoacoustic model coupled with single echo is the most appropriate encoding option to use when additional signal processing is imposed.

Watermark Security

For the methods described in this chapter, the watermarks adopted are sequences of alphanumeric characters. If secrecy of the embedded information is an important consideration, then established cryptographic techniques (Schneier, 1996) may be applied.

CONCLUSION

Three techniques for audio watermarking are described in this chapter that illustrate the nonblind watermarking scheme and the blind watermarking schemes with or without synchronisation information. For the time-frequency expansion-compression technique, the original signal is required to recover the watermark. For the peak-point method, watermark can be recovered from the watermarked signal itself. For the adaptive echo hiding methods, although the original signal is not required for the recovery of the watermark, synchronisation information is required.

The audio quality of the watermarked signals using the three techniques remains high. Although the rates of recovery of the watermark are affected by the various forms of attack, the signature and hence the ownership of the signals can be identified as long as there are sufficient number of embedded duplicate watermarks. Thus all three techniques are suitable for audio watermarking. The choice of method depends very much on the types of signal to be watermarked and the applications.

FUTURE RESEARCH DIRECTIONS

Many methods have been proposed for audio watermarking, these include nonblind watermarking schemes and the blind watermarking schemes with or without synchronisation information. The robustness of a watermarking scheme depends on the technique used, the types of audio signals, and the modes of attack. To preserve the quality of the watermarked signal, psychoacoustic models or perceptual coding scheme are popularly adopted.

The various schemes have their own merits and limitations. For example, the use of least significant bits has the advantage of simplicity and high data rate; however, the speech quality is affected. It is suggested that for future research, some hybrid schemes using a combination of established methods be investigated. The component methods may be chosen in such a way using the strengths of one technique to overcome the weaknesses arising from other techniques.

As psychoacoustic model plays an important role in preserving the quality of watermarked audio signals, it should be explored in any method investigated.

REFERENCES

Arnold, M. (2000). *Audio watermarking: Features, applications and algorithms.* Paper presented at the IEEE International Conference on Multimedia and Expo 2000 (Vol. 2, pp. 1013-1016).

Bassia, P., & Pitas, I.P. (1998, September 8-11). Robust audio watermarking in time domain. *EUSIPCO,* 25-28.

Boney, L., Tewfik, A., & Hamdy, K. (1996). *Digital watermarks for audio signals.* Paper presented at the International Conference on Multimedia Computing and Systems.

Coello, C.A., Veldhuizen, D.A., & Lamont, G.B. (2002). *Evolutionary algorithms for solving multi-objective problems*. Kluwer Academic Publishers.

Cox, I.J., Kilan, J., Leighton, T., & Shamoon, T. (1997). Secure spread spectrum watermarking for multimedia. *IEEE Transactions on Image Processing, 6*(12), 1673-1687.

Cox, I.J. (1998). Spread spectrum watermark for embedded signalling. *United States Patent 5,848,155.*

Cvejic, N., & Seppänen, T. (2005). Increasing robustness of LSB audio steganography by reduced distortion LSB coding. *Journal of Universal Computer Science, 11*(1), 56-65.

Foo, S.W., Yeo, T.H., & Huang, D.Y. (2001). An adaptive audio watermarking system. *IEEE Tencon.*, 509-513.

Foo, S.W., Ho, S.M., & Ng, L.M. (2004). *Audio watermarking using time-frequency compression expansion.* Paper presented at the IEEE International Symposium on Circuits and Systems (pp. 201-204).

Foo S.W., Xue, F. & Li M. (2005). *A blind audio watermarking scheme using peak point extraction*. Paper presented at the IEEE International Symposium on Circuits and Systems (pp. 4409-4412).

Garcia R.A. (1999) *Digital watermarking of audio signals using a psychoacoustic auditory model and spread spectrum theory.* Paper presented at the 107th Convention on Audio Engineering Society (preprint 5073).

Gruhl, D., Lu, A., & Bender, W. (1996). Echo hiding. In *Proceedings of the Information Hiding Workshop* (pp. 295-315).

Hsieh, C.T., & Tsou, P.Y. (2002). *Blind cepstrum domain audio watermarking based on time energy features.* Paper presented at the 14th International Conference on Digital Signal Processing (pp. 705-708).

Huang, J.W., Wang, Y., & Shi, Y.Q. (2002). *A blind audio watermarking algorithm with self-synchronization*. Paper presented at the International Symposium on Circuits and Systems 2002.

Kim, H.J., Choi, Y.H., Seok, J.W., & Hong, J.W. (2004). *Audio watermarking techniques: Intelligent watermarking techniques* (pp. 185-218).

Kiroski, D.K., & Malvar, H.M. *Embedding and detecting spread spectrum watermarks under the estimation attack*. Microsoft Research.

Lathi, B.P. (1998). *Modern digital and analog communication system* (3rd ed., pp. 728-737). Oxford University Press.

Li, W., Xue, X.Y., & Li, X.Q. (2003). Localized robust audio watermarking in regions of interest. *ICICS-PCM 2003.*

Neubauer, C., & Herre, J. (2000). *Advanced audio watermarking and its application*. Paper presented at the 109th AES Convention (AES preprint 5176).

Pohlmann, K.C. (1991). *Advanced digital audio.* Carmel.

Schneier, B. (1996). *Applied cryptography.* John Wiley & Sons.

Swanson, M.D., Zhu, B., Tewfik, A.H., & Boney L. (1998). Robust audio watermarking using perceptual masking. *Signal Processing, 66*(3), 337-355.

Vladimir, B., & Rao, K.R. (2001). An efficient implementation of the forward and inverse MDCT in MPEG audio coding. *IEEE Signal Processing Letters, 8*(2).

Wu, C.P., Su, P.C., & Kuo, C.J. (1999). *Robust audio watermarking for copyright protection.* Paper presented at the SPIE's 44th Annual Meeting on Advanced Signal Processing Algorithms, Architectures, and Implementations IX (SD39).

Xu, C., Wu, J., Sun, Q., & Xin, K. (1999). Applications of watermarking technology in audio signals. *Journal Audio Engineering Society, 47*(10).

Zwicker, E., & Fastl, H. (1990) *Psychoacoustics facts and models.* Berlin: Springer-Verlag.

Zwicker, E., & Zwicker, U.T. (1991) Audio engineering and psychoacoustics: Matching signals to the final receiver, the human auditory system. *J. Audio Eng. Soc., 39*, 115-126.

ADDITIONAL READING

Cassuto, Y., Lustig, M., & Mizrachy, S. (2001). *Real time digital watermarking systems for audio signal using perceptual masking* (Internal Report). Israel Institute of Technology.

Cedric, T., Adi, R., &Mcloughlin. I. (2000). Data concealment in audio using a nonlinear frequency distribution of prbs coded data and frequency-domain lsb insertion. In *Proceedings of the IEEE Region 10 International Conference on Electrical and Electronic Technology* (pp. 275-278). Kuala Lumpur, Malaysia.

Cox, I.J., & Miller, M.L. (1997) A review of watermarking and the importance of perceptual modeling. In *Proceedings of SPIE Human Vision and Electronic Imaging* (Vol. 3016, pp. 92-99).

Cox, I., Miller, M., & Bloom, J. (2003) Digital watermarking. San Francisco: Morgan Kaufmann Publishers.

Cvejic, N. (2007). *Algorithms for audio watermarking and steganography.* Retrieved March 20, 2007, from http://herkules.oulu.fi/isbn9514273842/isbn9514273842.pdf

Furon, T., Moreau, N., & Duhamel, P. (2000). Audio public key watermarking technique. In *Proceedings of the IEEE International Conference on Acoustics, Speech, and Signal Processing* (pp. 1959-1962). Istanbul, Turkey.

Ikeda, M., Takeda, K., & Itakura, F. (1999). Audio data hiding use of band-limited random sequences. In *Proceedings of the IEEE International Conference on Acoustics, Speech, and Signal Processing* (pp. 2315–2318). Phoenix, Arizona.

Kahr, M., & Branderburg, K. (1998). *Applications of digital signal processing to audio and acoustics.* Kluwer Academic Publishers.

Kirovski, D., & Malvar, H. (2001). Spread-spectrum audio watermarking: Requirements, applications, and limitations. In *Proceedings of the IEEE International Workshop on Multimedia Signal Processing* (pp. 219-224). Cannes, France.

Kirovski, D., & Malvar, H. (2003) Spread-spectrum watermarking of audio signals. *IEEE Transactions on Signal Processing, 51*(4), 1020-1033.

Kundur, D. (2001). Watermarking with diversity: Insights and implications. *IEEE Multimedia, 8*(4), 46-52.

Kundur, D., & Hatzinako, D. (1997). A robust digital image watermarking method using wavelet-based fusion. In *Proceedings of the IEEE International Conference on Image Processing* (Vol. 1, pp. 544-547).

Moore, B.J.C. (1997). *An introduction to the psychology of hearing.* Academic Press.

Neoliya, M. (2003). Digital audio watermarking using psychoacoustic model and spread spectrum theory. In *Proceedings of the UROP Congress,* Nanyang Technological University, Singapore.

Neubauer, C., & Herre, J. (2000). Advanced audio watermarking and its applications. In *Proceedings of the AES Convention, Audio Engineering*

Society preprint 5176 (pp. 311-319). Los Angeles, California.

Neubauer, C., & Herre, J. (1998). Digital watermarking and its influence on audio quality. In *Proceedings of the AES Convention on Audio Engineering Society* (preprint 4823) (pp. 225-233). San Francisco.

Oh, H.O., Seok, J., Hong, J., & Youn, D. (2001). *New echo embedding technique for robust and imperceptible audio watermarking.* Paper presented at the IEEE International Conference on Acoustic, Speech and Signal Processing (pp. 1341-1344). Salt Lake City, Utah.

Seok, J., & Hong, J. (2001) Audio watermarking for copyright protection of digital audio data. *Electronics Letters, 37*(1), 60–61.

Pan, D. (1995). A tutorial on mpeg/audio compression. *IEEE Multimedia, 2*(2), 60-74.

Wu, M., & Liu, B. (2003). *Multimedia data hiding.* New York: Springer Verlag.

Xu, C., Wu, J., Sun, Q., & Xin, K. (1999). Applications of watermarking technology in audio signals. *Journal Audio Engineering Society, 47*(10), 1995-2007.

Yardimci, Y., Cetin, A.E., & Anson, R. (1997). *Data hiding in speech using phase coding.* Paper presented at the ESCA, Eurospeech 97 (pp.1679-1682). Greece.

Chapter VI
Advanced Audio Watermarking Based on Echo Hiding:
Time-Spread Echo Hiding

Ryouichi Nishimura
Tohoku University, Japan

Yôiti Suzuki
Tohoku University, Japan

Byeong-Seob Ko
Samsung Electronics, Korea

ABSTRACT

This chapter introduces time-spread ***echo*** *hiding as an advanced audio* ***watermarking*** *method based on echo hiding. After reviewing some relevant researches, theoretical derivation of echo hiding and time-spread echo hiding is presented. As for embedding process, differences in the structure of echo* ***kernels*** *between the ordinary echo hiding and the time-spread echo hiding are schematically depicted to explain advantages of the new method. Several watermark-extracting methods are introduced for the decoding process to raise the detection rate for various conditions. Performance of the method in robustness against several* ***attacks*** *is evaluated in terms of d' because of its statistical preciseness. Results of computer simulations tell that the time-spread echo hiding show fairly better performance than the ordinary echo hiding.*

INTRODUCTION

Violation of copyrights is a serious concern in the field of sound development of digital media contents. Digital watermarking is an important technique for protecting copyrights of digital media contents (Czerwinski, Fromm, & Hodes, 1999; Podilchuk & Delp, 2001). Many novel techniques have been proposed for watermarking of digital audio contents over the past few years, for example, techniques based on **masking** (Boney, Tewfik, & Hamdy, 1996; Cvejic, Keskinarkaus, & Seppanen, 2001; Swanson, Zhu, Tewfik, & Boney, 1998), spread spectrum (SS) (Bender, Gruhl, Morimoto, & Lu, 1996; Cox, Kilian, Leighton, & Shamoon, 1995), phase coding (Bender et al., 1996), phase modulation (Nishimura, Suzuki, & Suzuki, 2001; Takahashi, Nishimura, & Suzuki, 2005), echo hiding (Gruhl & Bender, 1996; Oh, Seok, Hong, & Youn, 2001; Xu, Wu, Sun, & Xin, 1999), and data mining (Sasaki, Nishimura, & Suzuki, 2006).

In those techniques, echo hiding provides many benefits from various points of view, for example, robustness, imperceptibility, simple encoding, and decoding processes (Gruhl & Bender, 1996). In addition, detection rules for conventional single-echo hiding are lenient; anyone can detect information embedded in a host signal without any special key. This is an advantage of this technique, but simultaneously, considering the possibility of malicious tampering, it is also a point of vulnerability. This disadvantage is fatal from the viewpoint of protecting copyrights (Katzenbeisser & Petitcolas, 2000). Meanwhile, the robustness of watermark can be strengthened simply by increasing the amplitude of the echo. However, a high level echo causes coloration in timbre of a watermarked signal because of the ripple in its frequency response.

Several useful encoding processes for echo hiding have been proposed to cope with these problems. One is multiecho embedding (Xu et al., 1999). However, this echo embedding has limitations in allocation of delay time of echo and of time-slot for assigning multiple bits because preserving imperceptibility is difficult to realize in the case of simple multiple echoes. Moreover, in the case of a malicious attack using multiple encoding (Katzenbeisser & Petitcolas, 2000; Lee & Ho, 2000), echo hiding cannot ensure robustness against such an attack because no secret key exists to hide correct information from that of malicious tampering.

Another successful extension of echo hiding based on multiple echoes is a scheme with backward and forward kernels (Kim & Choi, 2003). Echo hiding, which utilizes the temporal forward masking to hide echoes, usually convolves an original signal with a kernel consisting of one impulse for the direct sound and successive one or more small impulses for echoes to embed watermarks. Meanwhile, a backward and forward kernel has an impulse before the one for the direct sound in addition to the usual one. Although it does not satisfy the causality, it could be unnoticeable thanks to the backward masking if the level of the echo and its temporal gap from the direct sound is sufficiently small. The most important result of this scheme is that it can drastically improve the watermark detection compared with a simple echo hiding.

In this chapter, a watermarking method, in which a time-spread echo is employed as an alternative to a single echo or a multiecho in echo hiding techniques, is introduced in detail. The basic algorithm of this method has been proposed in the articles (Ko, Nishimura, & Suzuki, 2002a, 2002b). By spreading an echo using **pseudo noise (PN) sequences**, the amplitude of each echo becomes small and its power spectrum becomes nearly flat in the mean time sense. This makes it difficult to decode the embedded data without the PN sequence used in the encoding process, and might result in good imperceptibility as well as natural and colorless sound quality after watermarking.

The method's decoding performance is evaluated as functions of following parameters: amplitude and length of PN sequences used in spreading an echo, length of the discrete Fourier transform (DFT) in the decoding process, and so on. The decoding performance and the frequency response of the method are also evaluated with those of conventional single echo watermarking. Moreover, robustness of the method against typical attacks, such as resampling, requantization, **MPEG 1 Layer III**, MPEG 2 AAC, digital to analog (D/A) and analog to digital (A/D), and so forth, is evaluated. The imperceptibility is also investigated via a listening test. Lastly, the refinement of the method to improve in watermark detection for the case of pitch scaling attack is discussed.

TIME-SPREAD ECHO KERNEL USING PN SEQUENCES

Figure 1 (a) shows that single echo hiding can be compared to a situation in which a direct sound and an echo (reflected sound) coexist. Accordingly, watermarks are embedded at a single constant displacement from the direct sound. However, as shown in Figure 1(b), sound in a real room includes both direct sound and many echoes heard as reverberation because of multiple reflections from the

Figure 1. Concept of single echo hiding and that of the time-spread echo method: (a) Single echo hiding and (b) time-spread echo method

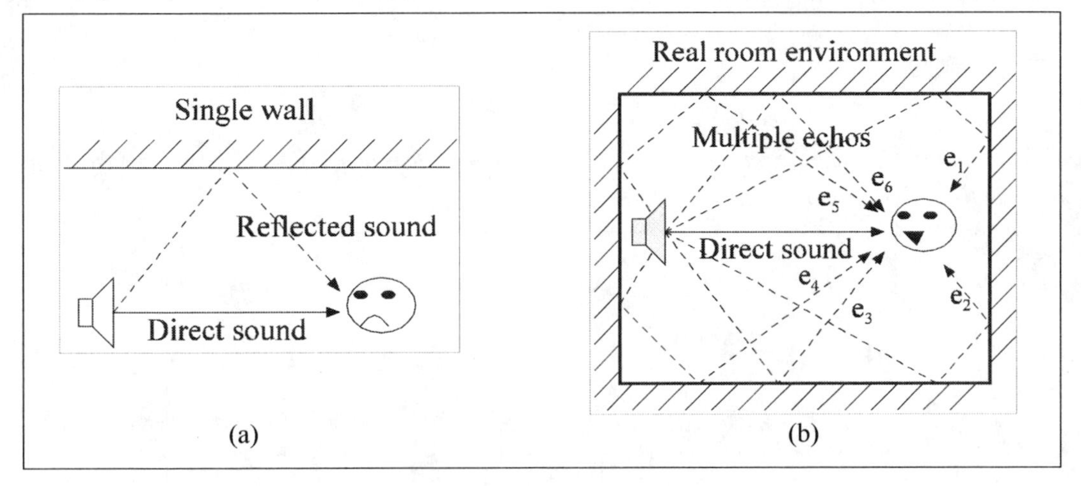

Figure 2. Single echo kernel and time-spread echo kernel using a PN sequence

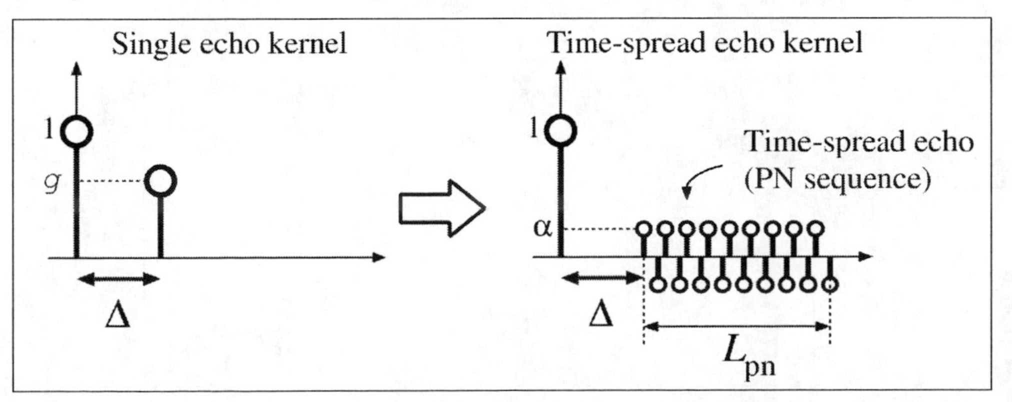

walls and other objects in the room. If watermarks can be embedded with various displacement from the direct sound, such complicated echoes can provide a more natural sound quality than only a single echo and several multiple echoes, even if those echoes are audible (Gardner, 1992; Ko & Kim, 1999). The time-spread echo method is based on this concept. Namely, to embed data into a host signal, the echo in Figure 1(a) is spread so that it becomes such as those in Figure 1(b).

Figure 2 illustrates the impulse responses of a single-echo kernel and a time-spread echo kernel. In the time-spread echo method, a single echo is temporally spread in the time domain using a PN sequence, which acts as a secret key to decode the embedded information from a watermarked signal. In Figure 2, g is the amplitude of the echo, α is the amplitude of the PN sequence, L_{PN} is the length of the PN sequence, and Δ is the time delay corresponding to watermark information (Gruhl & Bender, 1996).

The time-spread echo kernel in Figure 2 is constructed from a PN sequence such as:

$$k(n) = \delta(n) + \alpha p(n-\Delta),\ 0 < \alpha << 1, \qquad (1)$$

where $p(n)$ is an original PN sequence whose amplitude is ± 1, and $\delta(n)$ is the Dirac delta function.

Figure 3 shows a schematic explanation of how data are embedded into a host signal using the time-spread echo kernel. The watermarked signal is obtained by taking a linear convolution of the host signal and the time-spread echo kernel. In other words, the watermarked signal comprises a

Figure 3. Schematic explanation of how data are embedded into a host signal using the time-spread echo kernel (Ko, Nishimura, & Suzuki, 2005, p. 213)

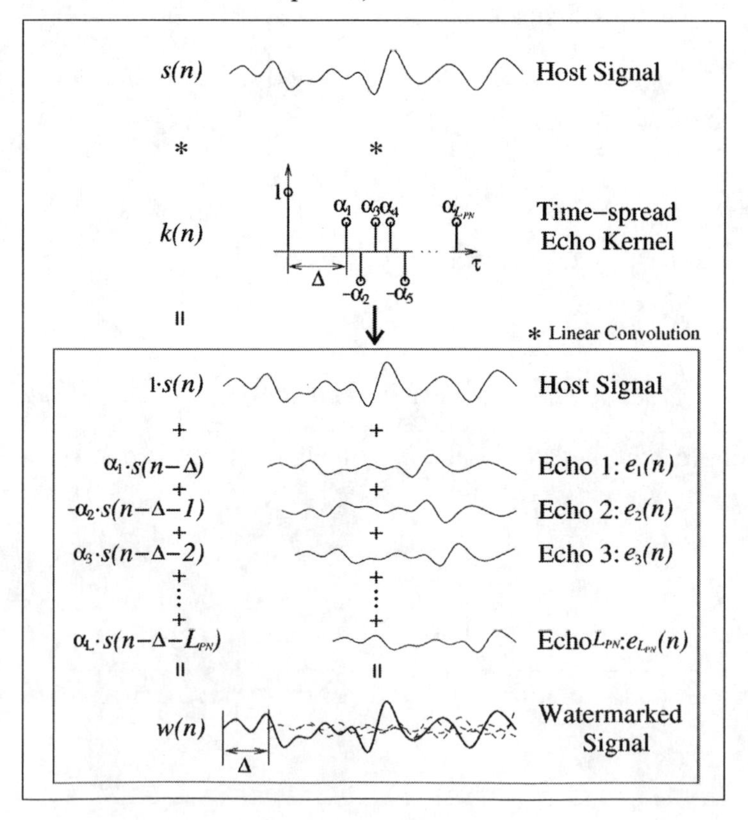

host signal and L_{PN} echoes generated by convolving the time-spread echo with the host signal.

Therefore, the watermarked signal is denoted as:

$$w(n) = s(n) * k(n), \tag{2}$$

where $s(n)$ is the host signal, $k(n)$ is the time-spread echo kernel, and * is a linear convolution.

DECODING PROCESS

Basic Idea

Figure 4 schematically shows the basic idea of the decoding process in the time-spread echo method. A watermarked signal is separable into a host signal component and echo signal components using a homomorphic process, for example, **cepstrum analysis** (Furui, 1992; Gruhl & Bender, 1996, Oppenheim & Schaffer, 1989). Namely, the peaks corresponding to the time-spread echo as well as the impulse at $\tau = 0$ are superposed linearly in the cepstrum domain, as shown at the bottom of Figure 4. In single-echo hiding, there is a large single peak corresponding to the echo in the cepstrum domain. For this reason, the embedded data might be decoded without any special information using only the cepstrum analysis. However, in the time-spread echo method, many peaks exist with very small amplitude in the cepstrum domain. Therefore, the embedded data are hardly detectable using cepstrum analysis alone. To decode the embedded data, we must despread the time-spread echo hidden in the cepstrum domain using the original PN sequence used in the encoding process, as shown at the bottom of Figure 4.

Formulation of the Decoding Process

Decoding based on the complex cepstrum: The decoding process shown in Figure 4 might be for-

Figure 4. Basic idea of the decoding process in the time-spread echo method (Ko, 2005, p. 213)

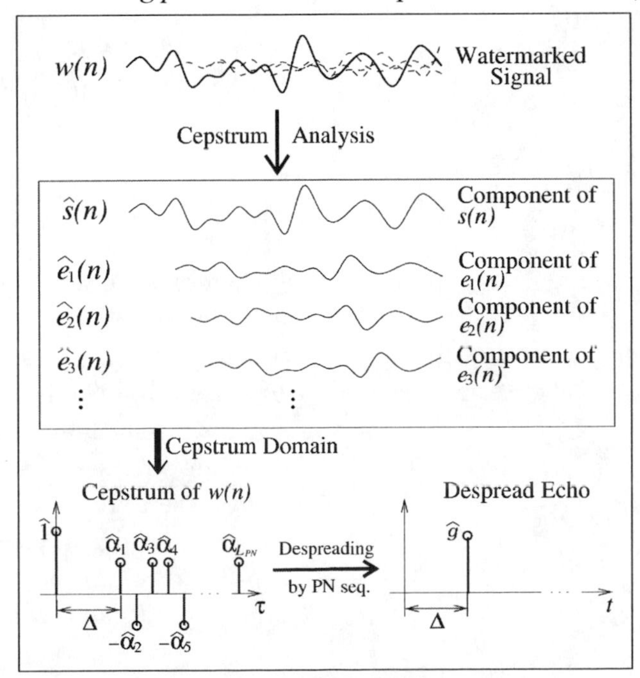

mulated as follows: Taking the cepstrum analysis of a watermarked signal $w(n)$ in equation (2), the complex cepstrum of $w(n)$ can be written as:

$$\hat{w}(n)=\hat{s}(n)+\hat{k}(n), \tag{3}$$

where $\hat{s}(n)=F^{-1}(\log(S(\omega)))$, $\hat{k}(n)=F^{-1}(\log(K(\omega)))$, $S(\omega)=F(s(n))$, $K(\omega)=F(k(n))$, $F(\cdot)$ is the Fourier transform, and $F^{-1}(\cdot)$ is the inverse Fourier transform. In equation (3), the complex cepstrum of a watermarked signal is represented as the addition of the complex cepstrum of a host signal and that of a time-spread echo kernel (Oppenheim & Schafer, 1989). To calculate the complex cepstrum of a time-spread echo kernel $k(n)$, taking the Fourier transform of $k(n)$ of equation (1), we get:

$$K(\omega)=1+\alpha e^{-j\omega\Delta}P(\omega), \tag{4}$$

where ω is radian frequency and $P(\omega)$ is the Fourier transform of $p(n)$. The logarithm of $K(\omega)$, $K_l(\omega)$, is expressed as:

$$K_l(\omega)=\log(K(\omega))=\log(1+\alpha e^{-j\omega\Delta}P(\omega)). \tag{5}$$

In equation (5), the second term in the logarithm might satisfy the condition:

$$|\alpha e^{-j\omega\Delta}P(\omega)|<1,$$

because the following equation stands for $0<\alpha<<1$ in equation (1):

$$0<\alpha<<\frac{1}{\max|P(\omega)|}, \tag{6}$$

where $\max|\cdot|$ is the maximum value. Equation (5) can then be expanded using Taylor series expansion as follows:

$$K_l(\omega)$$
$$=e^{-j\omega\Delta}P(\omega)-\frac{\alpha^2}{2}(e^{-j\omega\Delta}P(\omega))^2$$
$$+\frac{\alpha^3}{3}(e^{-j\omega\Delta}P(\omega))^3-\frac{\alpha^4}{4}(e^{-j\omega\Delta}P(\omega))^4+\cdots$$
$$=\alpha e^{-j\omega\Delta}P(\omega)+H.O.T.$$
$$\approx\alpha e^{-j\omega\Delta}P(\omega), \tag{7}$$

where H.O.T. represents higher-order terms. In addition, α is assumed to satisfy the condition of equation (6). Therefore, the higher-order terms in equation (7) can be omitted.

Next, taking the inverse Fourier transform of equation (7), the complex cepstrum of $k(n)$ becomes:

$$\hat{k}(n)=\alpha p(n-\Delta). \tag{8}$$

By substituting equation (8) into equation (3), we obtain:

$$\hat{w}(n)=\hat{s}(n)+\alpha p(n-\Delta). \tag{9}$$

The decoded signal is then obtained by taking cross-correlation between equation (9) and the original PN sequence $p(n)$. Consequently, the decoded signal using the complex cepstrum is denoted as:

$$d_c(n)=\hat{w}(n)\otimes p(n)$$
$$=n_{\hat{s}}(n)+\alpha p(n-\Delta)\otimes p(n), \tag{10}$$

where $\otimes$ is an operator of the cross-correlation and $n_{\hat{s}}(n)=\hat{s}(n)\otimes p(n)$ corresponds to the effect of the host signal as noise.

Figure 5 shows an example of the results of the decoding process: the decoded signal $d_c(n)$ when $\alpha=0.006$, $L_{PN}=1023$, $\Delta=1$ ms, and the length L_{FT} of DFT is 4096. Panel (a) shows the decoded signal when the host signal $s(n)$ is applied to the decoding process as the watermarked signal σ_{wn}, that is, there are no watermarks, and Panel (c) is when σ_{wn} is applied. Panels (b) and (d) respectively depict parts of (a) and (c). In Panel (c), the value of σ_{wn} is the standard deviation of $d_c(n)$ except

Figure 5. Decoded signal $d_c(n)$ when α=0.006, L_{PN} = 1023, L_{FT} = 4096, and Δ=1 ms (it corresponds to n≈44 samples when the sampling frequency F_S is 44.1 kHz)

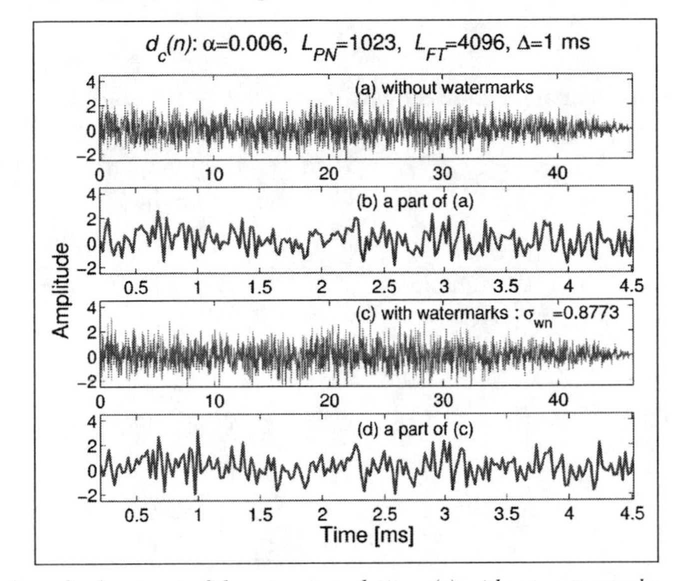

Note. The horizontal axis is time displacement of the cross-correlation: (a) without watermarks, (b) a part of (a), (c) with watermarks, and (d) a part of (c). (Ko et al., 2005, p. 214)

for the value at Δ because of the existence of a strong peak. In other words, the standard deviation corresponds to that of host-signal noise, $n_s(n)$. Therefore, smaller σ_{wn} is able to provide a clearer peak corresponding to the embedded information.

In Figures 5(a) and 5(b), no distinct peak is apparent in the decoded signal. The decoded result would be a false alarm if there were a distinct peak at Δ. In contrast, we can see a peak at 1 ms in Figures 5(c) and 5(d), though that peak is rather indistinct.

Decoding based on the real cepstrum: Because the complex cepstrum is composed of two components, that is, magnitude and phase, the real cepstrum, which has only a magnitude component, is generally used in many practical applications such as estimation of pitch and detection of echoes (Furui, 1992; Oppenheim & Schaffer, 1989). Here, we examine the decoding process using the real cepstrum instead of the complex cepstrum. The cepstrum (here, as usual, "the cepstrum" means the real cepstrum) of a watermarked signal $w(n)$ can be denoted as:

$$\tilde{w}(n)=\tilde{s}(n)+\tilde{k}(n), \tag{11}$$

where $\tilde{s}(n)$ and $\tilde{k}(n)$ are the cepstrums of $s(n)$ and $k(n)$, respectively. Then, $\tilde{k}(n)$ in equation (11) becomes:

$$\tilde{k}(n)=F^{-1}(\log|K(\omega)|). \tag{12}$$

To calculate $|K(\omega)|$, first, $p(n-\Delta)$ in equation (1) is expanded using the Dirac delta function $\delta(n)$ as follows:

$$p(n-\Delta)=\pm\delta(n-\Delta)\pm\delta(n-\Delta-1)\pm\cdots\pm\delta(n-\Delta-L_{PN}), \tag{13}$$

where each $\pm$ has random order. The Fourier transform of equation (13) becomes:

$$P_\Delta(\omega)=F(p(n-\Delta))=\pm e^{-j\omega\Delta}\pm e^{-j\omega(\Delta+1)}\pm\cdots$$
$$\pm e^{-j\omega(\Delta+L_{PN})}. \qquad (14)$$

Thus, the Fourier transform of the time-spread echo kernel $k(n)$ in equation (1) becomes:

$$K(\omega)=1+\alpha P_\Delta(\omega). \qquad (15)$$

Using equation (15), $|K(\omega)|$ is obtainable as:

$$|K(\omega)|=\frac{K(\omega)+K^*(\omega)}{2}$$
$$=\frac{1}{2}(1+\alpha P_\Delta(\omega)+1+\alpha P_\Delta^*(\omega))$$
$$=1+\frac{\alpha}{2}[\{\pm e^{-j\omega\Delta}\pm e^{-j\omega(\Delta+1)}\pm\cdots$$
$$\pm e^{-j\omega(\Delta+L_{PN})}\}$$
$$+\{\pm e^{j\omega\Delta}\pm e^{j\omega(\Delta+1)}\pm\cdots$$
$$\pm e^{j\omega(\Delta+L_{PN})}\}]$$
$$=1+\alpha[\pm\cos(\omega\Delta)\pm\cos(\omega(\Delta+1))\pm\cdots$$
$$\pm\cos(\omega(\Delta+L_{PN}))], \qquad (16)$$

where $K^*(\omega)$ and $P_\Delta^*(\omega)$ respectively denote the complex conjugates of $K(\omega)$ and $P_\Delta(\omega)$.

Using equation (16) and Taylor series expansion like in equation (7), $\log(|K(\omega)|)$ in equation (12) can be approximated as follows:

$$\log|K(\omega)|\approx\alpha[\pm\cos(\omega\Delta)\pm\cos(\omega(\Delta+1))\pm\cdots$$
$$\pm\cos(\omega(\Delta+L_{PN}))]. \qquad (17)$$

After calculating the inverse Fourier transform of equation (17), the cepstrum $\tilde{k}(n)$ in equation (12) can be expressed as follows:

$$\tilde{k}(n)=\alpha\ [\pm\frac{1}{2}(\delta(n-\Delta)+\delta(n+\Delta))$$
$$\pm\frac{1}{2}\{\delta(n-\Delta-1)+\delta(n+\Delta+1)\}\pm\cdots$$

Figure 6. Decoded signal d(n) when α=0.006, L_{PN} = 1023, L_{FT} = 4096, and Δ=1 ms: (a) Without watermarks, (b) a part of (a), (c) with watermarks, and (d) a part of (c).(Ko, 2005, p. 215)

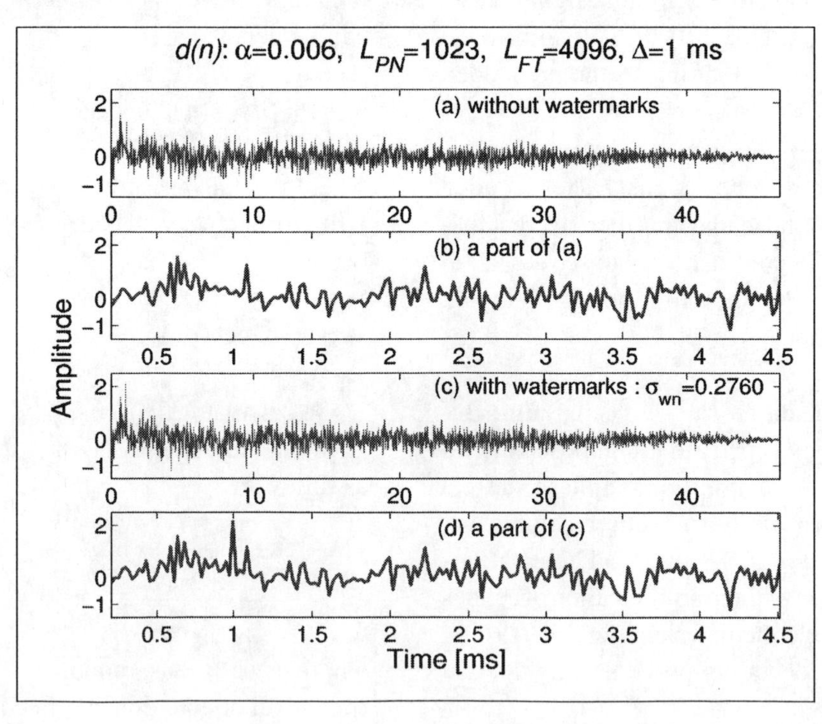

$$\pm\frac{1}{2}\{\delta(n-\Delta-L_{PN})+\delta(n+\Delta+L_{PN})\}]$$

$$=\frac{\alpha}{2}[p(n-\Delta)+p(-n+\Delta)], \tag{18}$$

Using equation (18), equation (11) becomes:

$$\tilde{w}(n)=\tilde{s}(n)+\frac{\alpha}{2}p(-n+\Delta)+\frac{\alpha}{2}p(n-\Delta). \tag{19}$$

Consequently, the decoded signal using the cepstrum can be rewritten as:

$$d(n)=\tilde{w}(n)\otimes p(n)$$

$$=n_s(n)+\frac{\alpha}{2}p(-n+\Delta)\otimes p(n)$$

$$+\frac{\alpha}{2}p(n-\Delta)\otimes p(n), \tag{20}$$

where $n_s(n)=\tilde{s}(n)\otimes p(n)$ corresponds to the noise by a host signal such as in equation (10). The third term $(\alpha/2)p(n-\Delta)\otimes p(n)$ in equation (20) is employed to detect the embedded watermarks because we are interested only in $n > 0$ in the decoding process; the other two terms have small contributions.

Figure 6 shows the decoded signal $d(n)$ when $\alpha = 0.006$, $L_{PN} = 1023$, $L_{FT} = 4096$, and $\Delta = 1$ ms. Panel (a) is when the host signal, $s(n)$, is applied to the decoding process instead of $w(n)$; Panel (c) is when $w(n)$ is applied; and Panels (b) and (d) are parts of (a) and (c), respectively. In Panel (c), the peak corresponding to the embedded data appears at the corresponding time delay Δ (1 ms). Comparing Figure 6(c) with Figure 5(c), the standard deviation σ_{wn} becomes smaller in the case when the cepstrum is used than when the complex cepstrum is used while the peak levels corresponding to the embedded watermarks in both cases are similar. Thus, the peak derived by use of the cepstrum is more distinct than that derived with the complex cepstrum.

Multiframe Average: One common method for reducing the standard deviation σ_{wn} of noise $n_s(n)$ is multiframe averaging. If N_s watermarks are

Figure 6: Decoded signal d(n) when α=0.006, LPN=1023, LFT=4096, and Δ=1 ms: (a) Without watermarks, (b) a part of (a), (c) with watermarks, and (d) a part of (c) (Ko, 2005, p. 215)

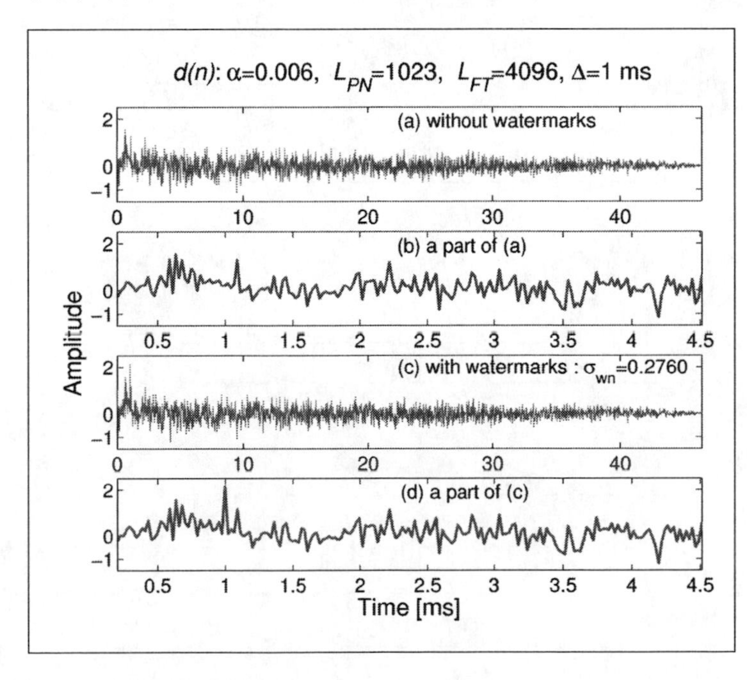

embedded into a host signal, the host signal can be split into N_s segments according to the number of the watermarks. Each watermark is embedded successively. Then to detect one watermark, one segment can be split into N_f frames according to the length of the segment and that of DFT, L_{FT}. The decoded signal with the multiframe average is denoted as:

$$d_i(n) = \frac{1}{N_f} \sum_{j=1}^{N_f} d_{f_j}(n), i = 1, 2, \cdots, N_s, \qquad (21)$$

where $d_{f_j}(n)$ is the decoded signal for the j-th frame and $N_f = \lfloor L_{seg}/L_{FT} \rfloor$, where L_{seg} is the length of each segment. This method is effective if the frames for calculating the cepstrum do not correlate with

Figure 7. Decoded signal for some numbers of N_f for a multiframe average with the same simulation conditions as in the case of Figure 6(c): That is, α=0.006, L_{PN} = 1023, L_{FT} = 4096, and Δ=1 ms (Ko, 2005, p. 216)

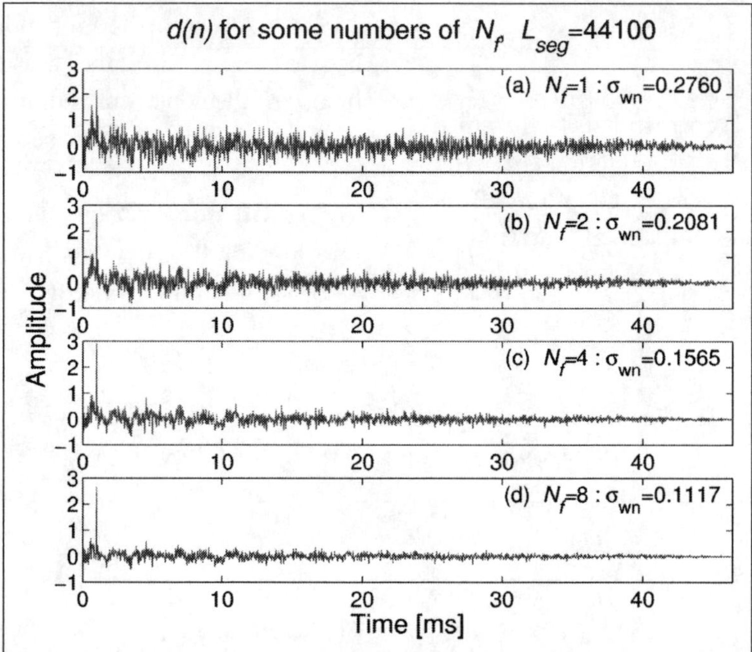

Figure 8. Block diagram of the decoding process based on the cepstrum analysis and the multiframe average shown in equation (21)

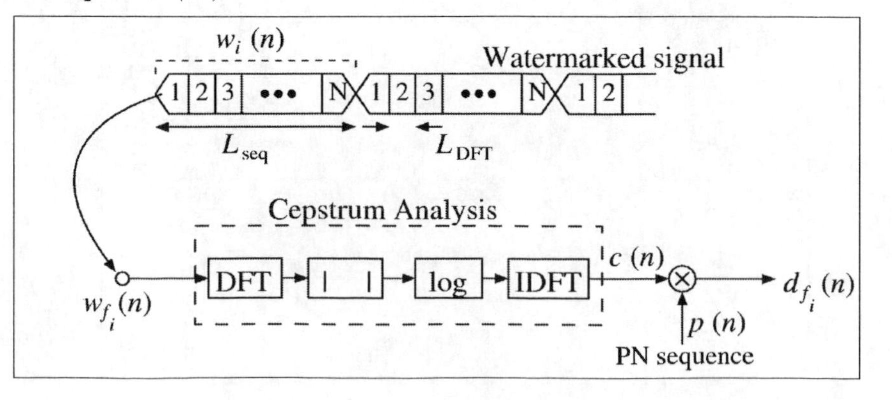

each other. It is noteworthy that no synchronizing process is necessary in the decoding process, that is, it is an auto-synchronization process.

Figure 7 shows the decoded signal for some numbers of N_f for a multiframe average with the same simulation conditions as in the case of Figure 6(c). It is readily apparent that the larger N_f is, the smaller the noise by a host signal, giving a clearer peak.

Figure 8 is a block diagram of the decoding process based on the cepstrum analysis and the multiframe average shown in equation (21). In equation (20) and Figure 8, the original PN sequence $p(n)$ used in the encoding process plays an important role as a secret key with which to decode the embedded data from the watermarked signal.

EVALUATION

In the time-spread echo method, several parameters exist, such as amplitude and length of PN sequences, length of DFT, and so on, which determine the decoding performance. Therefore, the decoding performance of the method should be evaluated as functions of these parameters and then be compared with conventional echo hiding. They are also compared in terms of the frequency responses of their kernels. In addition to these, the robustness against typical attacks and the sound quality of the watermarked signal are presented.

Table 1. Five host signals (five musical pieces) used for evaluation

Host Signal #	Genre
S01	American pops
S02	Japanese pops
S03	Brass ensemble
S04	Piano forte
S05	Finnish folk music

For simplicity of evaluation, L_{PN} was used as the maximum length of PN sequences (Boney et al., 1996; Cvejic et al., 2001) and 44.1 kHz of the sampling frequency of a host signal as the chip-rates of PN sequences. All host signals used for the evaluation were recorded from audio CD to PCM format (sampling frequency F_s: 44.1 kHz; quantization bits: 16 bits; stereo) using a personal computer. Table 1 shows a list of five pieces of music used as the host signal. In Table 1, S01 and S02 are popular vocal songs with impulsive sounds in the background. In addition, S03 and S04 are classical pieces performed by a brass ensemble and a piano, respectively, and S05 is Finnish folk music performed by violin, clarinet, accordion, and so on.

The discriminability or detectability index, denoted as d', is employed in signal detection theory as an index to evaluate the decoding performance and robustness (Boff, Kaufman, & Thomas, 1986; Green & Swets, 1998; Heeger, 1997; Wickens, 2002). The quantity d' is expected to provide an estimate of the strength of a signal. In echo hiding, a "signal" means a single bit of the embedded watermark corresponding to a peak/spike in the decoded signal $d(n)$ at the expected time delay Δ, as shown in Figures 5, 6, and 7. In this situation, is suitable to describe the performance or, in other words, the discriminability of the embedded watermark from the original signal. That is, the larger d' is, the stronger the signal, that is, the more distinct the peak/spike.

Distributions of responses both to "signal+noise" and to "noise" are needed to calculate d'. In this case, the decoded signal of watermarked signal is referred to as "signal+noise," and that of host signal is referred to as "noise." Responses to these decoded signals appear as values of these signals at the time displacement Δ. A blind detection is assumed here. Therefore, the two distributions of these responses might be regarded as independent. Thus, if these distributions are Gaussian distributions, d' is given as follows (Boff et al., 1986):

$$d' = \frac{\mu_s - \mu_n}{\sqrt{\sigma_s^2 + \sigma_n^2}}, \qquad (22)$$

where μ_s and μ_n are mean responses to "signal+noise" and "noise," respectively, and σ_s^2 and σ_n^2 are variances of responses to "signal+noise" and "noise," respectively.

Figure 9 shows histograms of $d(\Delta)$ for the actual data and their distributions calculated from the means (μ_s and μ_n) and the standard deviations (σ_s and σ_n) of $d(\Delta)$ in equation (22), assuming Gaussian distribution. In Figure 9, it is apparent that the increase of d' decreases the overlapped area between the two distributions of "signal+noise" and "noise." The hit rate increases and the false alarm rate decreases because the overlapped area corresponds to the false alarm and the miss when a criterion is set at the cross point of the distributions.

Table 2 shows the hit and false alarm rates calculated from actual data and d' assuming Gaussian distribution. Here, the criterion, that is, threshold, is set at the cross point of the distributions and $N_S = 1000$ in equation (21). The rates from d' are calculated from the following equations:

$$P_H = \Phi\left(\frac{\mu_s - c}{\sigma_s}\right)$$

$$P_F = 1 - \Phi\left(\frac{c - \mu_n}{\sigma_n}\right), \qquad (23)$$

where P_H and P_F respectively signify the hit and the false alarm rates, $\Phi(z)\ (= (1/\sqrt{2\pi})\int_{-\infty}^{z} e^{-x^2/2}dx)$ is the cumulative standard Gaussian distribution, and c is the value of a decision criterion. Assuming, the criterion c and d' in equation (22) are written as follows:

$$c = \frac{\mu_s + \mu_n}{2}$$

$$d' = \frac{\mu_s - \mu_n}{\sqrt{2}\sigma}. \qquad (24)$$

Using equation (24), equation (23) can be rewritten as:

$$P_H = \Phi\left(\frac{d'}{\sqrt{2}}\right)$$

Figure 9. Histograms of $d(\Delta)$ for the actual data and their distributions calculated from the means (μ_s and μ_n) and the standard deviations (σ_s and σ_n) of $d(\Delta)$ in equation (22) assuming a Gaussian distribution: (a) d'=1.6 and (b) d'= 3.1 (Ko et al., 2005, p. 217)

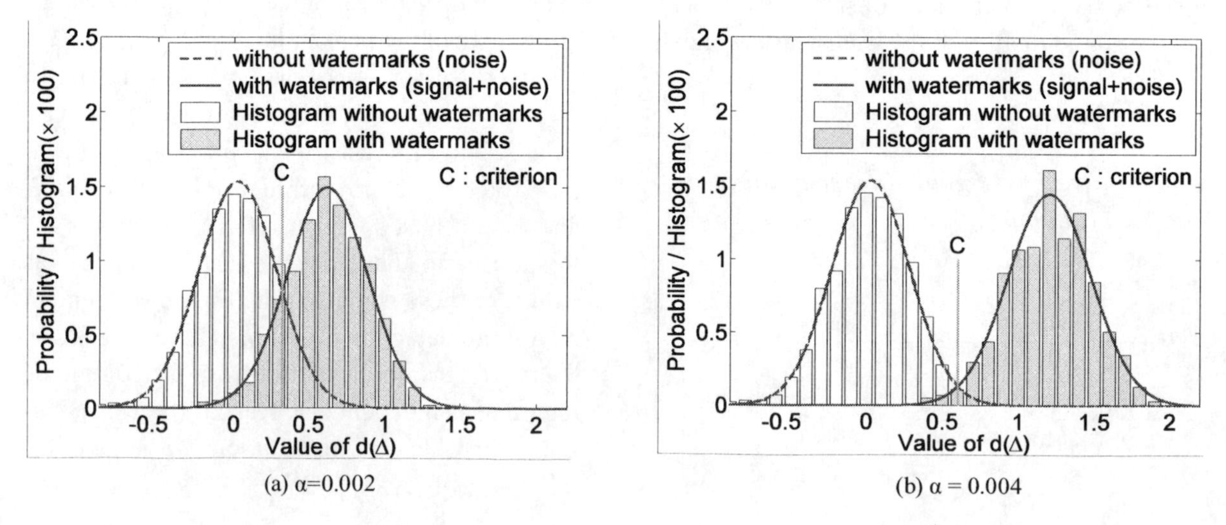

$$P_F = 1 - \Phi\left(\frac{d'}{\sqrt{2}}\right). \quad (25)$$

From Table 2, we might say that the actual hit and false alarm rates are nearly equivalent to those estimated from *d'*.

The quantity of *d'*, therefore, gives us the capability to estimate the hit and false alarm rates. Moreover, as shown in equation (22), *d'* is not dependent on the criterion, although the hit and false alarm rates are strongly dependent on the threshold. The above results indicate that the quantity *d'* is useful not as an optimal detection statistic to detect the embedded watermarks, but as an evaluating index to evaluate the performance of the method.

Decoding Performance

To evaluate the decoding performance, amplitude α and length L_{PN} of PN sequences, length L_{FT} of DFT, baud rate (Baud), and five host signals in Table 1 were set as parameters. Table 3 shows these parameters and their conditions. In Table 3, to avoid a time-aliasing error in the cepstrum analysis (Oppenheim & Schaffer, 1989), the length L_{FT} of DFT is set as $L_{FT}=k(L_{PN}+1)$ for each L_{PN}, where k = 1, 2, 4, and 8. The baud rate is employed instead of the bit rate because the bit rate for a given baud rate is changeable according to the adopted coding/signaling methods such as *M*-ary signaling (*M* bits are treated as one symbol) (Proakis, 2001; Sklar, 1998). The baud rate is 1 and F_S = 44.1 kHz. Therefore, the length L_{seg} of each segment in Figure 8 is 44,100 samples. Moreover, the numbers of frames, N_f, are given as $N_f=\lfloor L_{seg}/L_{FT}\rfloor$ for each L_{FT}.

Among the many parameters presented in Table 3, the amplitude α and length L_{PN} of the PN sequences can be considered as the major parameters that determine the decoding performance because they directly specify the time-spread echo kernel for embedding. For that reason, we investigated the effects of these parameters. Figure 10 shows *d'* for each α and L_{PN} averaged over all L_{FT} and host signals.

Table 2. Hit and false alarm rates calculated from real data and d'

d'	Decision	From *d'*	Actual
1.6	Hit	87.1	86.1
	False Alarm	12.9	13.3
3.1	Hit	98.6	98.2
	False Alarm	1.4	0.9

Table 3. Parameters and conditions for evaluating the decoding performance

Parameters		Conditions
Amplitude of p(n)	(α)	0.002, 0.004, ..., 0.018, 0.02
Length of p(n)	(L_{PN})	255, 511, 1023, 2047, 4095
Length of DFT	(L_{FT})	$L_{FT} = k(L_{PN}+1)$, k = 1, 2, 4, 8
Baud rate	(Baud)	1 (L_{seg} = 44,100 samples)
Number of host signals		Five kinds

Figure 10. d' For each α and L_{PN} averaged over all L_{FT} and host signals (Ko et al., 2005, p. 218)

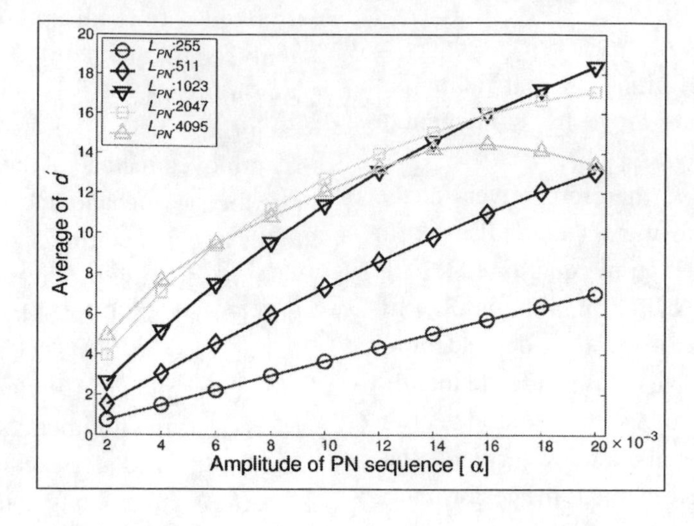

In Figure 10, it is apparent that the decoding performance increases almost proportionally as the amplitude α, and that the length L_{PN} of the PN sequences increases when α is small, for example, less than about 0.006, which realizes a sufficient decoding performance, as shown in Figure 7. From this result, we can say roughly that the decoding performance is proportional to amplitude α and length L_{PN} of the PN sequences.

Comparison with Single Echo Hiding

We next compare the decoding performance and the frequency response of the kernel of the time-spread echo method with those of single-echo hiding. Figure 11 shows d' for several amplitudes of PN sequence, α, and those of echo hiding, g, averaged over all host signals. From the viewpoint of the total power of a kernel, comparison with the following conditions would be fair:

$$g^2 \approx \alpha^2 L_{PN}. \quad (26)$$

Other parameters, namely, $L_{PN} = 1023$, and $L_{FT} = 2048$, are set. Figure 11 clarifies that the decoding performance of the time-spread echo method is better than that of single-echo hiding.

The frequency spectrum of a watermarked signal is specified by the product of the frequency spectrum of a host signal and that of a kernel. Therefore, the sound quality of a watermarked signal is influenced by the frequency response of the kernels (Oh et al., 2001). Figure 12 shows the frequency responses of the time-spread echo kernel and the single positive echo kernel. The amplitude of the PN sequence, α, and that of a single echo, g, applied in this investigation are indicated by the two arrows in Figure 11. These values are selected when the decoding performances of the time-spread echo method and the single echo hiding are nearly identical. Figure 12 shows that the time-spread echo kernel has a compound frequency characteristic with much smaller fluctuation than the single-echo kernel. That is, with the time-spread echo method, more natural and more colorless watermarked sound is obtainable than that which is obtainable using single-echo hiding.

Robustness

Digital contents can be easily edited and modified using software or other means. These manipulations, for example, compression/decompression,

Figure 11. d' For several amplitudes of PN sequence, α, and those of echo hiding, g, averaged over all host signals. Other parameters are set as $g=\alpha\sqrt{L_{PN}}$, $L_{PN} = 1023$, *and* $L_{FT} = 2048$.

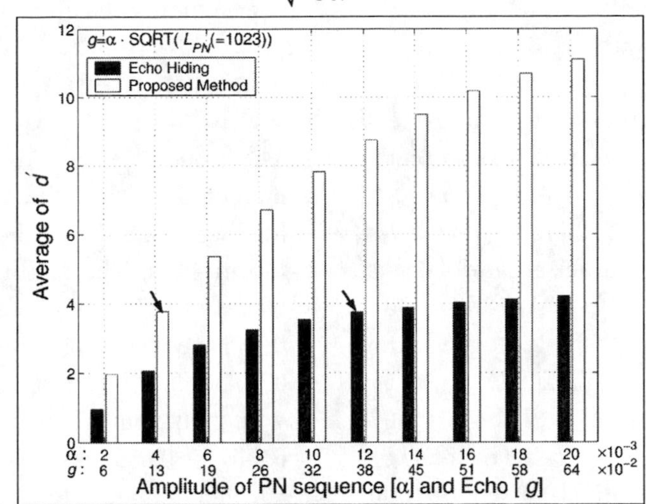

Note. The two arrows indicate the conditions that were selected for comparison (Ko, 2005, p. 218)

Figure 12. Frequency response of a single positive echo and a time-spread echo kernel: the upper panel shows that of a single positive echo kernel when g is about 0.3838; the bottom panel shows that for the time-spread echo kernel when α=0.004 and L_{PN} *= 1023 (Ko et al., 2005, p. 218)*

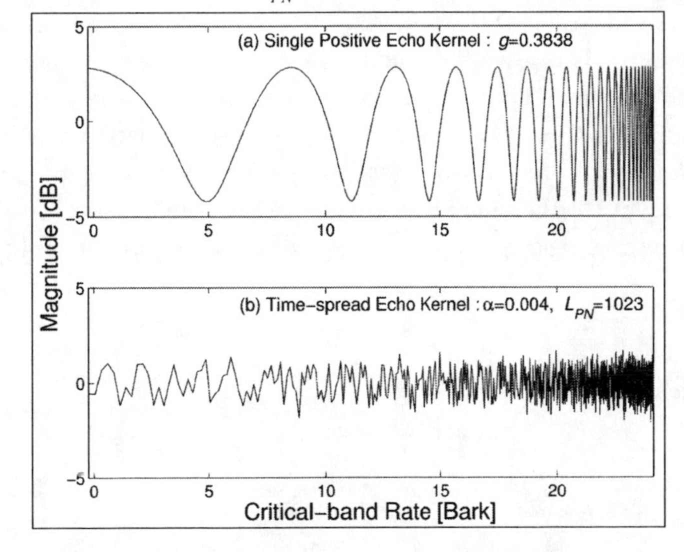

resampling, D/A and A/D conversions, and so forth, are considered as attacks to tamper with digital watermarking whether they are intentionally malicious or not. Digital watermarking, therefore, must be robust against such attacks.

Let us evaluate robustness of the time-spread echo method against the typical attacks presented in Table 4. We calculated *d*' after exposure to each attack in Table 4 to evaluate its robustness. Multi-attack (MUL) is an attack scenario, that is, "MP3" → "AAC" → "RES" → "AM8" → Audio CD, and analog output of CD player to the analog input of a personal computer (DAD).

Figure 13 shows normalized *d*' for S01, S02, S03, S04, S05, and "AVG" (averaged over all host signals) for each kind of attack, when α=0.006, L_{PN} = 1023, L_{FT} = 2048, L_{seg} = 44100, and N_f = 21 ($\lfloor L_{seg}/L_{FT} \rfloor$). In Figure 13, "NON" denotes the case in which the watermarked signal was not exposed to any attack. All performances were normalized by the case of "NON" in each panel. In Figure 13, it is readily apparent that *d*' is very low in the case of "MUL" and "PHI." Furthermore, the case of "PHI" has lower *d*' than that of "MUL" except for the case of Figure 13(d). This low *d*' was not improved, even though the amount of pitch scaling was very small, that is, ±1%.

As might be apparent in Figure 13(f), *d*' after attacks, except for the case of "MUL and PHI," is generally half of that in the case of no attack. Therefore, the time-spread echo method is expected to be able to maintain the watermark information to some extent even after typical attacks.

Listening Test

The imperceptibility of the time-spread echo method was investigated via a listening test. Two host signals, S01 and S04, were employed for the listening test. The length of the PN sequence, L_{PN}, was set to 511, 1023, or 2047. The amplitude of the PN sequence, α, was set to 0.002, 0.006, 0.01, 0.014, or 0.02 and the amplitude of single echo, *g*, was, on the other hand, set to 0.064, 0.192, 0.32, 0.448, or 0.635, using equation (26) when L_{PN} is 1023. Four male listeners and one female listener in their twenties with normal hearing participated in the listening tests.

The AXB paradigm was employed to investigate the difference of sound quality between a host signal and a watermarked signal. In these tests, A and B were always different from each other, that is, they were a host signal or a watermarked signal. Also, X was set randomly to either A or B. Listeners were asked to judge which of A or B was same as X. Because the chance level is 1/2, the discrimination limen is defined as α, which gives a correct response rate of 75% (Moore, 1997); listeners are considered to be able to distinguish the sound quality between a host and a watermarked signal beyond this value.

Table 4. Kinds of attacks and their conditions

Attack		Conditions
Resampling	(RES)	16 kHz
Requantization	(AM8)	8 bits
MPEG 1 Layer III	(MP3)	128 kbps
MPEG2 AAC	(AAC)	128 kbps
DAC and ADC	(DAD)	44.1 kHz, 16 bits
Multi-Attack	(MUL)	Including above all attacks
Time Scaling	(TUP)	+20% speed-up
Pitch Scaling	(PHI)	+20% higher-pitch

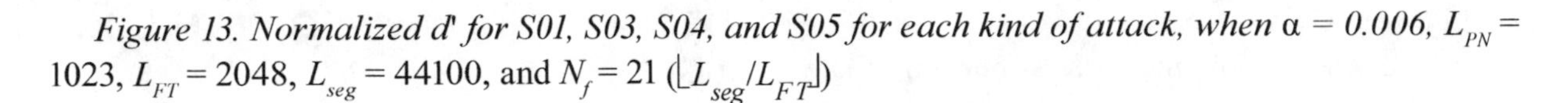

Figure 13. Normalized d' for S01, S03, S04, and S05 for each kind of attack, when $\alpha = 0.006$, $L_{PN} = 1023$, $L_{FT} = 2048$, $L_{seg} = 44100$, and $N_f = 21$ ($\lfloor L_{seg}/L_{FT} \rfloor$)

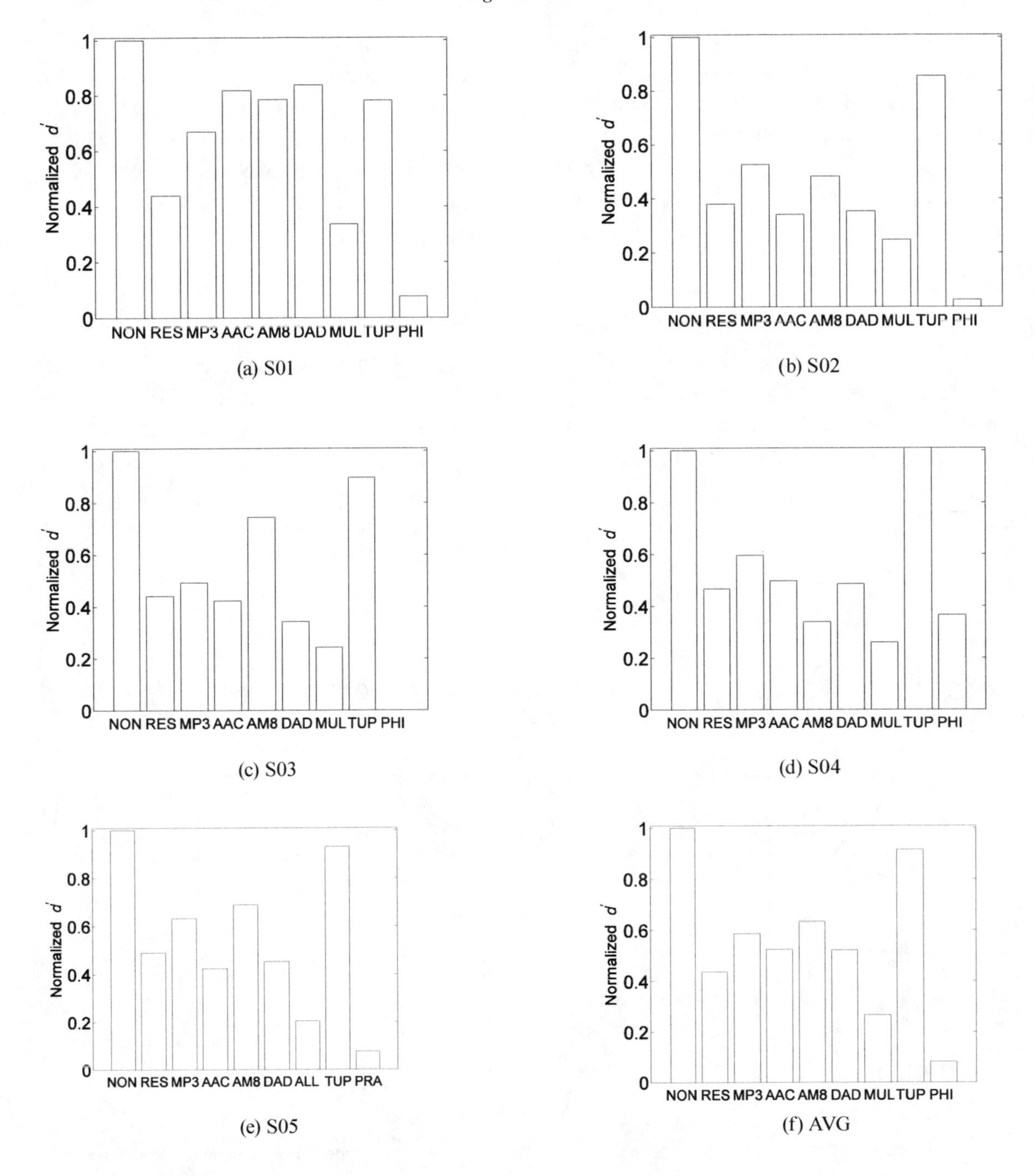

Note: "NON" is the case when the watermarked signal was not exposed to any attack. All performances were normalized by the case of "NON" in each panel: (a) S01, (b) S02, (c) S03, (d) S04, (e) S05, and (f) AVG (averaged over all host signals) (Ko et al., 2005, p. 219).

Figure 14 shows the results of the listening tests. The ordinate and the abscissa of each panel respectively exhibit the correct response rate [%] and α and g. Horizontal dashed lines indicate the correct response rate of 75%, which yields the discrimination threshold. For example, the discrimination threshold is about α = 0.01 for S01 and 0.014 for S04 when L_{PN} = 1023. Recall that the peak corresponding to Δ can be observed clearly when α=0.006 and L_{PN} = 1023, as shown in Figure 7. Moreover, we can see that the time-spread echo method provides the lower correct response rate than that of single echo as a whole, meaning that the time-spread echo method provides better imperceptibility than the single-echo method. Therefore, we can conclude that the time-spread echo method can realize sufficient decoding performance while maintaining good imperceptibility.

DECODING COMBINED WITH LOG SCALING

Pitch scaling is applied to audio signals as a special effect in certain applications. The term "pitch scaling" of an audio signal commonly indicates that the pitch of a signal is scaled by a scale factor κ (Garas & Sommen, 1998; Kahrs & Brandenburg, 1998). This operation is formulated simply as follows:

$$f' = \kappa f, \tag{27}$$

where κ (>0) is the scale factor, f is the original frequency, and f' is the κ-scaled frequency. If κ > 1, the pitch becomes higher and if κ<1, it becomes lower. Figure 15 shows an example of pitch scaling in the frequency domain.

Cepstrum analysis is carried out by taking the inverse Fourier transform of the logarithm of

Figure 14. Results of the listening test for the time-spread echo method and single-echo method as functions of L_{PN}, α and g: (a) S01 and (b) S04 (Ko et al., 2005, p. 220)

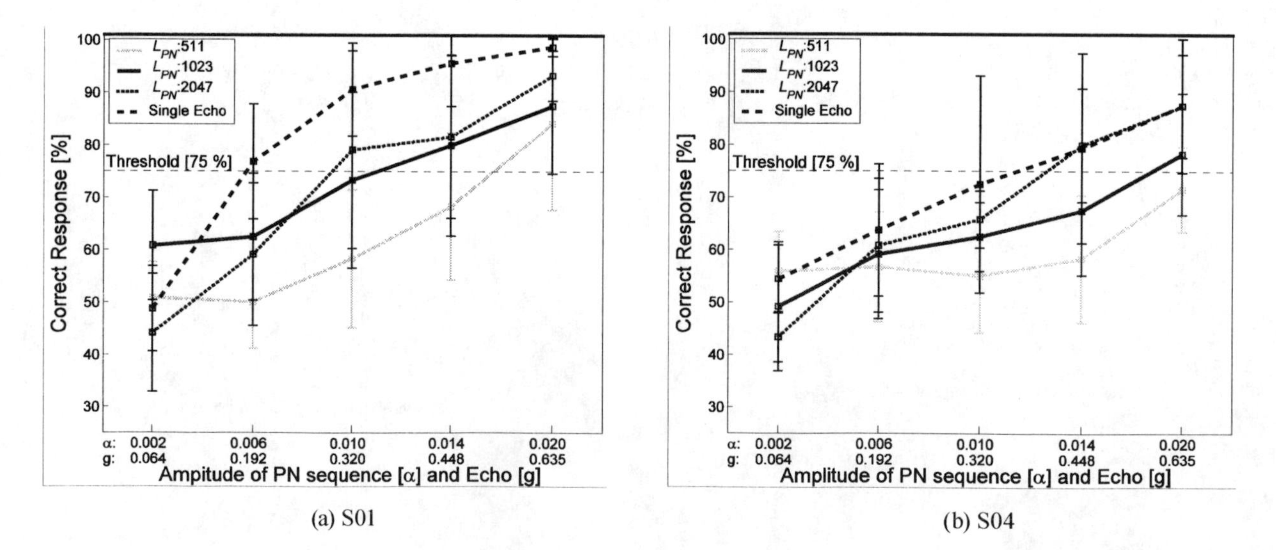

(a) S01

(b) S04

Figure 15. Example of pitch scaling in the frequency domain

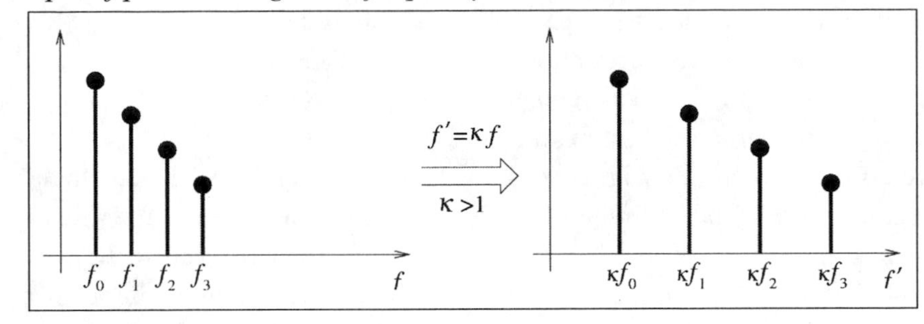

equation (27). Therefore, the effects of pitch scaling as shown in Figure 15 appear in the quefrency domain as follows:

$$\tau' = \beta\tau, \tag{28}$$

where β (=1/κ) is the pitch-scaling factor in the quefrency domain, τ is the original quefrency, and τ' is the scaled quefrency.

Accordingly, from equations (27) and (28), if the amount of pitch-scaling were known, the pitch-scaled secret key, that is, a PN sequence, would be scaled simply by $\tau = (1/\beta)\tau'$. However, the amount is generally unknown and is very difficult to estimate without the original signal. Consequently, a method to recover the synchronization is required for blind detection. A logarithm is widely exploited to convert the operation of multiplication into addition. Similarly, it converts the scaling (multiplying) process into a shifting (adding) process in the logarithm axis. That is, taking the logarithm of equation (28), the equation is rewritten as:

$$\log(\tau') = \log(\tau) + \log(\beta), \tag{29}$$

where log(·) is the common logarithm. In equation (29), we see that scaling by β is converted into shifting by log(β). This logarithm operation for the frequency or quefrency axis is referred to as "log scaling" hereafter.

Figure 16 shows a schematic explanation of how log scaling works in watermark detection from a pitch-scaled signal. If the original version,

Figure 16. Schematic explanation of the decoding process based on log scaling so as to be robust against pitch scaling

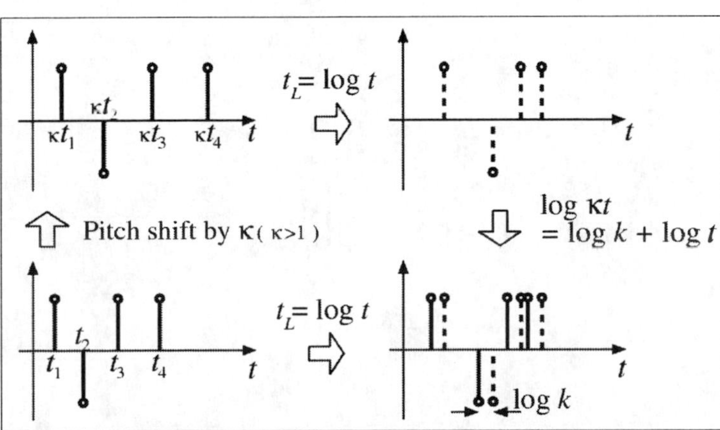

a signal without pitch scaling, and the pitch-scaled version are both converted using log scaling, the patterns of the two log-scaled versions are brought into accord even though the pitch-scaled version is shifted according to the amount of pitch scaling. Thus, correlation between the original and pitch-scaled version can be expected regardless of whether pitch scaling is used or not.

However, the log scaling in equation (29) cannot be directly implemented because discrete-time signal processing is basically used in the embedding and decoding process. That is, assuming that τ is either 1100 or 1101 sample points, for example, $\log(\tau)$ is then about 3.04139 or 3.04178 sample points, respectively, if the base 10 logarithm is used. These values, therefore, cannot be used directly because the point of each sample must be an integer.

To cope with this problem, the log scaling described in equation (29) is first scaled by a scale factor γ to obtain discrete sample points as follows:

$$\tau_L' = \lfloor \gamma \log(\tau') \rfloor, \tag{30}$$

where $\lfloor \cdot \rfloor$ rounds the element to the nearest integer towards minus infinity and γ is a log-scaling factor. Applying equation (30) to the preceding examples, 1100 and 1101 sample points of τ are scaled into 3041 and 3041 sample points, respectively, when γ = 1000, and 30413 and 30417 sample points, respectively, when γ = 10000. These examples indicate that the larger γ is, the better the log-scaling result. However, the number of samples increases dramatically when a large γ is applied, which might give rise to many problems such as long processing time and a huge amount of necessary memory. Hence, γ must be selected carefully. Linear interpolation is performed to obtain the samples between the log-scaled samples to obtain the complete log-scaled version of the signal.

Figure 17. Embedding and decoding process of the time-spread echo method

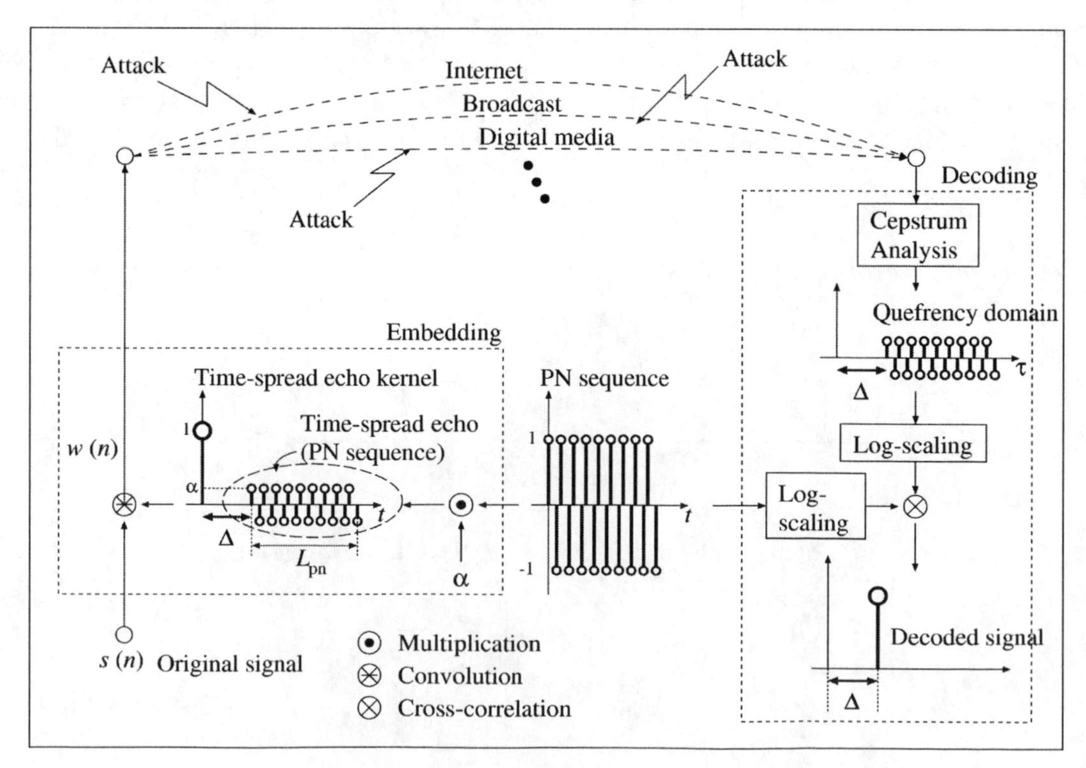

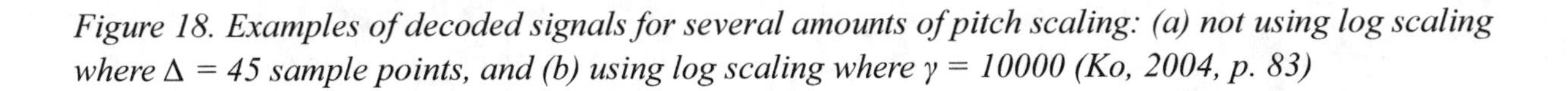

Figure 18. Examples of decoded signals for several amounts of pitch scaling: (a) not using log scaling where Δ = 45 sample points, and (b) using log scaling where γ = 10000 (Ko, 2004, p. 83)

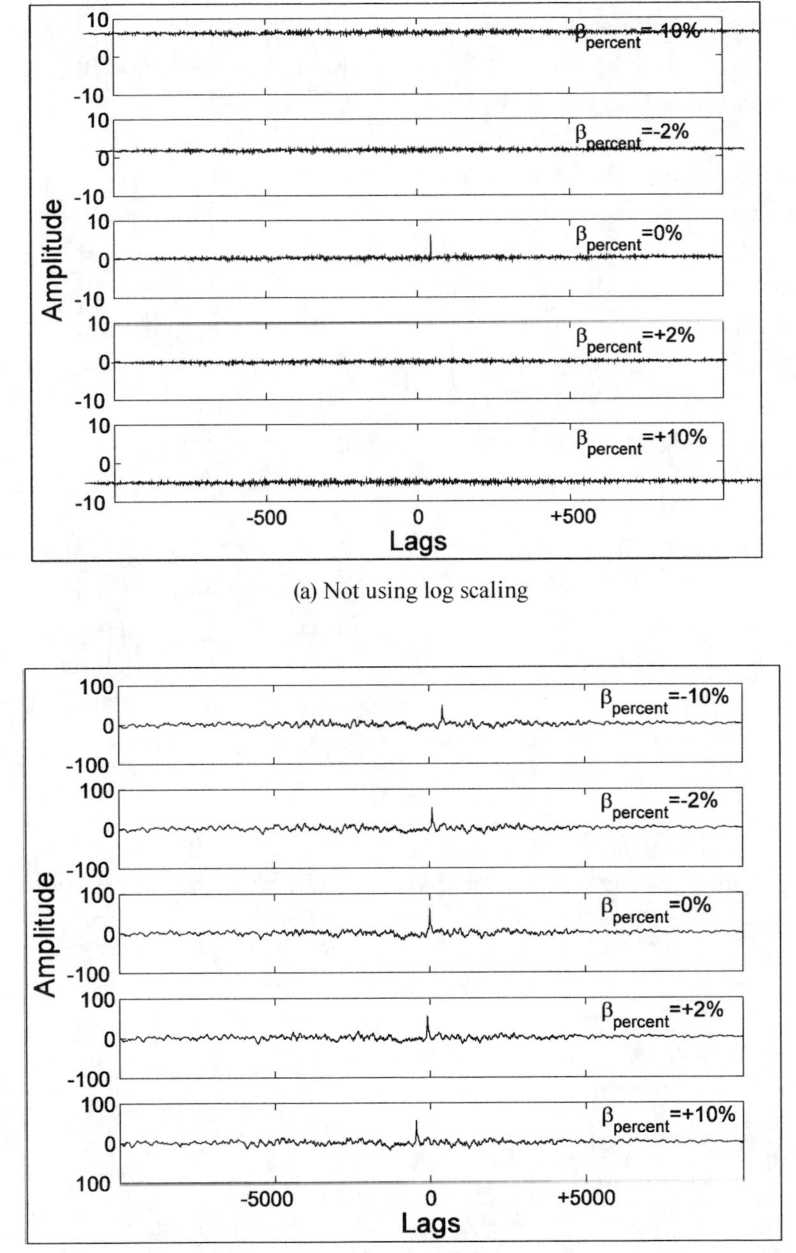

Figure 17 shows the embedding and decoding processes of the resultant time-spread echo method. Figure 17 shows a PN sequence used in the embedding and decoding processes; the embedded PN sequence is separated from the watermarked signal by cepstrum analysis in the quefrency domain. In the decoding process of Figure 17, if the watermarked signal is exposed to pitch scaling, correlation between the original and separated PN sequence cannot be performed because the pattern of the separated PN sequence is changed by pitch scaling.

Before employing the log-scaling method, we insert $\Delta-1$ zeros in front of the original PN sequence to eliminate the effect of the time delay Δ. In this process, note that the delay time Δ is assumed to be known, although the amount of pitch scaling is unknown. Because $\Delta-1$ zeros are inserted in front of the original PN sequence, the despread echo, that is, peak, in the decoded signal appears at t=0 when no pitch scaling is used. In contrast, the peak is shifted according to the amount of pitch scaling if there is pitch scaling.

Figure 18 shows examples of decoded signals for several degrees of pitch scaling. In this figure, $\beta_{percent}$ is the percentage of pitch scaling where $\beta_{percent} = 0\%$ corresponds to $\beta = 1$ in equation (28).

Figure 18(a) shows a case in which log scaling is not used and Figure 18(b) shows that when log scaling is used in the decoding process. In Figure 18(a), a distinct peak appears at Δ only when $\beta_{percent}=0\%$. In contrast, Figure 18(b) shows clearly that distinct peaks appear regardless of $\beta_{percent}$ and that the location of each peak shifts corresponding to the amount of pitch scaling.

The amount of peak shifting, $\varepsilon_{\pm}$, is given by the following equation:

$$\varepsilon_{\pm}=\gamma\log(\beta), \tag{31}$$

where γ is the log-scaling factor and β is the pitch-scaling factor. Here, $\varepsilon_{+}(<0)$ indicates the case of pitch scaling towards a higher pitch and $\varepsilon_{-}(>0)$ is for pitch scaling towards a lower pitch. From equation (31), the pitch-scaling factor β is estimated as:

$$\beta=10^{\varepsilon_{\pm}/\gamma}. \tag{32}$$

Figure 19, which is plotted by overlapping the results of Figure 18(b) with magnification, shows an actual example of the shifts of the peak according to the amount of pitch scaling. Table 5 shows the actual and calculated amounts of shifting, $\varepsilon_{\pm}$, and the pitch-scaling factor β. The actual $\varepsilon_{\pm}$ was

Figure 19. Actual example of the shifts of the peak according to the amount of pitch scaling

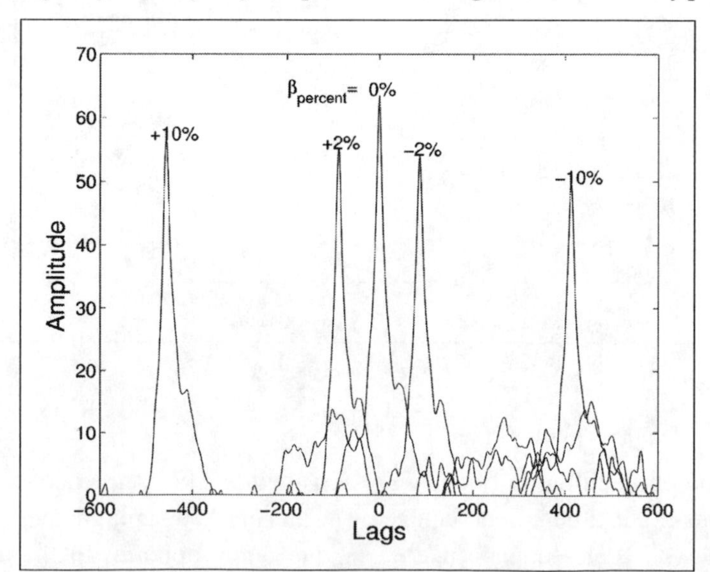

Note: This figure was plotted by magnifying the results of Figure 18(b) (Ko, Nishimura, & Suzuki, 2004a, p. 84)

measured from results of Figure 19. The actual β was calculated by substituting the actual $\varepsilon_{\pm}$ into equation (32). The calculated $\varepsilon_{\pm}$ was, on the other hand, computed using equation (31) based on the given β. Table 5 shows that the actual and calculated values have good coincidence. The results from Figure 19 and Table 5 show clearly that the amount of the peak shift corresponds exactly to the amount of pitch scaling.

DISCUSSION AND FUTURE WORK

The time-spread echo kernel should have characteristics that are more similar to the room reverberation, as schematized in Figure 1, than conventional methods with a single echo and multiple echoes. With such characteristics, this kernel is expected to provide more natural sound than a single echo kernel (Gardner, 1992; Ko & Kim, 1999). That is, if the amplitude of PN sequences is sufficiently small, listeners would be unable to even notice embedded "reverberation" in a host signal. In the worst case, even if the change in sound quality caused by watermarking were perceptible, it would not markedly degrade the musical sound quality in comparison with conventional echo-hiding techniques because of its reverberation-like nature and less coloration, as shown in Figure 12.

The algorithm of time-spread echo method offers several advantageous characteristics. One is that the decoding of embedded information from a watermarked signal is very difficult using only the cepstrum analysis and autocepstrum, which are the general decoding methods used in single echo hiding (Gruhl & Bender, 1996). This is true because the delay Δ, which corresponds to the embedded information in echo hiding, cannot be determined if only such signal processing is used, because the echo is spread over the time axis using a PN sequence. Figure 20 shows examples of the decoding results of single echo hiding and the time-spread echo method with the cepstrum analysis and the autocepstrum when $g = 0.3$, L_{PN}=1023, $\alpha = g/\sqrt{L_{PN}}$=0.0094, and time delay Δ= 1 ms. Panels (a) and (b) show results for single-echo hiding. A peak corresponding to the embedded data is clearly visible at about 1 ms. Furthermore, the decoding result of (b) with the autocepstrum is slightly clearer than the case of (a) with the cepstrum analysis. In contrast, no peak, such as that in the case of single echo hiding, is visible in Panels (c) and (d), which show the results of the time-spread echo method. This fact means that, with the time-spread echo method, it is very hard for malicious attackers to decode the embedded data from a watermarked signal with only the cepstrum analysis or the autocepstrum.

The multiple encoding process is useful for controlling multiauthorities or copies, though it is susceptible to malicious attacks (Lee & Ho, 2000). Using the time-spread echo technique, encoding with multiple watermarks is very easy.

Table 5. Actual and calculated and β

$\beta_{percent}$	Actual		Calculated	
	$\varepsilon_{\pm}$	β	$\varepsilon_{\pm}$	β
−10	413	1.10	413.9	1.10
−2	86	1.02	86.0	1.02
0	0	1	0	1
+2	−87	0.98	−87.7	0.98
+10	−457	0.90	−457.5	0.90

It can be achieved merely using uncorrelated PN sequences and can detect each piece of information by applying one of the PN sequences used for encoding the information.

It is necessary to consider the characteristics of a host signal in the encoding process to improve imperceptibility and performance of decoding. One example is a method in which a host signal is divided into several subbands and the amplitude of watermark for each subband is adjusted according to the masking level of each subband (Ko, Nishimura, & Suzuki, 2004). The decoding process is identical to the original time-spread echo method. Therefore, this method is easy to implement.

For more colorless watermarking and imperceptibility, it might be effective to apply exponential or linear decay in equation (1). However, this would probably result in problems, such as a decrease of the decoding performance. For stronger robustness and better imperceptibility, the use of other spreading kernels, for example, chaotic sequences (Kubin, 1995), time-stretching pulses (TSP) (Suzuki, Asano, Kim, & Sone, 1995), and so forth, should be considered in the future.

In addition to that, reflections in a real environment, such as an indoor room, and a collusion attack are two noticeable tough attacks to deal with for the time-spread echo method. In both **attacks**, inferring echoes are superimposed to the original multiple echos that represent the embedded watermarks, and strongly influence badly on the watermark detection rate. A pitch scaling attack is another one that attempts to desynchronize between the extracted PN sequence and the original PN sequence. Although one solution was presented in this chapter, that method requires much computational cost and needs further improvement for real use.

Figure 20. Examples of the decoding results of conventional echo hiding and the time-spread echo method with the cepstrum analysis and the autocepstrum when $g=0.3$, $L_{PN}=1023$, $\alpha=g/\sqrt{L_{PN}}=0.0094$, and time delay $\Delta=1$ ms. (a) Echo hiding with the cepstrum analysis, (b) echo hiding with the autocepstrum, (c) time-spread echo method with the cepstrum analysis, and (d) time-spread echo method with the autocepstrum. (Ko et al., 2005, p. 220)

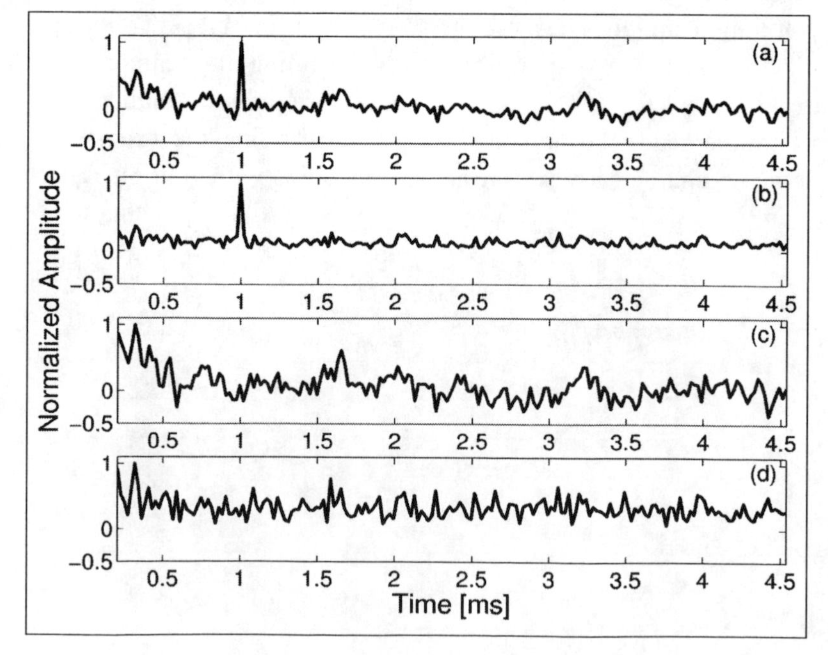

CONCLUSION

This chapter described a new digital audio watermarking method developed to improve the weak points of conventional echo-hiding techniques, such as a lenient decoding process, weakness against multiple encoding attacks, and so on. The basic idea of the method is to spread an echo using PN sequences in the time domain. Such a PN sequence acts as a secret key that is useful to decode the embedded information, resulting in secrecy superior to that of conventional echo hiding, which has no secret key for decoding. Results of a listening test showed that this method also has imperceptibility superior to that of conventional echo hiding while maintaining good decoding performance. Moreover, because spread echoes are comparable to natural reflections from walls, even if the embedded watermark is perceptible, this method can provide more natural sound quality, as confirmed by a computer simulation and a listening test. Furthermore, this method exhibited fair tolerance to typical attacks such as resampling and MPEG coding.

FUTURE RESEARCH DIRECTIONS

Research on watermarking hereafter will grow heated not only for the copyright protection but also for the universal design of multimedia information. Additional information which can serve for effective and efficient communication of information particularly for handicapped people will be highly required in the future. Rich site summary (RSS) of Web contents, which at present helps to provide timely information to the user, may be a foretaste of such idea. Watermarking is closely related to data hiding and therefore has a proper capability to serve for this purpose.

When an application of watermarking other than the copyright protection of digital audio data is concerned, the amount of information to be embedded and high resistance to degradation in sound quality during data transmission are principal two considerations. Some applications may permit certain degradation in sound quality to gain the bitrate of watermark. Other applications may require the watermark to survive after transmission through an analogue communication path, for example air transmission path. This point of view will open up new vistas in the research on watermarking since only a few have been done for these applications.

Many researchers from several fields, such as social system designers, information system designers, experts of human hearing, audio engineers and system developers, lawyers for copyright ownership, specialist in cryptography for assessment of safety, and those engaged in data compression and transmission, should be involved in developing a new watermarking method. Good relationship with others sometimes helps to create a novel idea. Hence, cooperation with colleagues in other research fields is highly appreciated. This is usually true in any research areas but seems to the authors particularly important in realizing practical and affordable watermarking technologies.

As for copyright protection, techniques to tamper watermarks must be investigated more vigorously. Performance evaluation of watermarks in robustness has usually been done assuming attacks of common signal processing, such as audio compression, resampling, re-quantization, and addition of white noise. However, considering an intentional deletion of watermark, several other approaches can be taken. One typical attack among them is the collusion attack. Many watermarking methods having been developed so far are fragile against this type of attack. Components of content relevant to serving for the human perception are common among individual contents while watermarks embedded in them are different. Therefore, the collusion attack can easily weaken the watermark preserving the perceptual quality of content just like the averaging taken in signal measurement to gain the signal to noise ratio.

Finally, the ultimate goal is to develop sound and safe watermarking methods that keep effect even after uncovering the algorithm. Watermarking methods currently used in practice conceal their algorithms to protect the safety and reliability of them. However, it in turn implies that the method is not safe once when the algorithm is opened. No one can ensure the perpetual secrecy of the algorithm. Hence, a truly safe watermarking method must be one which algorithm can be opened. This goal is actually difficult to achieve but therefore highly challenging.

ACKNOWLEDGMENT

A part of study in this chapter was supported by a Grant-in Aid for Development of Innovative Technologies by the Ministry of Education, Culture, Sports, Science, and Technology of Japan, as well as by the Strategic Information and Communications R&D Promotion Programme (SCOPE) by the Ministry of Internal Affairs and Communications of Japan.

REFERENCES

Bender, W., Gruhl, D., Morimoto, N., & Lu, A. (1996). Techniques for data hiding. *IBM Systems Journal, 35*(3/4).

Boff, R.K., Kaufman, L., & Thomas, D.J. (1986). *Handbooks of perception and human performance* (Vol. I). John Wiley & Sons.

Boney, L., Tewfik, H.A., & Hamdy, N.K. (1996). Digital watermarks for audio signals. In *Proceedings of the International Conference on Multimedia Computing and Systems* (pp. 473-480).

Cox, J.I., Kilian, J., Leighton, T., & Shamoon, T. (1995). *Secure spread spectrum watermarking for multimedia* (Tech. Rep. No. 95-10). Princeton, NJ: NEC Research Institute.

Cvejic, N., Keskinarkaus, A., & Seppanen, T. (2001). *Audio watermarking using m-sequences and temporal masking.* Paper presented at the IEEE Workshop on Applications of Signal Processing to Audio and Acoustics (pp. 227-230).

Czerwinski, S., Fromm, R., & Hodes, T. (1999). *Digital music distribution and audio watermarking* (Project Report UCB 1S 219).

Furui, S. (1992). *Digital speech processing, synthesis, and recognition.* Marcel Dekker Inc.

Garas, J., & Sommen, P. (1998). Time/pitch scaling using the constant-Q phase vocoder. In *Proceedings of the STW's 1998 Workshops CSSP98 and SAFE98* (pp. 173-176).

Gardner, G.W. (1992). *The virtual acoustic room.* Unpublished master's thesis, MIT Computer Science and Engineering.

Green, M.D., & Swets, A.J. (1998). *Signal detection theory and psychophysics.* Peninsula Publishing.

Gruhl, D., & Bender, W. (1996). Echo hiding. In *Proceedings of the Information Hiding Workshop* (pp. 295-315).

Heeger, D. (1997). Signal detection theory. *Teaching handout.* Department of Psychology, Stanford University.

Kahrs, M., & Brandenburg, K. (Eds.). (1998). *Applications of digital signal processing to audio and acoustics.* Kluwer Academic Publishers.

Katzenbeisser, S., & Petitcolas, P.A.F. (Eds.). (2000) *Information hiding techniques for steganography and digital watermarking.* Artech House.

Kim, H.J., & Choi, Y.H. (2003). A novel echo-hiding scheme with backward and forward kernels. *IEEE Transactions on Circuits and Systems for Video Technology, 13*(8), 885-889.

Ko, B., & Kim, H. (1999). Modified circulant feedback delay networks (MCFDNs) for artificial reverberator using a general recursive filter and CFDNs. *J. Acoustical Society of Korea, 18*(4E), 31-36.

Ko, B., Nishimura, R., & Suzuki, Y. (2002a). *Proposal of an echo-spread watermarking method using PN sequence*. Paper presented at the 2002 Spring Meeting of The Acoustical Society of Japan (pp. 535-536).

Ko, B., Nishimura, R., & Suzuki, Y. (2002b). Time-spread echo method for digital audio watermarking using PN sequences. *Proceedings of ICASSP 2002, 2*, 2001-2004.

Ko, B., Nishimura, R., & Suzuki, Y. (2004a). Log-Scaling watermark detection in digital audio watermarking. *Proceedings of ICASSP 2004, 3*, 81-84.

Ko B., Nishimura R., & Suzuki Y. (2004b). Robust watermarking based on time-spread echo method with subband decomposition. *IEICE Transactions on Fundamentals, E87-A*(6), 1647-1650.

Ko, B., Nishimura, R., & Suzuki, Y. (2005). Time-spread echo method for digital audio watermarking. *IEEE Transactions on Multimedia, 7*(2), 212-221.

Kubin, G. (1995). *What is a chaotic signal?* Paper presented at the IEEE Workshop on Nonlinear Signal and Image Processing (pp. 141-144).

Lee, S., & Ho, Y. (2000). Digital audio watermarking in the cepstrum domain. *IEEE Transactions on Consumer Electronics, 46*(3), 744-750.

Moore, J.C.B. (1997). *An Introduction to the psychology of hearing* (4th ed.). Academic Press.

Nishimura, R., Suzuki, M., & Suzuki, Y. (2001). Detection threshold of a periodic phase shift in music sound. In *Proceedings of the 17th International Congress on Acoustics.*

Oh, O.H., Seok, W.J., Hong, W.J., & Youn, H.D. (2001). New echo embedding technique for robust and imperceptible audio watermarking. *Proceedings of ICASSP 2001, 3*, 1341-1344.

Oppenheim, V.A., & Schaffer, W.R. (1989). *Discrete-time signal processing.* NJ: Prentice Hall.

Podilchuk, I.C., & Delp, J.D. (2001). Digital watermarking: Algorithms and applications. *IEEE Signal Processing Magazine, 18*(4), 33-46.

Proakis, G.J. (2001). *Digital communications* (4th ed.). McGraw-Hill.

Sasaki, N., Nishimura, R., & Suzuki, Y. (in press). Audio watermarking based on association analysis. *Proceedings of ICSP 2006.*

Sklar, B. (1998). *Digital communications fundamentals and applications.* NJ: Prentice-Hall.

Suzuki, Y., Asano, F., Kim, H., & Sone, T. (1995). An optimum computer-generated pulse signal suitable for the measurement of very long impulse response. *J. Acoustical Society of America, 97*(2), 1119-1123.

Swanson, D.M., Zhu, B., Tewfik, H.A., & Boney, L. (1998). Robust audio watermarking using perceptual masking. *Signal Processing, 66*(3), 337-355.

Takahashi, A., Nishimura, R., & Suzuki, Y. (2005). Multiple watermarks for stereo audio signals using phase-modulation techniques. *IEEE Transactions on Signal Processing, 53*(2), 806-815.

Xu, C., Wu, J., Sun, Q., & Xin, K. (1999). Applications of digital watermarking technology in audio signals. *J. Audio Engineering Society, 47*(10), 805-812.

Wickens, D.T. (2002). *Elementary signal detection theory.* Oxford University Press.

ADDITIONAL READING

Anderson, R. (1996, May/June). Information hiding. In *Proceedings of the First International Workshop,* Cambridge, UK.

Arttameeyanant, P., Kumhom, P., & Chamnongthai, K. (2002). Audio watermarking for Internet. In *Proceedings of the IEEE ICIT'02* (pp. 976-979).

Aucsmith, D. (1998, April). Information hiding. In *Proceedings of the Second International Workshp, IH'98*, Portland, Oregon.

Bender, W., Gruhl, D., & Morimoto, N. (1995). Techniques for data hiding. In *Proceedings of SPIE,* (2420), 164-173.

Boney, L., Tewfik, A.H., & Hamdy, K.N. (1996). Digital watermarks for audio signals. In *Proceedings of the IEEE Multimedia '96* (pp. 473-480).

Choi, K-P., & Lee, K-Y. (2002, June). An efficient audio watermarking by using spectrum warping. *IEICE Transactions on Fundamentals, 85*(6), 1257-1264.

Ejima, M., & Miyazaki, A. (2000, March). A wavelet-based watermarking for digital images and video. *IEICE Transactions on Fundamentals, 83*(3), 532-540.

Garcia. R.A. (1999, September). Digital watermarking of audio signals using a psychoacoustic auditory model and spread spectrum theory. In *Proceedings of the AES 107th Convention* (Preprint 5073 (N-3)).

Hartung, F., & Kutter, M. (1999, July). Multimedia watermarking techniques. In *Proceedings of the IEEE, 87*(7), 1079-1107.

Hartung, F., & Ramme, F. (2000, November). Digital rights management and watermarking of multimedia content for m-commerce applications. *IEEE Communications Magazine*, 38(11), 78-84.

Ikeda, M., Takeda, K., & Itakura, F. (2000, October). Robust audio data hiding by sse of trellis coding. Proc. *WESTPRAC VII.*

Kaewkamnerd, N., & Rao, K. R. (2000, February). Wavelet based image adaptive watermarking scheme. *Electronics Letters, 36*(4), 312-313.

Kim, H.J., & Choi, Y.H. (2003, August). A novel echo-hiding scheme with backward and forward kernels. *IEEE Transactions on Circuits and Systems for Video Technology, 13*(8), 885-889.

Kirovski, D., & Malvar, H.S. (2003, April). Spread-spectrum watermarking of audio signals. *IEEE Transactions on Signal Processing, 51*(4), 1020-1033.

Ko, B-S., Nishimura, R., & Suzuki, Y. (2004). Robust watermarking based on time-spread echo method with subband decomposition. *IEICE Transactions on Fundamentals, E87-A*(6), 1647-1650.

Ko, B-S., Nishimura, R., & Suzuki, Y. (2005). Time-spread echo method for digital audio watermarking. *IEEE Transactions on Multimedia, 7*(2), 212-221.

Larbi, S.D., & Jaidane-Saidane, M. (2005, February). Audio watermarking: A way to stationarize audio signals. *IEEE Transactions on Signal Processing, 53*(2), 816-823.

Lemma, A., Aprea, J., Oomen, W., & van de Kerkhof, L. (2003, April). A temporal domain audio watermarking technique. *IEEE Transactions on Signal Processing, 51*(4), 1088-1097.

Liu, J., Zhang, X., & Sun, J. (2003). *A new image watermarking scheme based on DWT and ICA.* Paper presented at the IEEE International Conference on Neural Networks and Signal Processing (pp. 1489-1492).

Lu, Z.-M., Xing, W., Xu, D.-G., & Sun, S.-H. (2003, December). Digital image watermarking method based on vector quantization with labeled codewords. *IEICE Transactions on Inf. & syst., 86*(12), 2786-2789.

Neubauer, C., & Herre, J. (2000, February). Audio watermarking of MPEG-2 AAC bit streams. In *Proceedings of the AES 108th Convention* (Preprint 5101 (F-2)).

Stefan Katzenbeisser, F., & Petitcolas (2000). *Information hiding: Techniques for steganography and digital watermarking.* Artech House Publishers.

Swanson, M.D., Kobayashi, M., & Tewfik, A.H. (1998, June). Multimedia data-embedding and watermarking technologies. *Proceedings of the IEEE, 86*(6), 1064-1087.

Takahashi, A., Nishimura, R., & Suzuki, Y. (2005, February). Multiple watermarks for stereo audio signals using phase-modulation techniques. *IEEE Transactions on Signal Processing, 53*(2), 806-815.

Takahashi, H., Nishimura, R. & Suzuki, Y. (2005). Time-spread echo digital audio watermarking tolerant of pitch shifting. *Acoustical Science and Technology, 26*(6), 530-532.

Chapter VII
Analysis-by-Synthesis Echo Watermarking

Wen-Chih Wu
Wu-Feng Institute of Technology, Taiwan

Oscal Chen
National Chung Cheng University, Taiwan

ABSTRACT

In this chapter, detailed explanations would be given on the role of echo hiding playing in audio watermarking in terms of background, functions, and applications. Additionally, a section is dedicated to discuss the various approaches proposed in the past to solve the flaws of echo hiding. Lastly, the proposed analysis-by-synthesis echo watermarking scheme based on interlaced kernels is introduced. Comparisons in audio quality and robustness performance are also looked at the proposed and conventional echo watermarking schemes.

INTRODUCTION

The concept of echo hiding was first proposed by Bender et al. in 1996 (Gruhl, Lu, & Bender, 1996). As a new technology, it is remarkably easy to materialize. In comparison to other audio watermarking techniques, echo hiding with many outstanding features is extensively adopted in various audio watermark applications. The conventional scheme of embedding watermarks on audio signals is to add random noises on the host audio signals. In particular, Boney Tewfik, and Hamdy (1996) presented the robust spread spectrum watermarking scheme that exploits temporal and frequency masking. With this scheme widely used in some of the high-quality audio compression algorithms, it is actually quite difficult to allow embedded signals to be higher than the truncation threshold of the compression algorithms, yet at the same time be lower than the masking threshold. Hence, the drawback of adding random noises on the host audio signals is that the random noises are easily removed by lossy data compression algorithms. By comparison, echo signals generated by the echo hiding technique have the same statistic and hearing qualities as the host signals. In fact, the echo is perceived as added resonance and is robust for lossy data compression algorithms. Additionally, the echo hiding technique does not need host signals to extract the embedded data, which is a blind watermarking technique.

As expected, there are also some disadvantages in using the echo hiding technique (Kim & Choi, 2003). First, more complicated computation is required for echo detection. Second, echo hiding is also prone to inevitable mistakes, such as the echo from the host signal itself may be treated as the embedded echo. Third and last, if the echo added has smaller amplitude, then the cepstrum peak would be covered by the surrounding peaks to make the echo detection an arduous task to perform. A larger echo may increase the accuracy rate of detection but it also easily exposes the system to deliberate attacks, which then affects the sound quality.

As for the watermark embedding system, there are a few factors to consider: first, the embedded watermark should not, if possible, affect the listening quality of the host signals. To meet this condition, Foo, Yeo, and Huang (2001) had proposed the adaptive and content-based audio watermarking system. By analyzing the psycho-acoustic model of the audio signals to adjust the amplitude of echoes, the embedded echoes would not exceed the limit of which the human ears can perceive. While this scheme can effectively guarantee the audio quality, it may also affect the accuracy of watermark restoration due to insufficient amplitudes of echoes embedded. Second, embedded watermarks must be able to endure the common signal processing attacks, such as filtering, lossless/lossy compression, resampling, digital-to-analog and analog-to-digital conversion, and adding noises, and so forth. To enhance the robustness under various attacks, many schemes have been developed by modifying the embedding mechanism in echo hiding (Kim & Choi, 2003; Oh, 2001). Third, embedded watermarks must ensure the safety of the system so that the interested parties cannot detect the embedding manner and deliberately destroy the existing watermark data. In general, to detect and remove the echo without knowledge of the host signal or the parameters of the embedded echo, this problem known as blind echo cancellation is a hard task (Petitcolas, Anderson, & Kuhn, 1998). Nevertheless, there are many researchers who devoted themselves to devise new approaches to increase security of echo hiding.

THEORETICAL FOUNDATION OF ECHO WATERMARKING

Embedding

By adding echoes, the echo hiding scheme hides data in the host audio signals. The hidden data can be adjusted by the two parameters shown in Figure 1: amplitude and offset that represent the magnitude and time delay for the embedded echo, respectively. The embedding process uses two echoes with different offsets, one to represent the

Figure 1. Manner of echoes being embedded

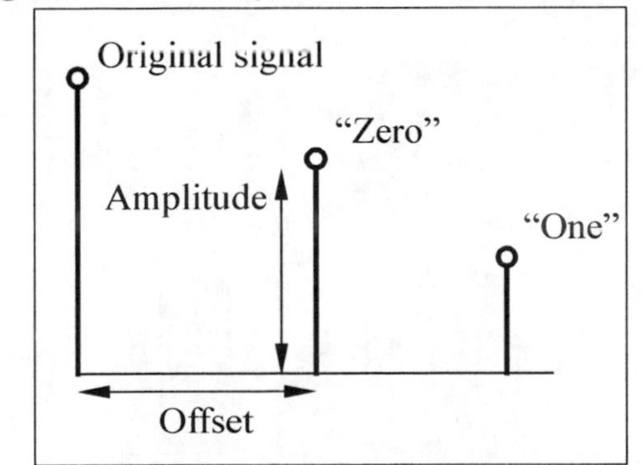

binary datum "Zero" and the other to represent the binary datum "One." With the echo amplitude set at a value lower than the audibility threshold, the distortion caused by adding the echoes would not be detected.

Assuming data, I, hidden in the host audio is expressed in binary codes, where $I \in \{0, 1\}$. The embedding process of hidden data can be considered as a system with two different kernels, impulse responses, that represents the data type to be added as 0 and 1. To be more simplified, the kernel $h[n]$ is only denoted by two impulses, one to represent the host audio signals, and the other to generate the echo, as expressed below:

$$h[n] = \delta[n] + \alpha\,\delta[n-d], \qquad (1)$$

where $\delta[n]$ is the Kronecker delta function and α (< 1) represents the amplitude of the embedded echo. The offset d of the echo signal is determined by the values of the hidden data. The kernel in Figure 2(a) is used to embed a binary zero, whereas the kernel in Figure 2(b) is used to embed a binary one. The difference between these two kernels is that the offsets of their echoes are different from each other. Here, d_a and d_b denote the offsets of the echoes for the "Zero" and "One" kernels, respectively. Additionally, α_a and α_b represent the corresponding echo amplitude values, which are generally set at an equal value, $\alpha_a = \alpha_b$. Here, the subscripts a and b are merely used to denote the positions of embedded echoes instead of the watermark data. Because the same position can be used to represent different watermark data in the proposed interlaced kernels, such notation is adopted to consider consistency in derivation. After being processed by one of the two kernels, the host signals can then be converted to watermarked

Figure 2. Impulse responses of echo kernels: (a) "Zero" kernel. (b) "One" kernel

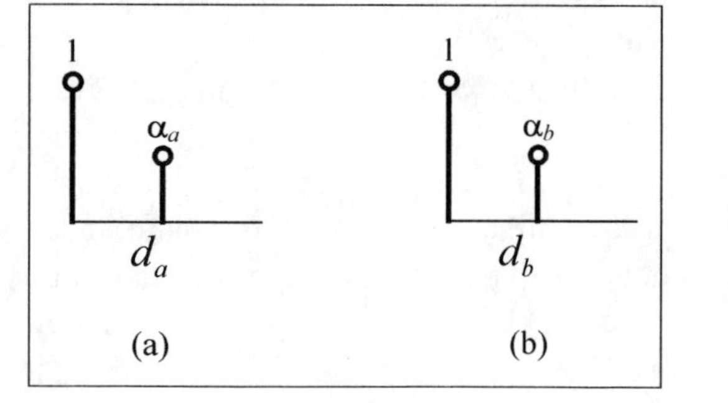

Figure 3. Echo embedding process (Gruhl et al., 1996)

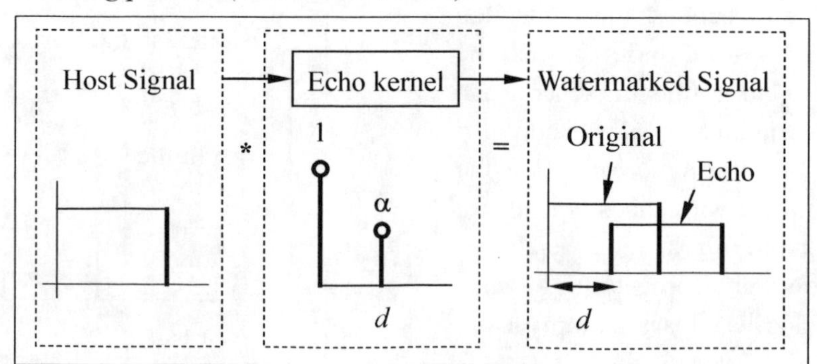

signals by the convolution operation depicted in Figure 3. That is, the watermarked audio signal $y[n]$ can be obtained by convolving the echo kernel $h[n]$ and the host signal $x[n]$:

$$y[n] = h[n] * x[n]. \quad (2)$$

In order to hide many watermark data, host audio signals are partitioned into multiple small audio segments. One binary bit datum can be added in each audio segment. To combine all echoes in the end, the windowing technique can be applied to allow a smooth transition between the neighboring segments.

EXTRACTING

As explained in the previous section, watermark data hidden in the host audio signals are achieved by embedding echoes at different offsets. Extracting watermark data from watermarked audio signals would involve the position detection for the echoes. Out of all the existing detection techniques available, the cepstrum computation is considered as one of those more significant techniques. It can be viewed as conducting the deconvolution process to separate the signals and echoes. The cepstrum of the watermarked audio signals $y[n]$ is defined as follows (Oppenheim & Schafer, 1989):

$$c_y[n] = F^{-1}(\ln F(y[n])), \quad (3)$$

where F and F^{-1} represent the Fourier transform and the inverse Fourier transform, respectively. Using equation (2), the cepstrum of the watermarked signals is written as:

$$\begin{aligned} c_y[n] &= F^{-1}(\ln H(e^{j\omega})) + F^{-1}(\ln X(e^{j\omega})) \\ &= c_h[n] + c_x[n]. \end{aligned} \quad (4)$$

Notably, the cepstrum of the watermarked audio signals comprises two parts: one is the cepstrum of the kernel $c_h[n]$ and the other is the cepstrum of the host signals $c_x[n]$. Hence, the performance of the echo hiding scheme would be affected by the host audio signals.

Using the power series expansion, the cepstrum of the kernel in equation (1) can be rewritten as follows:

$$\begin{aligned} c_h[n] &= F^{-1}(\ln(1 + \alpha e^{-j\omega d})) \\ &= F^{-1}\left(\alpha e^{-j\omega d} - \frac{\alpha^2}{2} e^{-2j\omega d} + \frac{\alpha^3}{3} e^{-3j\omega d} - \cdots\right) \\ &= \alpha\,\delta[n-d] - \frac{\alpha^2}{2}\delta[n-2d] + \frac{\alpha^3}{3}\delta[n-3d] - \cdots. \end{aligned} \quad (5)$$

The cepstrum of the watermarked audio signals is illustrated in Figure 4. The cepstrum of the kernel at the offset d is the value α which corresponds to the amplitude of the embedded echo. However, there are a number of pulses that repeatedly appear for every delay time d. Furthermore, the amplitudes of the impulses representing the echoes may be small relative to the host signals. Consequently, it is difficult to detect the positions of the echoes. Bender et al. solved this problem by considering the signal power at each delay using the autocorrelation of the cepstrum which is defined as follows (Gruhl et al., 1996).

$$Mag(x) = F^{-1}\left(\ln_{complex}(F(x))^2\right). \quad (6)$$

However, either the cepstrum or the cepstrum autocorrelation may have the similar performance of identifying the watermark data. In order to conduct the consistent analyses, only the ceptrum is adopted in this work.

Perception of Echoes

There are two types of perception we have on echoes: one is the echo itself, and the other is coloration. When the offset of the echo is at about 50ms, human ears can clearly hear one single echo.

Figure 4. Cepstrum of the watermarked audio signals

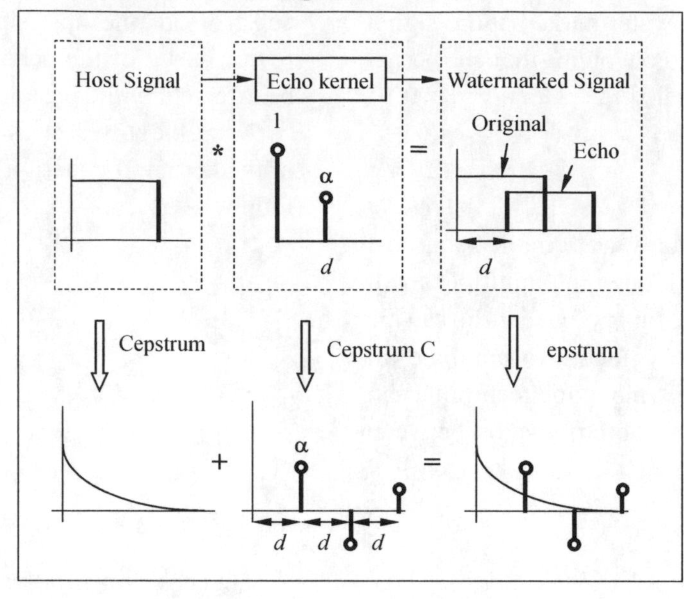

If several echoes of this type occur continuously, then it would sound like a clutter of echoes. On the contrary, if an echo has a short delay, at about 2ms, it would be perceived as changes in timbre, commonly known as coloration. In human ear model, signals entering the ears are divided into several frequency bands which correspond to ears' critical bands. These critical bands are caused by cochlear filters. Depending on the critical bands, cochlear filters have a 5 to 20 ms duration memory that is approximately inversely proportional to their bandwidth. Pulses with long delays do not deviate at all under different frequencies because the memory of ear filters is shorter than the time difference between pulses, which then ensures the nature of echoes being properly retained. Under short delays, the frequency content of pulses is retained but their time information is lost. With the duration of cochlear filters larger than the time difference between pulses, pulses will interact with each other in the cochlea (Oh, 2001). In temporal masking, stronger signal pulses create pre and postmasking regions in the time period during which the weaker signals beneath the audibility thresholds produced by a masker will not be perceived. The postmasking will exponentially decay during 50 to 200 ms, depending upon the strength and duration of the masker (Painter & Spanias, 1997). In considering temporal masking, adequately small values should be selected for the offsets of echo kernels so that the embedded echoes would be positioned in the coloration region. However, offsets that are too small may induce a distortion in host signals, which then makes it difficult to use the cepstrum technique to detect the positions of echoes. Hence, selecting the appropriate offset values is not an easy task.

The robustness of the echo hiding can be enhanced by increasing amplitudes of echo signals, but audio quality would be degraded. Moreover, other parameters such as offsets, numbers of embedded echoes, frame size, and overlap size also pose influences on audio quality. Oh et al. (2001) addressed a positive single echo kernel by using experiments to categorize the relationship between audio quality and echo kernel parameters. In their conclusion, an embedded echo whose offset value ranges from 0.9 to 3.4 msec and amplitude value is below 0.31, is imperceptible. If the amplitude is not increased beyond 0.47, the embedded echo can be perceived but not annoying. These phenomena can be referred to select adequate values of offset and amplitude of the embedded echo to preserve audio quality. With consideration of the frame size, a long frame size naturally increases the accuracy rate of detection, but the number of embedded bits would then decrease. Therefore, the frame size is typically set at a value larger than 45msec (Oh, 2001).

CONVENTIONAL ECHO WATERMARKING SCHEMES

The echo hiding concept proposed by Bender et al. has a significant influence on audio watermarking. They not only introduced the technique of using backward echo kernel to embed watermark information in host signals, they also addressed on how to use computation similar to cepstrum to detect the position of an echo so that the embedded watermark information can be restored. Additionally, all kinds of schemes are later proposed by many literatures in an attempt to resolve the issue of making a choice between the imperceptibility and robustness of echo hiding to enhance its performance. The attention is mainly focused on utilizing the perception of human hearing so that the embedded echo is not easily perceived. Furthermore, the amplitude of the embedded echo is lowered as much as possible but yet it is able to maintain the high watermark recovery accuracy rate.

In the literature, Xu, Wu, Sun, and Xin (1999) proposed a multiple echo technique. Instead of embedding one large echo into the host audio signal, they use multiple echoes with different offsets. Under the condition of maintaining the same recovery accuracy, the use of multiple-echo kernel would decrease the amplitudes of echoes, which improves the audio quality. Concurrently, an increase in the number of echoes also brings on additional unknown parameters. This change would decrease the possibilities of malicious users in detecting the echo positions and removing them.

In another work, Ko, Nishimura, and Suzuki (2005) went further to propose the time-spread echo kernel. With the use of pseudo-noise (PN) sequence, an echo is spread out as numerous little echoes in a time region. When the embedded data of watermarked audio signals are extracted, the PN sequence functions like a secret key. Without obtaining the PN sequence used in the embedding process, extracting the embedded data would be more difficult. Due to the increase in the number of echoes, amplitude of each echo then becomes smaller. Hence the distortion caused by the embedded watermark would be less severe, while it still maintains good recovery accuracy.

In addition, Oh et al. (2001) introduced the positive-negative echo hiding scheme. Their echo kernels comprise positive and negative echoes at nearby locations. Since the frequency response of a negative echo is the inversed shape having similar ripples as that of a positive echo, the frequency response of the positive and negative echoes has the smooth shape in the low frequency band. By employing positive and negative echoes, one can thus embed multiple echoes to enable that the host audio quality is not apparently deteriorated.

Recently, Kim and Choi (2003) presented an echo hiding scheme with backward and forward kernels. The theoretically-derived results show that the amplitude of the cepstrum coefficient at the echo position from the backward and forward kernels is larger than that from the backward kernel only when the embedded echoes are symmetric. Hence, the backward and forward kernels can improve the robustness of echo hiding scheme.

In the following, the properties of cepstrum coefficients in the conventional echo hiding schemes are investigated and the corresponding problems are discussed. The echo hiding scheme with positive and negative echo kernels introduced by Oh et al. comprises positive and negative pulses with a small difference in their offsets,

$$h[n]=\delta[n]+\alpha_{PB}\delta[n-d_{PB}]-\alpha_{NB}\delta[n-d_{NB}], \quad (7)$$

where d_{PB} and d_{NB} denote the offsets of the positive backward and negative backward echoes, respectively, and α_{PB} and α_{NB} denote the amplitudes of the positive backward and negative backward echoes, respectively. The echo hiding scheme with positive and negative kernels can reduce noise perceptibility and improve the robustness performance considerably. Subsequently, in addition to the positive and negative kernels, Kim et al. (2003) presented the echo hiding scheme with backward and forward kernels. The echo kernel is given as follows:

$$h[n]=\delta[n]+\alpha_{PB}\delta[n-d_{PB}]+\alpha_{PF}\delta[n+d_{PF}]$$
$$-\alpha_{NB}\delta[n-d_{NB}]-\alpha_{NF}\delta[n+d_{NF}], \quad (8)$$

where α_{PF}, and α_{NF} denote the amplitudes of the positive and the negative forward echoes, respectively. Additionally, the offsets of the positive forward echo and the negative forward echo are denoted as d_{PF}, and d_{NF}, respectively. In the case of symmetric impulse response, that is, $d_{PB}=d_{PF}=d_P$, $d_{NB}=d_{NF}=d_N$, $\alpha_{PB}=\alpha_{PF}=\alpha_P$, and $\alpha_{NB}=\alpha_{NF}=\alpha_N$, Kim et al. theoretically analyzed that the values of the cepstrum coefficients at $n=d_P$ and $n=d_N$ are as follows:

$$\begin{cases} c_h[d_P]=\dfrac{\alpha_P}{1-\alpha_P^2} \\ c_h[d_N]=-\dfrac{\alpha_N}{1-\alpha_N^2} \end{cases}. \quad (9)$$

For $\alpha < 1$, the value $\alpha/(1-\alpha^2)$ is larger than α. Therefore, the amplitudes of echoes can be reduced accordingly when embedding echoes in pairs. For simplicity, the same value is used for the amplitudes of positive and negative echoes, that is, $\alpha_P=\alpha_N=\alpha$. The impulse responses of the "Zero" and "One" kernels are depicted in Figure 5. For the watermark data "Zero", the embedded echoes would be placed at the position "*a*" where the offsets of the positive and negative echoes are in the d_{a_P} and d_{a_N}, respectively. But for the watermark data "One", the embedded echoes would be placed at the position "*b*" where the offsets of the positive and negative echoes are in the d_{b_P} and d_{b_N}, respectively. If the embedded watermark datum is "Zero", the "Zero" kernel shown in Figure 5(a) is used. The watermarked audio signals can then be written as:

$$y[n]=x[n]+\alpha x[n-d_{aP}]-\alpha x[n-d_{aN}]$$
$$+\alpha x[n+d_{aP}]-\alpha x[n+d_{aN}]. \quad (10)$$

Figure 5. Impulse responses of the positive, negative, backward and forward kernels. (a) "Zero" kernel. (b) "One" kernel

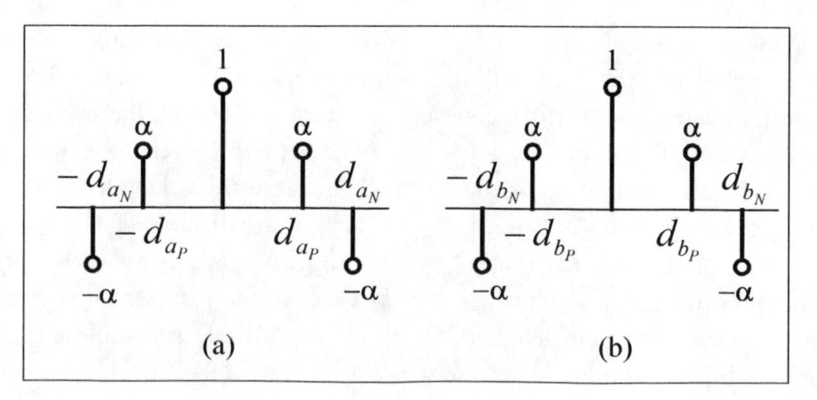

In the extracting process, the cepstrum of the watermarked signals after attacks, $\hat{y}$, is given by:

$$c_{\hat{y}}[n]=c_x[n]+\frac{\alpha}{1-\alpha^2}\delta[n-d_{ap}]$$
$$-\frac{\alpha}{1-\alpha^2}\delta[n-d_{aN}]+\cdots+c_A[n], \qquad (11)$$

where $c_A[n]$ denotes the cepstrum component that is resulted from the attacks exerted on the watermarked audio signals. The cepstrum value of the watermark data "Zero" $c_{\hat{y}\_Z}$ can be obtained as follows, (see equation (12)), where c_{x_ab} and c_{A_ab} represent $(c_x[d_{ap}]-c_x[d_{aN}])-(c_x[d_{bp}]-c_x[d_{bN}])$ and $(c_A[d_{ap}]-c_A[d_{aN}])-(c_A[d_{bp}]-c_A[d_{bN}])$, respectively. Similarly, if the embedded watermark datum is "One," the "One" kernel shown in Figure 5(b) is used. The cepstrum value of the watermark data "One" $c_{\hat{y}\_O}$ would be:

$$c_{\hat{y}\_O}=-\frac{2\alpha}{1-\alpha^2}+c_{x\_ab}+c_{A\_ab}. \qquad (13)$$

Therefore, in order to extract the embedded data correctly, the following condition should be satisfied:

$$\frac{2\alpha}{1-\alpha^2}+c_{x\_ab}+c_{A\_ab}\geq 0, \quad if\ I=0$$
$$-\frac{2\alpha}{1-\alpha^2}+c_{x\_ab}+c_{A\_ab}<0, \quad if\ I=1. \qquad (14)$$

In equation (14), the amplitude of the embedded echo is affected by the values of c_{x_ab} and c_{A_ab}. Since either type of watermark data can occur, the amplitude of embedded echo can decrease when the cepstrum components resulting from the host signal and the malicious attacks are as small as possible. That is:

$$c_{x\_ab}+c_{A\_ab}=0. \qquad (15)$$

Generally, the cepstrum difference caused by the various attacks is uncorrelated with that caused by the host signals. Thus, this condition is seldom to hold. In addition, the amplitude of the embedded echo is usually fixed in the conventional schemes. However, in equation (14), the term c_{x_ab} is related to the characteristic of the host signal and varied for different segments. Therefore, it may be redundant to embed the echoes with a fixed amplitude for certain segments. If the

Equation 12.

$$\begin{aligned}
c_{\hat{y}_Z} &= (c_{\hat{y}}[d_{ap}]-c_{\hat{y}}[d_{aN}])-(c_{\hat{y}}[d_{bp}]-c_{\hat{y}}[d_{bN}]) \\
&= \left\{\left(c_x[d_{ap}]+\frac{\alpha}{1-\alpha^2}+c_A[d_{ap}]\right)-\left(c_x[d_{aN}]-\frac{\alpha}{1-\alpha^2}+c_A[d_{aN}]\right)\right\}- \\
&\quad \{(c_x[d_{bp}]+c_A[d_{bp}])-(c_x[d_{bN}]+c_A[d_{bN}])\} \\
&= \frac{2\alpha}{1-\alpha^2}+\{(c_x[d_{ap}]-c_x[d_{aN}])-(c_x[d_{bp}]-c_x[d_{bN}])\}+ \\
&\quad \{(c_A[d_{ap}]-c_A[d_{aN}])-(c_A[d_{bp}]-c_A[d_{bN}])\} \\
&= \frac{2\alpha}{1-\alpha^2}+c_{x_ab}+c_{A_ab},
\end{aligned}$$

echo's amplitude is adapted according to the host signals in the encoding process, the detection accuracy would be maintained even with small echo amplitude. Furthermore, the amplitude of the embedded echoes would contemplate the deviation in the cepstrum difference caused by the various attacks, c_{A_ab}, to give the watermark data recovered accurately in the extraction process.

PROPOSED ECHO WATERMARKING SCHEME

Among the echo hiding scheme and its modifications, the efforts are focused on improving the performance of the echo hiding by modifying the kernels. However, they have not taken the characteristics of host audio signals into account. Certainly, there is no consideration in connection with the attacks on the watermarked audio signals. In this chapter, we present an echo hiding scheme based on the analysis-by-synthesis approach (Wu, Chen, & Wang, 2003; Wu & Chen, 2004, 2006). The amplitude of the embedded echo signal is adapted during the embedding process by considering not only the properties of the host signals but also the situations when the watermarked audio signals have suffered various attacks. Moreover, the interlaced kernels are developed to enable that each audio segment is divided into two parts in which the echoes' positions for embedding "Zero" and "One" are exchanged alternately. As compared to the conventional schemes, the proposed scheme demonstrates a superior performance on the robustness and perceptual quality.

Figure 6. Block diagram of the proposed watermark embedding process

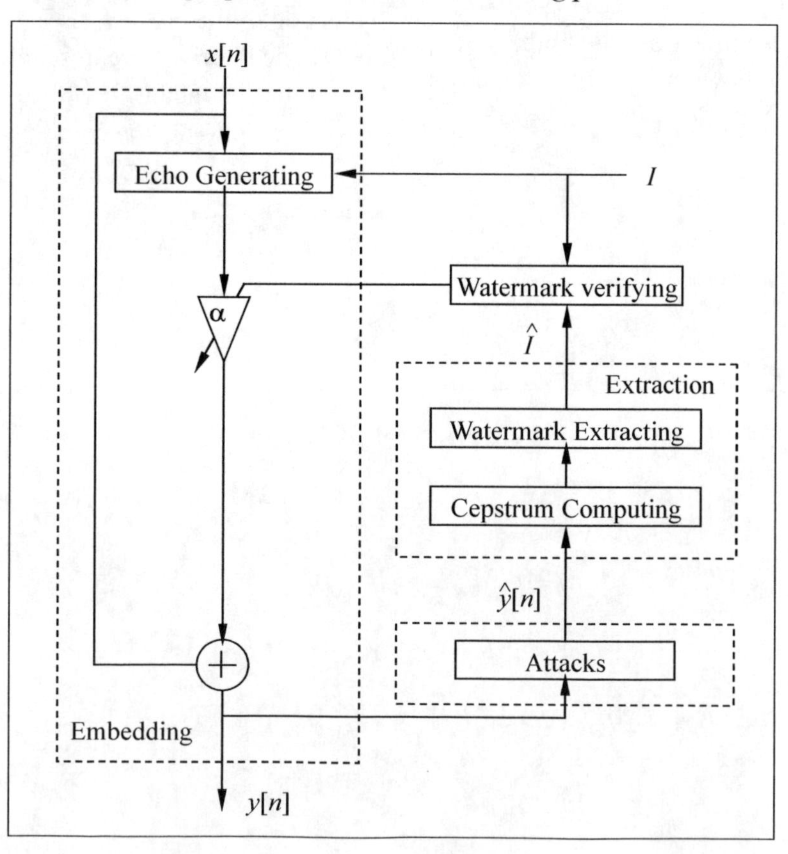

The analysis-by-synthesis approach is adopted in the proposed echo hiding scheme. During the watermark embedding process, these embedded data are extracted from the watermarked audio signals to see if they can be accurately recovered in the decoding process. If the watermark data extracted are not consistent with the added watermark data, the amplitudes of the embedded echoes are increased gradually. According to the characteristics of the host audio signals, the amplitudes of the embedded echoes are adapted appropriately so as to ensure that the embedded data are recovered accurately after various common attacks and have the impacts on the host audio signals as small as possible. The embedding process of the proposed echo hiding scheme using the analysis-by-synthesis approach is shown in Figure 6.

In general, the values of high-order cepstrum coefficients are smaller than those of low-order cepstrum coefficients (Lee & Ho, 2000; Oppenheim & Schafer, 1989). Furthermore, the cepstrum coefficients of the host signals at two offsets may not be equal. If the offset of the embedded echo is fixed for each type of watermark datum "Zero" or "One," it will lead to a problem that an audio segment is likely for embedding one type of watermark datum but improper for the other type of watermark datum. In this study, we develop the interlaced kernels to resolve this problem. For the interlaced kernels, the host audio signals are partitioned into many segments without overlapping each other and then each segment is further separated into two parts. In the front part of a segment, the echo is placed at the offsets of d_a and d_b for watermark data "Zero" and "One," respectively. Contrarily, the echo positions are interchanged in the rear part of this segment. That is, the echo is placed at the offsets of d_a and d_b to represent the watermark data "One" and "Zero," respectively. The impulse responses of the "Zero" and "One" kernels are depicted in Figure 7. For the watermark data "Zero," the embedded echoes in the front part of a segment are placed at the position "a" where the offsets of the positive and negative echoes are in the d_{a_P} and d_{a_N}, while the embedded echoes in the rear part of a segment are placed at the position "b" where the offsets of the positive and negative echoes are in the d_{b_P} and d_{b_N}. But for the watermark data "One," the embedded echoes in the front part of a segment are placed at the position "b" where the offsets of the positive and negative echoes are in the d_{b_P} and d_{b_N}, while the embedded echoes in the rear part of a segment are placed at the position "a" where the offsets of the positive and negative echoes are in the d_{a_P} and d_{a_N}. In the watermark extracting process, each audio segment is also partitioned into two parts. The cepstrum value of each part is calculated individually. The cepstrum value of a segment is obtained by adding the cepstrum values of these two parts at the corresponding positions indicating the same watermarks. Finally, the cepstrum values that correspond to the watermark data "One" and "Zero" are compared to determine the embedded watermark data.

If the embedded watermark datum is "Zero," the watermarked audio signal can be generated by using the kernel "Zero" shown in Figure 7(a) and written as follows, (seed eqatuion (16)), where $x_f[n]$ and $x_r[n]$ denote the host audio signals in the front part and rear part of a segment, respectively.

Next, the cepstrum of the watermarked audio signals that have gone through the malicious attacks are obtained, (see equation (17)) where $c_{A_f}[n]$ and $c_{A_r}[n]$ denote the cepstrum components that are resulted from the various attacks exerted on the front and rear part of the watermarked audio signals. The cepstrum value of the watermark data "Zero" $c_{\hat{y}\_Z}$ can then be derived at, (see equation (18)), where $c_{x_f\_ab}$, $c_{x_r\_ab}$, $c_{A_f\_ab}$ and $c_{A_r\_ab}$ represent $(c_{x_f}[d_{ap}]-c_{x_f}[d_{aN}])-(c_{x_f}[d_{bp}]-c_{x_f}[d_{bN}])$ $(c_{x_r}[d_{ap}]-c_{x_r}[d_{aN}])-(c_{x_r}[d_{bp}]-c_{x_r}[d_{bN}])$ $(c_{A_f}[d_{ap}]-c_{A_f}[d_{aN}])-(c_{A_f}[d_{bp}]-c_{A_f}[d_{bN}])$ and $(c_{A_r}[d_{ap}]-c_{A_r}[d_{aN}])-(c_{A_r}[d_{bp}]-c_{A_r}[d_{bN}])$, respectively. Similarly, if the embedded watermark

Figure 7. Impulse responses of the interlaced kernels. (a) "Zero" kernel. (b) "One" kernel

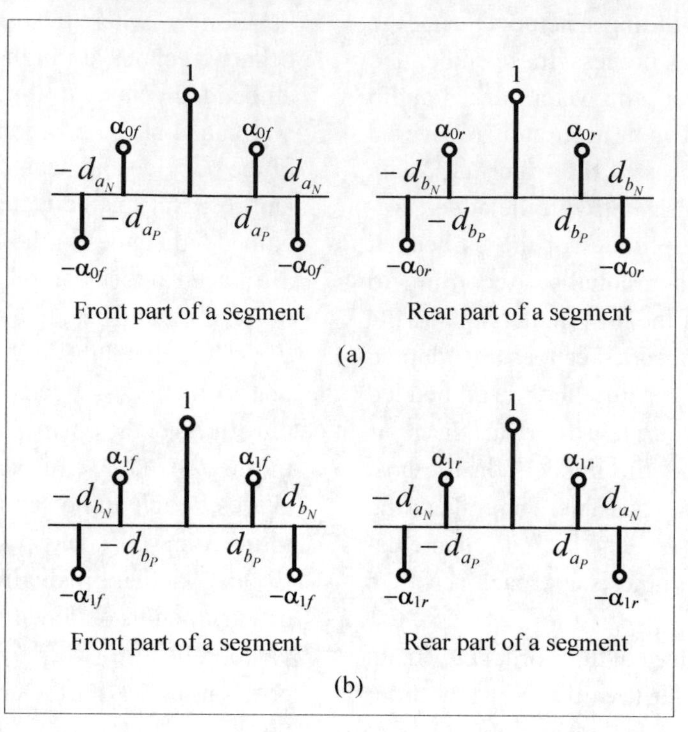

Equation 16.

$$\begin{aligned} y[n] = {} & x_f[n] + \alpha_{0f} x_f[n - d_{aP}] - \alpha_{0f} x_f[n - d_{aN}] + \alpha_{0f} x_f[n + d_{aP}] - \alpha_{0f} x_f[n + d_{aN}] \\ & + x_r[n] + \alpha_{0r} x_r[n - d_{bP}] - \alpha_{0r} x_r[n - d_{bN}] + \alpha_{0r} x_r[n + d_{bP}] - \alpha_{0r} x_r[n + d_{bN}], \end{aligned}$$

Equation 17.

$$\begin{aligned} c_{\hat{y}}[n] & \equiv c_{\hat{y}_f}[n] + c_{\hat{y}_r}[n] \\ & = c_{x_f}[n] + \frac{\alpha_{0f}}{1-\alpha_{0f}^2}\delta[n - d_{aP}] - \frac{\alpha_{0f}}{1-\alpha_{0f}^2}\delta[n - d_{aN}] + \cdots + c_{A_r}[n] + \\ & \quad c_{x_r}[n] + \frac{\alpha_{0r}}{1-\alpha_{0r}^2}\delta[n - d_{bP}] - \frac{\alpha_{0r}}{1-\alpha_{0r}^2}\delta[n - d_{bN}] + \cdots + c_{A_f}[n], \end{aligned}$$

datum is "One," the "One" kernel shown in Figure 7(b) is used. The cepstrum value of the watermark data "One" $c_{\hat{y}\_o}$ would be:

$$c_{\hat{y}\_o} = -\left(\frac{2\alpha_{1f}}{1-\alpha_{1f}^2} + \frac{2\alpha_{1r}}{1-\alpha_{1r}^2}\right) + (c_{x_f\_ab} - c_{x_r\_ab}) + (c_{A_f\_ab} - c_{A_r\_ab}). \quad (19)$$

Accordingly, to extract the embedded data correctly, the following condition should be satisfied, (see equation (20)).

To reduce the amplitude of embedded echo, the values of the cepstrum components resulting from the host signal and the malicious attacks need be minimized and thereby the best condition is derived at:

$$(c_{x_f\_ab} - c_{x_r\_ab}) + (c_{A_f\_ab} - c_{A_r\_ab}) = 0. \quad (21)$$

Equation (20) shows that the cepstrum components resulting from the host audio signals and various attacks are counteracted in the front and rear parts of a segment. For a short segment, the host audio signals generally have similar properties and the influences generated by attacks can be also resembled in the front and rear parts of a segment. Therefore, the cepstrum value resulting from the host signals and various attacks is reduced in equation (20). Even though the condition in equation (21) is not feasible, the proposed analysis-by-synthesis approach would modify the echo amplitude to ensure that the condition in equation (20) is satisfied. However, if the interlaced kernels are not applied, the condition to extract the hidden datum correctly is satisfied in equation (14). In general, the variance of $(c_{x_f\_ab} - c_{x_r\_ab}) + (c_{A_f\_ab} - c_{A_r\_ab})$ is smaller than that of $c_{x\_ab} + c_{A\_ab}$. Therefore, the interlaced kernels can enhance the robustness of the echo hiding under small echo amplitudes.

Equation 18.

$$\begin{aligned} c_{\hat{y}\_z} &\equiv \left\{(c_{\hat{y}_f}[d_{ap}] - c_{\hat{y}_f}[d_{aN}]) - (c_{\hat{y}_f}[d_{bp}] - c_{\hat{y}_f}[d_{bN}])\right\} \\ &\quad + \left\{(c_{\hat{y}_r}[d_{bp}] - c_{\hat{y}_r}[d_{bN}]) - (c_{\hat{y}_r}[d_{ap}] - c_{\hat{y}_r}[d_{aN}])\right\} \\ &= \left(\frac{2\alpha_{1f}}{1-\alpha_{1f}^2} + \frac{2\alpha_{1r}}{1-\alpha_{1r}^2}\right) + (c_{x_f\_ab} + c_{x_r\_ba}) + (c_{A_f\_ab} + c_{A_r\_ba}) \\ &= \left(\frac{2\alpha_{1f}}{1-\alpha_{1f}^2} + \frac{2\alpha_{1r}}{1-\alpha_{1r}^2}\right) + (c_{x_f\_ab} - c_{x_r\_ab}) + (c_{A_f\_ab} - c_{A_r\_ab}), \end{aligned}$$

Equation 20.

$$\begin{aligned} &\left(\frac{2\alpha_{1f}}{1-\alpha_{1f}^2} + \frac{2\alpha_{1r}}{1-\alpha_{1r}^2}\right) + (c_{x_f\_ab} - c_{x_r\_ab}) + (c_{A_f\_ab} - c_{A_r\_ab}) \geq 0, \quad \text{if } I = 0 \\ -&\left(\frac{2\alpha_{1f}}{1-\alpha_{1f}^2} + \frac{2\alpha_{1r}}{1-\alpha_{1r}^2}\right) + (c_{x_f\_ab} - c_{x_r\_ab}) + (c_{A_f\_ab} - c_{A_r\_ab}) < 0, \quad \text{if } I = 1. \end{aligned}$$

In the analysis-by-synthesis approach, whether the embedded data can be accurately recovered from the watermarked audio signals is considered during the watermark embedding process. Since the watermarked audio signals may undergo various malicious attacks before the extracting process, these attack conditions should be considered in the watermark embedding process. Because the characteristics of the common signal processing and the malicious attacks are so diverse, the actual influences on the audio signals are unknown *a priori*. Therefore, the amplitudes of the embedded echoes can be adjusted such that the absolute value of the cepstrum coefficient difference at echo positions that correspond to the watermark data "Zero" and "One" is above a specific threshold value, c_{th}. The robustness performance can be improved by increasing the threshold value, albeit deteriorating the perceptual quality.

Comparison of Echo Watermarking Scbemes

In the following, ten audio pieces with different styles are utilized as the host signals to test the performances of the proposed scheme. Each host signal has 4,096,000 samples. The audio signals are sampled at 44.1 kHz where each sample is quantized by 16 bits. In the embedding process, host signals are divided into many segments without overlapping each other, each of which has 4,096 samples. The segmentation of the host signals is shown in Figure 8. The offsets of the embedded echoes are set as follows: $d_{a_P} = 100$, $d_{a_N} = 103$, $d_{b_P} = 110$, and $d_{b_N} = 113$. The Hanning window with 4,096 points is employed to smooth the boundaries of the segments of the echo signal. The symmetric coefficients of the Hanning window are formulated as follows, (see quation (22)), where n equals to 4,096.

The echo hiding scheme using the positive and negative echo kernels (EHPN) and the echo hiding scheme, using the positive, negative, backward, and forward echo kernels (EHPNBF) are two conventional approaches used for comparison in this chapter (Oh, 2001; Kim & Choi, 2003). Since the embedded echoes are different in these schemes, the equivalent echo amplitude for each scheme is employed to do fair comparison. The equivalent echo amplitude α_e for the EHPN scheme is defined as:

Equation 22.

$$w[k+1] = 0.5\left(1 - \cos\left(\frac{2\pi k}{n-1}\right)\right) \qquad k = 0, \cdots, n-1$$

Figure 8. Segmentation of the host signals

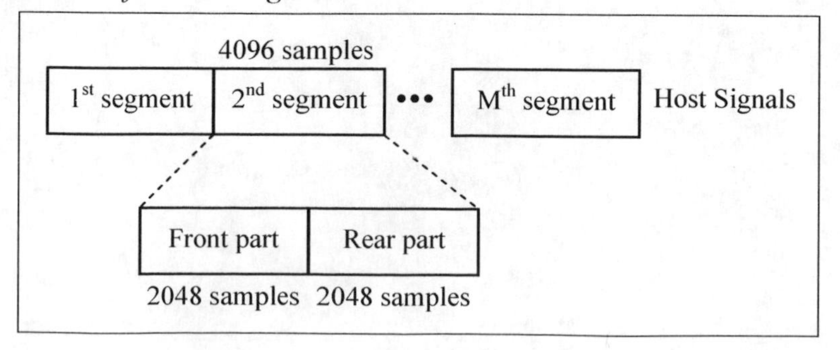

$$\alpha_e = (\alpha_{PB} + \alpha_{NB})/2. \tag{23}$$

The equivalent echo amplitude for the EHPN-BF scheme is defined as follows:

$$\alpha_e = (\alpha_{PB} + \alpha_{NB} + \alpha_{PF} + \alpha_{NF})/2. \tag{24}$$

In the proposed scheme, the embedded echoes are divided into two parts in each audio segment. Thus, the following equation is used to represent the equivalent echo amplitude:

$$\alpha_e = (\alpha_{0f} + \alpha_{0r} + \alpha_{1f} + \alpha_{1r}). \tag{25}$$

In general, a lower echo amplitude implies that the host audio signals are less degraded. Accordingly, the equivalent echo amplitude can be adopted to evaluate the performance on audio quality. Additionally, the signal-to-noise ratio (SNR) of watermarked audio signals is also given as a reference. A higher SNR reveals that the watermarked audio signals are much closer to the host signals. However, it should be mentioned that these metric data are used to compare the relative audio quality of the proposed and conventional echo hiding schemes rather than subjective audio quality. For the robustness comparison, the bit error rate (BER) is utilized and defined as follows:

$$BER = 1 - \frac{\text{number of bits correctly extracted}}{\text{number of bits embedded}} \tag{26}$$

Herein, the lower is the bit error rate, the higher is the recovery accuracy.

The robustness of the proposed and conventional schemes is compared at the following cases:

1. **Closed-loop:** The watermark data are extracted from the watermarked audio signals without any attacks.
2. **MP3 attack:** Encoding/decoding was conducted on the watermarked audio signals by using the MPEG-1 audio layer III at a bit rate of 64kbps.
3. **Quantization attack:** Each sample of the watermarked audio signals is quantized to 8 bits.
4. **Adding random noises:** After adding random noises, the watermarked audio signals would have a signal-to-noise ratio of 20 dB.
5. **Re-sampling attack:** The watermarked audio signals are down sampled by a factor of 2, and then up-sampling.
6. **Band-pass filtering attack:** A 10th order Butterworth band-pass filter which has 100 Hz and 10 kHz cutoff frequencies is applied to the watermarked audio signals.
7. **Echo addition attack:** An echo, that has an amplitude of 0.3 and a delay offset of 100, is added to the watermarked signals.
8. **De-synchronization attack:** Audio signals are shifted left 500 samples.

First of all, the advantage of the interlaced kernels is discussed. In the conventional watermarking schemes, the echo position represents a definite identification tag. Occasionally, there is the case that the cepstrum coefficient value at one position is much smaller than that of the other. However, the embedded watermark data should be realized by inserting the echo signal just at the position whose cepstrum coefficient is small. In this case, increasing the echo amplitude at that position is hardly to change the relationship of cepstrum coefficients for these two echo positions. To solve this problem, the technique of the interlaced kernels is developed. Each segment is separated into two parts in which the identification tags defined for the echoes are interchanged. From equation (20), the cepstrum coefficients in the front and rear parts of a segment counteract each other. Therefore, the influence on the echo amplitude resulted from the host audio signals is reduced. Furthermore, the cepstrum coefficients generated by the various attacks are also opposed to each other in the front and rear part of a segment. Hence, the degradation on recovery accuracy caused by various attacks can be lessened. This phenomenon becomes apparent particularly for the attack of echo addition. Because the actual effects of various attacks on the watermarked audio signals are unknown *a priori*, the amplitudes of the embedded echoes are adjusted such that the absolute value of the cepstrum coefficient difference at echo positions that correspond to the watermark data "Zero" and "One" is above a specific threshold value. This threshold value is utilized as the buffer range for attacks that are not considered in advance. The simulation results of the proposed scheme with and without the interlaced kernels under the equivalent echo amplitude smaller than or equal to 0.4 are listed in Table 1. The robustness is highly improved by using the interlaced kernels even though its equivalent echo amplitude is smaller than that without interlaced kernels. In other words, the interlaced kernels can improve the robustness and the audio quality concurrently. Furthermore, interchanging the embedded echoes' positions would make the cepstrum be likely independent of the type of embedded data. Therefore, the positions of the embedded echoes are difficult to detect by the third parties and the security of the proposed embedding scheme against intended attacks is improved

Finally, the performance of the proposed scheme is compared to the conventional approaches. The equivalent echo amplitude of the proposed scheme is set to be smaller than or equal to those of the conventional schemes for each segment. Such setting in the amplitudes of the embedded echoes considers mainly for fair comparison in the audio quality. The performances of the proposed and the conventional echo hiding schemes, with the maximum equivalent echo amplitudes 0.2 and 0.4, are summarized in Tables 2 and 3, respectively. Simulation results reveal the high recovery accuracy rate of the proposed scheme under various attacks, which is much better than those of the conventional schemes. Accordingly, the robustness performance has much improved in the proposed scheme. Moreover, the mean values of the equivalent echo amplitude of the proposed scheme are smaller than those of the conventional schemes. Since the maximum amplitudes of embedded echoes used for the proposed scheme are limited by the fixed value used in the conventional schemes for each segment, the proposed scheme can yield better audio quality than the conventional schemes. The SNR values also confirm this statement. The simulation results show that the SNR values of the proposed scheme are higher than those of the conventional schemes. Additionally, the distribution of the equivalent echo amplitudes for the proposed scheme under the limit of the maximum equivalent echo amplitude 0.4 is shown in Figure 9. In proposed scheme, there are 75% segments that have equivalent echo amplitude smaller than that in the conventional schemes. Therefore, it is concluded that the proposed scheme can improve the robustness with even embedded echoes' amplitudes smaller than those of the conventional schemes.

Table 1. Effects of the proposed schemes with and without interlaced kernels

Performances \ Proposed Schemes		Without interlaced kernels	With interlaced kernels
Robustness	Closed loop	0.040	0.006
	MP3	0.055	0.014
	Quantization	0.080	0.065
	Noise	0.124	0.110
	Re-sampling	0.091	0.019
	Band-pass filtering	0.056	0.013
	Echo addition	0.416	0.017
	Desynchronization	0.040	0.006
Audio quality	SNR(dB)	24.3	24.9
	Mean of α_e	0.27	0.20

CONCLUSION

In this chapter, the role of echo hiding in the audio watermarking has been described in the range of background, functions, and applications. The imperfections of the echo hiding and the relevant solutions in the past are also presented. However, the characteristics of the host signals are not well addressed in the conventional schemes. The tradeoff between robustness and imperceptibility can then be improved by the proposed analysis-by-synthesis approach echo hiding. In the watermark encoding process, the amplitudes of echoes are adapted by using an analysis-by-synthesis approach to take the advantage of the properties of host signals and then minimize the amplitudes of the echo signals. Moreover, the interlaced kernels are proposed to eliminate the

Table 2. Comparisons of the proposed and conventional schemes with equivalent echo amplitudes smaller than and equal to 0.2

Performances \ Schemes		EHPN $\alpha_P = 0.2$ $\alpha_N = 0.2$	EHPNBF $\alpha_{PB} = \alpha_{NB} = 0.1$ $\alpha_{PF} = \alpha_{NF} = 0.1$	Proposed ($c_{th} = 0.2$) $\alpha_{0f}, \alpha_{1f} \leq 0.1$ $\alpha_{0r}, \alpha_{1r} \leq 0.1$
Robustness	Closed loop	0.086	0.080	0.027
	MP3	0.146	0.140	0.048
	Quantization	0.178	0.166	0.097
	Noise	0.270	0.261	0.153
	Re-sampling	0.156	0.154	0.059
	Band-pass filtering	0.114	0.113	0.039
	Echo addition	0.207	0.199	0.054
	Desynchronization	0.086	0.080	0.027
Audio quality	SNR(dB)	23.7	26.8	28.1
	Mean of α_e	0.20	0.20	0.16

Table 3. Comparisons of the proposed and conventional schemes with equivalent echo amplitudes smaller than and equal to 0.4

Performances \ Schemes		EHPN $\alpha_P = 0.4$ $\alpha_N = 0.4$	EHPNBF $\alpha_{PB} = \alpha_{NB} = 0.2$ $\alpha_{PF} = \alpha_{NF} = 0.2$	Proposed ($c_{th} = 0.2$) $\alpha_{0f}, \alpha_{1f} \leq 0.2$ $\alpha_{0r}, \alpha_{1r} \leq 0.2$
Robustness	Closed loop	0.031	0.024	0.006
	MP3	0.058	0.047	0.014
	Quantization	0.088	0.071	0.065
	Noise	0.163	0.145	0.110
	Re-sampling	0.064	0.052	0.019
	Band-pass filtering	0.044	0.037	0.013
	Echo addition	0.066	0.051	0.017
	Desynchronization	0.031	0.024	0.006
Audio quality	SNR(dB)	17.7	20.8	24.9
	Mean of α_e	0.40	0.40	0.20

Figure 9. Distribution of equivalent echo amplitudes in the proposed scheme

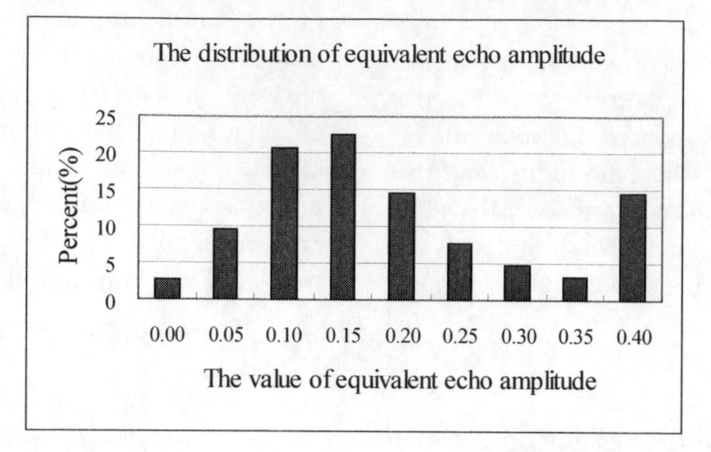

influences of the host signals and various attacks on the watermarked data. The simulation results reveal that the proposed scheme is superior to the conventional schemes both in the perceptual quality of the watermarked audio signals and the robustness for various attacks.

FUTURE RESEARCH DIRECTIONS

The success of the watermark detection relies greatly on the allocation of the audio signal segments. Losing synchronization may result in a false detection. There are two synchronization problems: one is to align the starting point of the audio signal segments and the other is the time-scale and frequency-scale modifications which are the most serious attacks by malicious attackers. Though the echo hiding is rather robust against certain type of synchronization, there is still room for improvement.

One of the advantages of the echo hiding is that the echo detection technique is easy to implement. However, it would be a serious drawback from the viewpoint of the copyright protection. The possibility of malicious tampering by the third parties is greatly increased. Thus, an adequate secure mechanism should be considered (Petitcolas et al., 1998; Craver, Wu, Liu, Stubblefield, Swartzlander, & Wallach, 2001). Generally, there exists a peak value in the ceptrum right at a delay offset for an embedded echo. Such a peak may allow the embedded echoes to be easily detected and removed by malicious attackers. Accordingly, it is necessary to devise a scheme to smooth cepstrum peaks and preserve detection accuracy.

In this work, the analysis-by-synthesis approach considers whether these embedded data can be accurately recovered from the watermarked audio signals during the watermark embedding process. The amplitudes of the embedded echoes can be adjusted such that the absolute value of the cepstrum coefficient that correspond to the watermark data "Zero" and "One" is above a specific threshold value. This threshold value is used as the buffer range for the malicious attacks that are not considered in advance. However, the corresponding relationship between the threshold value and the robustness performance is not so straightforward. It is difficult to decide the suitable value for the threshold. If all attacks can be known in advance, the proposed scheme can consider these attacks during the analysis-by-synthesis process. Such approach does not need to decide the threshold value. Apparently, to know all attacks in advance and to implement these attacks in the analysis-by-synthesis process of the proposed scheme is not feasible. Perhaps, it can be an easy way to build a model to characterize the malicious attacks. If these malicious attacks can be appropriately modeled, the analysis-by-synthesis approach will modify the amplitudes of the embedded echoes by considering only this model on the watermarked audio signals in advance during the embedding process.

REFERENCES

Boney, L., Tewfik, A., & Hamdy, K. (1996). *Digital watermarks for audio signals.* Paper presented at the IEEE International Conference on Multimedia Computing and Systems (pp. 473-480).

Craver, S.A., Wu, M., Liu, B., Stubblefield, A., Swartzlander, B., & Wallach, D. S. (2001). Reading between the lines: Lessons from the SDMI challenge. In *Proceeding of the 10th USENIX Security Symposium.*

Foo, S.W., Yeo, T.H., & Huang, D.Y. (2001). An adaptive audio watermarking system. In *Proceeding of the IEEE International Conference on Electrical and Electronic Technology* (Vol. 2, pp. 509-513).

Gruhl, D., Lu, A., & Bender, W. (1996). Echo hiding. In *Proceeding of the Information Hiding Workshop* (pp. 295-315). University of Cambridge, UK.

Kim, H.J., & Choi, Y.H. (2003). A novel echo-hiding scheme with backward and forward kernels. *IEEE Transactions on Circuits and Systems for Video Technology, 13,* 885-889.

Ko, B.S., Nishimura, R., & Suzuki, Y. (2005). Time-spread echo method for digital audio watermarking. *IEEE Transactions on Multimedia, 7,* 212-221.

Lee, S.K., & Ho, Y.S. (2000). Digital audio watermarking in the cepstrum domain. *IEEE Transactions on Consumer Electronics, 46,* 744-750.

Oh, H.O., Kim, H.W., Seok, J.W., Hong, J.W., & Youn, D.H. (2001). Transparent and robust audio watermarking with a new echo embedding technique. In *Proceeding of the IEEE International Conference on Multimedia and Expo* (pp. 433-436).

Oh, H.O., Seok, J.W., Hong, J.W., & Youn, D.H. (2001). New echo embedding technique for robust and imperceptible audio watermarking. In *Proceeding of IEEE International Conference Acoustic, Speech, and Signal Processing* (Vol. 3, pp. 1341-1344).

Oppenheim, A.V., & Schafer, R.W. (1989) *Discrete-time signal processing.* Englewood Cliffs, NJ: Prentice Hall.

Painter, T., & Spanias, A. (1997). A review of algorithms for perceptual coding of digital audio signals. In *Proceeding of the International Conference on Digital Signal Processing* (pp. 179-205).

Petitcolas, F.A.P., Anderson, R.J., & Kuhn, M.G. (1998). *Attacks on copyright marking systems.* Paper presented at the International Workshop on Information Hiding (pp. 218-238).

Wu, W.C., & Chen, O.T.-C. (2004). An analysis-by-synthesis echo watermarking method. In *Proceeding of IEEE International Conference on Multimedia and Expo* (Vol. 3, pp. 1935-1938).

Wu, W.C., & Chen, O.T.-C (2006). Analysis-by-synthesis echo hiding scheme using mirrored kernels. In *Proceeding of IEEE International Conference on Acoustic, Speech & Signal Processing,* II 325-II 328.

Wu, W.C., Chen, O.T.-C, & Wang, Y.H. (2003). An echo watermarking method using an analysis-by-synthesis approach. In *Proceeding of the 5th IASTED International Conference on Signal and Image Processing* (pp. 365-369).

Xu, C., Wu, J., Sun, Q., & Xin, K. (1999). Applications of digital watermarking technology in audio signals. *Journal of the Audio Engineering Society, 47*(10).

ADDITIONAL READING

Arnold, M. (2000). Audio watermarking: Feature, applications and algorithms. In *Proceeding of the IEEE International Conference on Multimedia and Expo* (Vol. 2, pp. 1013-1016).

Arnold, M., Wolthusen, S.D., & Schmucker, M. (2003). *Techniques and applications of digital watermarking and content protection.* Norwood, MA: Artech House Inc.

Barni, M., & Bartolini, F. (2004). *Watermarking systems engineering.* New York: Marcel Dekker Inc.

Bassia, P., Pitas, I., & Nikolaidis, N. (2001). Robust audio watermarking in the time domain. *IEEE Transactions on Multimedia, 3,* 232-241.

Bender, W., Gruhl, D., Morimoto, N., & Lu, A. (1996). Techniques for data hiding. *IBM System Journal, 35,* 313-336.

Cox, I.J., Kilian, J., Leighton, F.T., & Shamoon, T. (1997). Secure spread spectrum watermarking for multimedia. *IEEE Transactions on Image Processing, 6,* 1673-1687.

Cox, I.J., Miller, M.L., & Bloom, J.A. (2002). *Digital watermarking: Principles & practice.* San Francisco: Morgan Kaufmann Publishers.

Craver, S., Memon, N., & Yeung, M.M. (1998). Resolving rightful ownerships with invisible watermarking techniques: Limitations, attacks, and implications. *IEEE Journal on Selected Areas in Communications, 16*(4), 573-586.

Cvejic, N., Keskinarkaus, A., & Seppanen, T. (2001). Audio watermarking using m-sequences and temporal masking. In *Proceeding of the IEEE Workshop on Applications of Signal Processing to Audio and Acoustics* (pp. 227-230).

Cvejic, N., & Seppanen, T. (2003). Robust audio watermarking in wavelet domain using frequency hopping and modified patchwork method. In *Proceeding of the International Symposium on Signal Processing* (pp. 251-255).

Furht, B., & Kirovski, D. (2006). *Multimedia watermarking techniques and applications.* Boca Raton, FL: Auerbach Publications.

Furht, B., Muharemagic, E., & Socek, D. (2005). *Multimedia encryption and watermarking.* New York: Springer-Verlag.

Johnson, N.F., Duric, Z., & Jajodia, S. (2000). *Information hiding: Steganography and watermarking – attacks and countermeasures.* Norwell, MA: Kluwer Academic Publishers.

Katzenbeisser, S., & Petitcolas, F.A.P. (2000). *Information hiding techniques for steganography and digital watermarking.* Norwood, MA: Artech House Inc.

Kipper, G. (2003). *Investigator's guide to steganography.* Boca Raton, FL: CRC Press.

Kirovski, D., & Malvar, H. (2001). Robust spread-spectrum audio watermarking. In *Proceeding of the IEEE International Conference on Acoustic, Speech, and Signal Processing* (pp. 1345-1348).

Lemma, A.N., Aprea, J., Oomen, W., & Kerkhof, L. (2003). A temporal domain audio watermarking technique. *IEEE Transactions on Signal Processing, 51*(4), 1088-1097.

Liu, Z., & Inoue, A. (2003). Audio watermarking techniques using sinusoidal pattern based on pseudorandom sequence. *IEEE Transactions on Circuits and Systems for Video Technology, 13*(8), 801-812.

Lu, C.S. (2005). *Multimedia security: Steganography and digital watermarking techniques for protection of intellectual property.* Hershey, PA: Information Science Publishing.

Mansour, M.F., & Tewfik, A.H. (2001). Audio watermarking by time-scale modification. In *Proceedings of IEEE International Conference on Acoustic, Speech, and Signal Processing* (Vol. 3, pp. 1353-1356).

Pan, J.S., Huang, H.C., & Jain, L.C. (2004). *Intelligent watermarking techniques.* River Edge, NJ: World Scientific Publishing Co.

Petitcolas, F.A.P., Anderson, R.J., & Kuhn, M.G. (1999). Information hiding: A survey. *Proceedings of the IEEE, 87*(7), 1062-1078.

Podilchuk, C.I., & Delp, E.J. (2001). Digital watermarking: Algorithms and applications. *IEEE Signal Processing Magazine, 18*(4), 33-46.

Seitz, J. (2005). *Digital watermarking for digital media.* Hershey, PA: Information Science Publishing.

Sencar, H.T., Ramkumar, M., & Akansu, A.N. (2004). *Data hiding fundamentals and applications: Content security in digital multimedia.* San Diego: Elsevier Academic Press.

Seok, J., Hong, J., & Kim, J. (2002). A novel audio watermarking algorithm for copyright protection of digital audio. *ETRI Journal, 24*(3), 181-189.

Swanson, M.D., Zhu, B., & Tewfik, A.H. (1999). Current state of the art, challenges and future directions for audio watermarking. In *Proceeding of the IEEE International Conference on Multimedia Computing and Systems* (Vol. 1, pp. 19-24).

Swanson, M.D., Zhu, B., Tewfik, A.H., & Boney, L. (1998). Robust audio watermarking using perceptual masking. *Signal Processing, 66*(3), 337-355.

Wu, M., Craver, S.A., Felten, E.W., & Liu, B. (2001). Analysis of attacks on SDMI audio watermarks. In *Proceeding of IEEE International Conference on Acoustic, Speech, and Signal Processing* (pp. 1369-1372).

Wu, M., & Liu, B. (2002). *Multimedia data hiding.* New York: Springer-Verlag.

Yeo, I.K., & Kim, H.J. (2003). Modified patchwork algorithm: A novel audio watermarking scheme. *IEEE Transactions on Speech and Audio Processing, 11*(4), 381-386.

Chapter VIII
Robustness Analysis of Patchwork Watermarking Schemes

Hyoung-Joong Kim
Korea University, Korea

Shijun Xiang
Korea University, Korea

In-Kwon Yeo
Sookmyung Women's University, Korea

Subhamoy Maitra
Indian Statistical Institute, India

ABSTRACT

The patchwork watermarking scheme is investigated in this chapter. The performance of this algorithm in terms of imperceptibility, robustness, and security has been shown to be satisfactory. Robustness of the patchwork algorithm to the curve-fitting attack and blind multiple-embedding attack is presented also in this chapter. Robustness against jitter attack, which is a natural enemy of this watermarking algorithm, is also studied.

INTRODUCTION

Watermarking algorithm must satisfy at least two constraints: imperceptibility and robustness. Embedded audio watermarks should be almost inaudible. Data can be embedded with psycho-acoustic analysis that will make the watermarked signal far much less audible. Also, the algorithm should be robust enough to withstand attempts such as removal or alteration of inserted watermarks. Intriguing attacks threat watermarking techniques. The algorithms we analyze in this chapter are robust against many attacks such as lossy compression, noise addition, filtering, quantization, and so on. Needless to say, these two constraints, imperceptibility and robustness, seem to be contradictive. However, both constraints must be satisfied. Other requirements include blind detection, capability, and high embedding capacity. The patchwork algorithm is blind; a nonwatermarked signal is not necessary for detecting hidden message. Embedding capacity of the patchwork is relatively high since each subset consists of a small number of populations.

Original patchwork algorithm (Bender, Gruhl, Morimoto, & Lu, 1996) is refreshingly novel among many watermarking methods. The algorithm has been proposed as an image watermarking scheme at the outset. This algorithm has inserted information into the time-domain signal. Moreover, the population of each subset was very large. It was not adaptive to the signal, but it added or subtracted constant d independently of the signal strength. Nonetheless, it has provided a solid base as an excellent tool for information hiding. Recently, Arnold (2000) has tried to improve the performance of the original patchwork algorithm. Arnold's algorithm is a landmark in the area of watermarking research, especially for patchwork algorithm.

Yeo and Kim (2003a) have focused on the improvement of the previous patchwork algorithms. They have derived mathematical formulations that help to improve robustness. The core idea of the improved scheme is called the modified patchwork algorithm (MPA) (Yeo & Kim, 2003a) which can enhance the power of the original patchwork algorithm considerably. The MPA is an additive watermarking scheme such as:

$$x(n) = s(n) \pm d(n), \qquad (1)$$

where $s(n)$ is an nonwatermarked audio sample at time instance n, $d(n)$ is an watermark signal, and $x(n)$ is an watermarked sample. Yeo and Kim (2003b) have generalized the patchwork algorithm by introducing multiplicative watermarking scheme such as:

$$x(n) = s(n)\,[1 \pm d(n)]. \qquad (2)$$

However, note that the multiplicative watermarking scheme is not so effective in case of image watermarking. Thus, they have combined the additive and multiplicative schemes with suitable parameters such as:

$$x(n) = k_1[s(n) \pm d(n)] + k_2\, s(n)\,[1 \pm d(n)], \qquad (3)$$

where $k_1 + k_2 = 1$. The algorithm in equation (3) is called the generalized patchwork algorithm (GPA) (Yeo & Kim, 2003b). Unfortunately, the best performance was achieved when $k_2 = 0$ in case of image watermarking. In other words, it implies the additive scheme only is the best in case of image watermarking.

Performance of the MPA has been thoroughly studied by Yeo and Kim (2003a), Cvejic and Seppänen (2003), and Cvejic and Tujkovic (2004). It has been concluded that the MPA is sufficiently robust. Recently, Das, Kim, and Maitra (2005) have performed cryptanalysis against the GPA for image watermarking based on the curve-fitting attack (Das & Maitra, 2004). The experimental results are based on the assumption that "the GPA is public, and the attackers have the information of how to divide watermarked images." It is impractical in real world since an image owner can divide the image into blocks of random size in such a manner that it is difficult to access the information of image block size for an attacker. That is to say, the cryptanalytic method (Das & Maitra, 2004; Das et al., 2005) is ineffective to the watermarking scheme once the size of image block for watermarking is saved as a secret key. At the same time, we note that this cryptanalysis against MPA audio watermarking (Yeo & Kim, 2003a) has not been performed yet.

Figure 1. Two sets sampled badly from two rectangular squares (left) and well randomly (right)

Thus, in this chapter we will survey the patchwork algorithms and their robustness as the audio and image watermarking schemes. In the next section, security of GPA image watermarking is investigated. Robustness of the MPA as an audio watermarking against the curve-fitting attack and blind multiple-embedding attack will be presented next. Their robustness against jitter attack will also be studied. Finally, the conclusion is drawn and future works will be considered.

SECURITY EVALUATION OF GENERALIZED PATCHWORK ALGORITHM

Here we survey a cryptanalytic method of the generalized patchwork algorithm under the assumption that the attacker possesses only a single copy of the watermarked image (Das et al., 2005). In the scheme, the watermark is inserted by modifying randomly chosen discrete cosine transform (DCT) values in each block of the original image (Yeo & Kim, 2003b). Towards the attack, the authors fit the low degree polynomials (which minimize the mean square error) on the data available from each block of the watermarked content. Then, corresponding DCT data of the attacked image is replaced by the available data from the polynomials to construct an attacked image.

The technique nullifies the modification achieved during watermark embedding. Experimental results show that recovery of the watermark becomes difficult after the attack under the assumption that the size of image blocks is leaked.

Take an image of size $M \times M$. Consider the image in DCT domain, where the discrete cosine transform has been done using block size $N \times N$. There is a secret key K_1 and a secret watermarking signal w. The watermarking signal $w = [w_1, \ldots, w_t]$ is $t = \dfrac{M \times M}{N \times N}$ bits binary pattern.

Bit Embedding

Consider the case each DCT block where a bit is inserted. The key K_1 is used as a seed of pseudo random number generator to get random index values from $[Z_1, Z_2]$, $1 \leq Z_1 \leq Z_2 \leq N \times N$, where the index values are matched with orders of the JPEG-like zigzag pattern in the DCT block. Now two index sets I_0 and I_1, each containing $2n$ elements, are generated using the secret key K_1 as follows:

$$I^0 = I^{0+} \cup I^{0-},\ I^{0+} = \{I_1^0, \ldots, I_n^0\},\ I^{0-} = \{I_{1+n}^0, \ldots, I_{2n}^0\},$$
$$I^1 = I^{1+} \cup I^{1-},\ I^{1+} = \{I_1^1, \ldots, I_n^1\},\ I^{1-} = \{I_{1+n}^1, \ldots, I_{2n}^1\}.$$

I^0 and I^1 are used to embed the watermark bits 0 and 1, respectively. Define $A^j = \{A_1^j, \ldots, A_n^j\}$ as the DCT coefficients corresponding to I^{j+} and $B^j = \{B_1^j, \ldots, B_n^j\}$ as the DCT coefficients corresponding to I^{j-} for $j = 0, 1$. Suppose that bit 0 is the watermark signal. Then, the means and the sample variances are calculated by:

$$\overline{A^0} = \frac{1}{n}\sum_{i=1}^{n} A_i^0,$$

$$\overline{B^0} = \frac{1}{n}\sum_{i=1}^{n} B_i^0,$$

$$S_{A^0}^2 = \frac{1}{n-1}\sum_{i=1}^{n} (A_i^0 - \overline{A^0})^2,$$

$$S_{B^0}^2 = \frac{1}{n-1}\sum_{i=1}^{n} (B_i^0 - \overline{B^0})^2,$$

respectively. The embedding function is as follows, (see equation (4)), where:

$$S_{E0} = \sqrt{\frac{\sum_{i=1}^{n} (A_i^0 - \overline{A_0})^2 + \sum_{i=1}^{n} (B_i^0 - \overline{B_0})^2}{n(n-1)}}.$$

Thus, in total $4n$ DCT coefficients are chosen for inserting the watermark and out of them $2n$ coefficients are modified.

Bit Extraction

For decoding and detecting the watermark, the following strategy is used. Note that here the authors work on the watermarked image and the scheme is oblivious, so the original image is not required. Only the secret key K_1 needs to be known to generate the index sets. Take:

$$T_0^Q = \beta \max\{\frac{S_{A^0}^2}{S_{B^0}^2}, \frac{S_{B^0}^2}{S_{A^0}^2}\} + (1-\beta)\frac{(\overline{A^0} - \overline{B^0})^2}{S_{E0}^2},$$

$$T_1^Q = \beta \max\{\frac{S_{A^1}^2}{S_{B^1}^2}, \frac{S_{B^1}^2}{S_{A^1}^2}\} + (1-\beta)\frac{(\overline{A^1} - \overline{B^1})^2}{S_{E1}^2}. \quad (5)$$

If $P_1 = 0$ and $P_2 > 0$, then $\beta = 0$. On the other hand, if $P_1 > 0$ and $P_2 = 0$, then $\beta = 1$. If $T_0^Q > T_1^Q$, then to extract bit 0 else to extract bit 1.

The basic idea behind the bit extraction is as follows:

- **Case 1:** Consider a specific DCT block and the case $P_1 > 0$ and $P_2 = 0$, that is, $\beta = 1$. If 0 has been inserted, then the values in DCT coefficients A^0 and B^0 will change, but A^1 and B^1 will not change. Thus, the sample variances $S_{A^1}^2$ and $S_{B^1}^2$ of the watermarked image will be the same as the sample variances of the original image and statistically the values will be equal (as they both are sample variances). So T_1^Q will stay close to 1. On the other hand, though in the original image $S_{A^0}^2$ and $S_{B^0}^2$ are statistically equal (both are sample variances), in the watermarked image one of those two values will increase and the other will decrease. Thus, T_0^Q will be away from (considerably larger than) 1.
- **Case 2:** Consider a specific DCT block and how the Das et al. (2005) explain the case $P_1 = 0$ and $P_2 > 0$, that is, $\beta = 0$. If 0 has been inserted, then the values in DCT coefficients

Equation 4.

$$A_i^{0*} = (1 + sign(S_{A^0}^2 - S_{B^0}^2)P_1)A_i^0 + sign(\overline{A_0} - \overline{B_0})\sqrt{P_2}\frac{S_{E0}}{2},$$

$$B_i^{0*} = (1 - sign(S_{A^0}^2 - S_{B^0}^2)P_1)B_i^0 - sign(\overline{A_0} - \overline{B_0})\sqrt{P_2}\frac{S_{E0}}{2},$$

$\overline{A^0}$ and $\overline{B^0}$ will not change, but $\overline{A^1}$ and $\overline{B^1}$ will change. Thus, $(\overline{A^0} - \overline{B^0})^2$ will be away from 0. On the other hand, $\overline{A^1}$ and $\overline{B^1}$, both being the sample mean, $(\overline{A^1} - \overline{B^1})^2$ will be close to 0. Thus, T_0^Q will be away from (considerably larger than) 0 and T_1^Q will be close to 0.

Cryptanalysis

As discussed above, due to the embedding of the bits one of T_0^Q and T_1^Q becomes much larger than 1 (Case 1) or 0 (Case 2) (due to the modification of DCT coefficients), and the other one stays close to 1 (Case 1) or 0 (Case 2) (as it is sample data from the DCT values of the original image). Towards the attack, it needs to modify the DCT data such that both T_0^Q and T_1^Q becomes close to 1 (Case 1) or 0 (Case 2), that is, the effect of the modification in some specific DCT coefficients can be nullified.

The main tool for implementing this attack is fitting a polynomial over the DCT data of the watermarked image. The strategy of polynomial fitting in DCT domain has been used by Das and Maitra (2004) for cryptanalysis of the CKLS scheme (Cox, Kilian, Leighton, & Shamoon, 1997). However, in that case the DCT data had been sorted first and then the polynomial fitting has been attempted. In this case the polynomial fitting is done without sorting the data and further it has been done on every $N \times N$ DCT block as the watermark insertion algorithm (Yeo & Kim, 2003b} works over each DCT block to insert one bit of watermark.

For each $N \times N$ DCT block, the authors take the DCT coefficients in the index range $[Z_1, Z_2]$. Corresponding to the data the authors fit a polynomial of degree d, such that the mean square error is minimized. Then the authors replace the DCT values in the index range $[Z_1, Z_2]$ by the estimated values available from the polynomial. This provides the basis of our attack. Since the DCT coefficients are approximated by a polynomial on the watermarked image, the effect of modifying the original DCT data will be approximated and nullified depending on the degree of the approximating polynomial.

It is clear that as increasing the degree d of the polynomial, the mean square error is less and the quality of the image will stay good, but the effect of nullification of the watermark will be less. On the other hand, if the degree is decreased, the mean square error will be more and the extracted image will not have good image quality, but the removal of watermark information will be achieved in a much better way. Thus, it needs to make a compromise over this based on the DCT data.

The reason why the authors do not sort the DCT data before fitting the DCT polynomial is as follows. If the data get sorted then it becomes monotonically increasing or decreasing (according to the sorting). Hence the data can be very well modeled with low degree polynomial with very small mean square error. Thus, when one gets back to the DCT coefficients from the polynomial, the modified data (and also the data that are not modified) are not disturbed much and consequently the watermark is not removed.

Experimental Results

The authors consider the parameters used by Yeo and Kim (2003b) for a concrete description of the attack. Image size of 512×512 has been considered. During the embedding procedure $N \times N$ DCT blocks are taken, where $N = 64$. The index range in each block is chosen in $[Z_1 = 200, Z_2 = 700]$. The value of n has been taken to be 30. The watermark is of length $\frac{512 \times 512}{64 \times 64} = 64$ bits.

The attacker buys a single copy and then modifies it in such a manner such that the buyer-specific watermark cannot be extracted from the attacked images. Thus, the main motivation of the attacker is removing the watermark completely. Towards that direction we in the chapter made extensive experiments. It is found that in the event that the information of block size divided in the embedding is accessed by the attacker, that is, the same size of image block is deployed for such a attack, we can use polynomials of degrees 3, 4, or 5 by which the watermark is removed completely and the attacked image quality also stays good (>30 dB PSNR with respect to the attacked image). However, the watermark can be extracted at a low bit error rate if we use the different size of image block as that in the embedding. It means that in the watermarking embedding, the size of blocks can be saved as key for improvement of security against fitting-curve attacks.

The authors (Das et al., 2005) present the experimental results listed in Table 1 for three different 512×512 images. The result is agreeing with our testing well under the assumption that the attacker has accessed the information of image block size. The attack process is presented as follows. We insert a 64 bit watermark considering two cases:

1. $P_1 = 0.5$, $P_2 = 0$ and extract the watermark taking $\beta = 1$
2. $P_1 = 0$, $P_2 = 9$ and extract the watermark taking $\beta = 0$

with the proper key after the attack and calculate how many bits are matching with the original watermark. That around half of the bits are matching (around 32 out of 64) demonstrates the success of the attack. The correlation can be calculated as $\frac{x-(64-x)}{64}$, when x bits are matching. That is, if $x = 32$, then the correlation is 0. In our experimentation, the maximum absolute value of correlation after the attack is 0.125, when $x = 36$ or 28. This clearly indicates that the cryptanalysis presented here is successful if the size of blocks is leaked.

In the testing, we provide the PSNR of the attacked image with respect to the watermarked image to present the image quality too. One can check that we get PSNR (of the attacked image with respect to the watermarked image) values greater than 30 dB. We present the "number of bits matching the PSNR value in dB" in each cell corresponding to the degree of the polynomials which we take as 3, 4, and 5.

To improve the PSNR of the attacked image with respect to the watermarked image further, the authors recommended to modify the attack in the following way. Instead of fitting one polynomial in the range of $[Z_1, Z_2]$, the authors fit number of polynomials in the range. For the experimental purpose, where $[Z_1 = 201, Z_2 = 700]$, the authors fit five polynomials in the intervals [201, 300], [301, 400], [401, 500], [501, 600], and [601, 700]. We have checked that this increases the PSNR by an average of 1 dB than what presented in Table 1.

MODIFIED PATCHWORK ALGORITHM AND ITS ROBUSTNESS

Basically, the elements of two sets should be taken randomly to make sure of good performance. If a set A is sampled, for example, from one rectangle just below the chin of a Korean actress, and an-

Table 1. Experimental results

	$P_1 = 0.5, P_2 = 0$			$P_1 = 0, P_2 = 9$		
	3	4	5	3	4	5
Lena	31, 30.11	30, 30.13	30, 30.14	31, 30.29	30, 30.31	29, 30.34
Goldhill	32, 30.13	30, 30.14	33, 30.16	36, 30.18	31, 30.27	34, 30.42
Fishboat	28, 30.56	33, 30.68	31, 30.86	29, 30.58	31, 30.64	33, 30.8

other sample B from another rectangle as shown in Figure1, it is difficult to ensure $\overline{a} - \overline{b} = 0$. Note that here are two ways to achieve mathematically $\overline{a} - \overline{b} = 0$. The first approach is to make the two sets A and B large in number of elements. However, due to the high cross-correlation between neighboring samples, especially in image, making larger sets usually contributes less to goodness of the sets. In other words, this approach seldom succeeds. The second approach is to set elements randomly. Random samples (see the right figure) marked by small circles and rectangle nearly ensures $\overline{a} - \overline{b} = 0$ even though the number of samples are small. That is why randomness is important in statistical image processing, especially in the patchwork algorithm.

As mentioned before, the patchwork scheme is one of the most viable solutions for digital audio and image watermarking. Basically the patchwork schemes choose four pairs of sample subsets randomly, and then modify the sample values of each subset in different manners. For example, a constant d is added to all sample values of one subset conceptually while d is subtracted from other subset values as long as the patchwork scheme is additive. In other words, it makes the large samples larger and the small samples smaller. Then accordingly, sample means of each subset are changed or location of mean is shifted according to the value d and its sign. Thus, the "location-shift" scheme stresses the fact of shifting sample means due to additive scheme (see Figure 2). Most of the previous patchwork schemes (Bender et al., 1996; Pitas & Kaskalis, 1995; Wang, 2004; Yeo & Kim, 2003a) are additive. Additive patchwork scheme detects watermark by computing difference between mean values of two subsets and applying hypothesis test. Theoretically, the difference will be $2d$ for the watermarked signal and 0 for nonwatermarked signal.

On the other hand, multiplicative patchwork scheme changes the sample variances by multiplying $(1+d)$ or $(1-d)$ by sample values. Accordingly, the sample variance value of one subset increases while another subset decreases. In other words, multiplicative scheme scales the variance values up or down. Similarly, the "scale-shift" scheme stresses the fact of scaling sample variances due to multiplicative scheme (Yeo & Kim, 2003b). The ratio of resulting variances between two subsets is a clue to detect embedded bit information.

The additive and multiplicative patchwork schemes can be combined appropriately in order to increase robustness and reduce perceptibility. The patchwork scheme can be applied to both audio and image. It can be applied in the DCT domain or in the DWT domain or in whatever domain.

Figure 2. Concepts of "location-shift" (above) and "scale-shift" (below) in patchwork algorithm

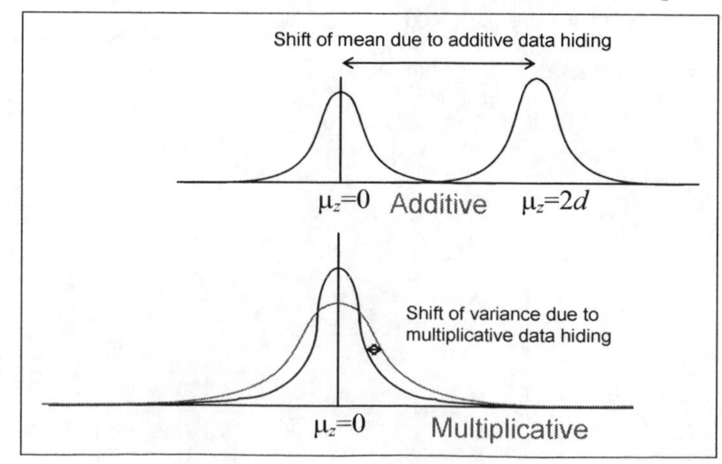

Transformation is more useful than embedding in the time-domain in terms of psychoacoustic analysis and high probability to have almost equal sample means for both subsets such as $\bar{a} - \bar{b}$. This property is important to reduce false alarm rate at the detection process. Since the patchwork scheme modifies a relatively small number of samples, it is desirable in terns of imperceptibility. Note that Bender et al. (1996), and Pitas and Kaskalis (1995) have modified samples across the whole image by dividing it into two subsets. Yeo and Kim have shown that around 30 samples are sufficient to hide one bit of data as long as the samples are randomly chosen.

Modified Patchwork Algorithm

In the embedding step, two subsets, A and B, are necessary. Each subset has n elements such as $A = \{a_1, \ldots, a_n\}$ and $B = \{b_1, \ldots, b_n\}$. In order to embed data the sample means:

$$\bar{a} = \frac{1}{n}\sum_{i=1}^{n} a_i \text{ and } bar\{b\} = \frac{1}{n}\sum_{i=1}^{n} b_i, \tag{6}$$

of each subset, and the pooled sample standard error,

$$S = \sqrt{\frac{\sum_{i=1}^{n}(\bar{a} - a_i)^2 + \sum_{i=1}^{n}(\bar{b} - b_i)^2}{n(n-1)}} \tag{7}$$

have to be computed. The embedding function is given as follows:

$$a_i^* = a_i + sign(\bar{a} - \bar{b})\sqrt{C}\frac{S}{2}$$

and

$$b_i^* = b_i - sign(\bar{a} - \bar{b})\sqrt{C}\frac{S}{2}. \tag{8}$$

For the detecting watermark two subsets $A_1 = \{a_{11}, \ldots, a_{1n}\}$ and $B_1 = \{b_{11}, \ldots, b_{1n}\}$, or $A_0 = \{a_{01}, \ldots, a_{0n}\}$ and $B_0 = \{b_{01}, \ldots, b_{0n}\}$ are obtained according to the hidden bit 1 or 0, respectively. Then, the test statistics:

$$T_0^2 = \frac{(\bar{a}_0 - \bar{b}_0)^2}{S_0^2} \text{ and } T_1^2 = \frac{(\bar{a}_1 - \bar{b}_1)^2}{S_1^2} \tag{9}$$

Figure 3. Distributions of $\bar{a} - \bar{b}$

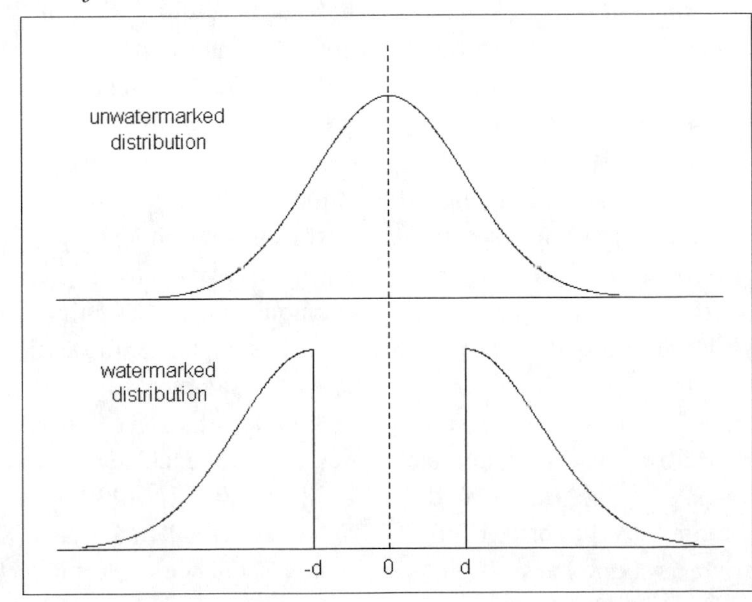

are calculated based on each sample means and pooled sample variances for subsets A_0, B_0, A_1, and B_1. Here, T^2 is defined to be as follows:

$$T^2 = \max\{T_0^2, T_1^2\}. \quad (10)$$

At the end of the watermark detection process, value T^2 is compared with the threshold M and a "watermark is embedded" decision is brought if $T^2 > M$. Only when $T^2 > M$, a zero bit is detected if $|T_0| > |T_1|$, and otherwise 1. For the detail algorithm, refer to Yeo and Kim (2003a). Figure 3 shows the probability distribution of $\bar{a} - \bar{b}$ for nonwatermarked signal and watermarked signal. Sample values for two sets are chosen at $[Z_L, Z_U]$. The lower bound and upper bounds are decided to minimize the audibility of noise and maximize the embedding capacity.

Robustness Analysis

Performance of the MPA has been thoroughly studied by Yeo and Kim (2003a), Cvejic and Seppänen (2003), and Cvejic and Tujkovic (2004) in terms of robustness and inaudibility. It has been concluded that the MPA is sufficiently robust and inaudible. Recently, Das et al. (2005) have performed cryptanalysis against the GPA for image watermarking based on the curve-fitting attack (Das & Maitra, 2004) under the assumption that an attacker can get the information of how to segment image blocks. Note that this cryptanalysis against audio watermarking has not been performed yet. In this section, the curve-fitting attack is applied to the audio samples for the cryptanalysis. In addition, the blind multiple-embedding attack is applied under the assumption that the embedding position is unknown while the modified patchwork algorithm is known to public.

The curve-fitting attack has two forms. The original attack (Das & Maitra, 2004) transforms the time-domain samples, sorts the transformed coefficients, and fits the curve by the sorted coefficients to modify the data as slightly as possible. The rationale is simple. If the data gets sorted, it becomes monotonically increasing or decreasing (according to the sorting). Consequently, low degree polynomials are enough to model the sorted data with small mean square error. Then, the attack can not disturb the image much but remove watermarks considerably as a consequence. The first method (Das & Maitra, 2004) was verified effective to some watermarking techniques, but almost ineffective to the patchwork algorithm. The second approach (Das et al., 2005) transforms the time-domain samples, and applies curve-fitting to the unsorted transformed coefficients. It is noted that the DCT data remains unsorted to disturb more for the image patchwork algorithm as long as the information of image block size is leaked. Figure 4 shows an example of unsorted samples and sorted samples by magnitudes of the DCT coefficient.

The fitted coefficients are replaced with the values corresponding to the fitted polynomial. In case of audio watermarking, such an attack is not so effective since the information of how to segment audio clips may be kept as a secret key. Extensive testing shows that under the condition that the length of segments is secret, it can not remove hidden messages while its audio quality degrades very much. In the event that an attacker has got the information of how to segment the clips, the watermark by patchwork algorithm (Yeo & Kim, 2003b) will be destroyed. In real world, it is impractical to get the information of how to segment audio since the length of audio segments can be selected randomly in a huge space.

In our experiments, the audio clips (for experiments, we choose a "Japanese pop music" as example clip) are basically chopped by multiples of 4,410 samples for the audio sampled at the rate of 44,100 samples per second. For each segment of samples, the DCT is first performed. The index range in each segment is chosen randomly in $[Z_L = 500, Z_U = 1000]$ as a key K_1. The value of n has been taken to be 25. The watermark is a binary sequence of 30 bits. The signal-to-noise

Figure 4. Concept of the curve-fitting attack against unsorted (above) and sorted data (below)

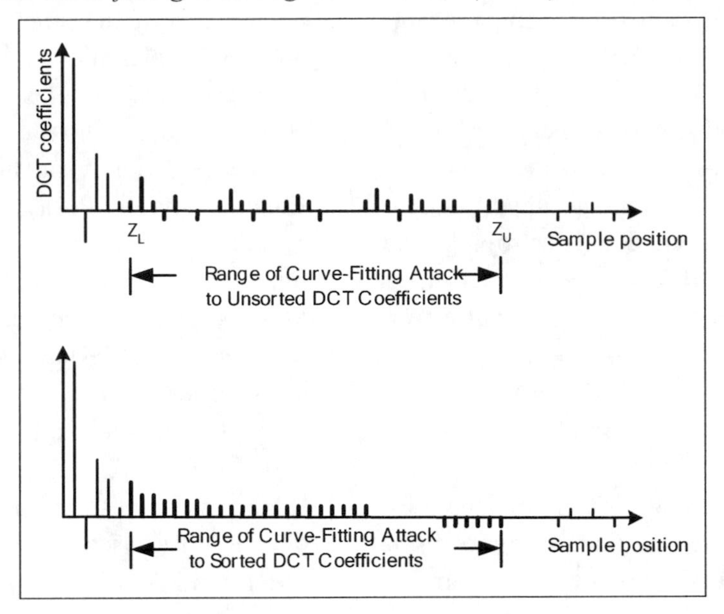

rate (SNR) of the watermarked audio is around 46.56 dB. Each bit of watermark is embedded five times for correction of error. About the details of embedding process and used parameters, we refer to the previous work of Yeo and Kim (2003b). Our motivation is to investigate whether the method (Das et al., 2005) is acting on the patchwork audio watermarking algorithm. Towards this direction we made extensive tests mainly including Test I: suppose that the attacker has known all parameters except the key K_1, and Test II: the attacker has the information of watermarking embedding scheme but the length of segments 4,410 (as the key K_2) and the key K_1 are unknown. We have optimized the polynomials with the degrees 3, 4, and 5 by the curve-fitting tool in the MATLAB. The polynomial type is described as follows:

$$f(x) = a_n x^n + a_{n-1} x^{n-1} + \cdots + a_1 x + a_0.$$

The attack test results are tabulated in Table 2. Corresponding to Table 2, we have the observations blow:

1. The SNR values of the attacked audio with respect to the watermarked audio decrease quickly from around 18 dB to 2 dB as the length of segments K_2 increases from 2,205 to 15,435 with the step of 2,205.
2. It is noted that when K_2=4,410, that is, K_2 as a secret key is accessed by the attacker, the embedded messages can not be recovered. Experiments show in this case the bit error rate (BER) is around 50%. When K_2 is not known by an attacker, the watermark is successfully extracted at a low error bit, and even the audio quality has been degraded a lot.

As a conclusion, the fitting-curve attacks degrade the audio quality a lot, and the audio watermarking based on patchwork algorithm is secure to polynomial fitting attacks as long as the information of segmenting audio for watermarking is not leaked.

A blind multiple-embedding attack is an attack that embeds data N times into the audio clip already watermarked with the MPA. Since the position of n samples is not known to attackers, what they can do is to embed data using the same watermarking algorithm N times in the hope of destroying watermarks. However, the probability of hitting m positions (or, equivalently, samples) exactly among n positions from the M populations is very low. Thus, multiple embedding by N times will increase the chance of destroying more numbers of watermarked samples. Experiments show this blind attack is to be shown ineffective. In the experiments, a 30-bit message was hidden. The detail embedding method can be found in the literature by Yeo and Kim (2003a). The test results are listed in Table 3.

As the number of blind embedding increases, the number of undetected bits also increases as a sequence. Of course, audio quality degrades quickly as N increases. In case of Japanese pop music, all 30 bits are detected successfully at low error rate even after 100 times of multiple embedding. Regarding other different types of music, the simulation results are similar.

Of course, the patchwork algorithm has a natural enemy: the jitter attack. Such an attack deleting samples randomly will cause the position of watermarked samples to shift (Xiang, Huang, & Yang, 2006). This position alteration incurs serious detection error. Figure 5 illustrates the situation vividly. Position of the watermarked samples is marked by 'o' in the Figure. Due to the intentional deletion of the forth sample, a detector locates the wrong positions since it does not know whether the samples have been attacked or not. As a conclusion, the jitter attack is a very effective attack to the patchwork algorithm because the MPA hides data into number of audio segments. If the position of samples is changed by jitter attack, it will cause the desynchronization of the watermark. As a result, the watermark will be difficult to be recovered. The test results of jitter attacks with different strength are tabulated in Table 4. Here 1/700 means that one sample is deleted out of each 700 samples.

Table 2. The test results based on the cryptanalytic attack (Das, 2005)

Length of Segments (K_2)	2205	4410	6615	8820	11025	13230	15435
BER (%) (if degree = 3)	0	60	16.67	0	3.33	3.33	3.33
BER (%) (if degree = 4)	0	50	13.33	0	3.33	3.33	10
BER (%) (if degree = 5)	0	26.67	13.33	0	0	3.33	7
SNR (dB) (if degree = 3)	18.56	15.66	12.08	9.24	7.90	6.46	3.00
SNR (dB) (if degree = 4)	18.39	15.46	11.89	9.08	7.88	6.28	2.65
SNR (dB) (if degree=53)	18.13	15.29	11.61	8.88	7.90	6.09	2.20

Table 3. Number of bits detected after blind multiple embedding attacks

Number of embeddings (N)	5	10	20	30	40	60	80	100
Number of undetected bits	0	1	1	2	1	2	2	3
SNR (dB)	39.25	36.14	33.00	31.02	27.56	26.04	24.78	27.20

Table 4. The BER values after blind jitter attack for Japanese pop song

Jitter attack Strength	1/700	1/900	1/1100	1/1300	1/1500	1/1700	1/1900
BER (%)	58.62	62.07	51.72	65.52	48.28	62.07	68.97

Unfortunately, this blind attack is to be shown ineffective. In the experiments, a 50-bit message was hidden. The detail embedding method can be found in the literature by Yeo and Kim (2003a). As the number of blind embeddings increases, the number of undetected messages also increases accordingly. Performance depends on the target music. In case of Japanese pop music, all 50 bits are detected successfully even after 30 times of multiple embedding. However, audio quality degrades quickly as N increases.

CONCLUSION

The patchwork algorithms, the generalized patchwork algorithm (GPA), and the modified patchwork algorithm (MPA), are investigated briefly as the image and audio watermarking schemes in this chapter. The performance of this algorithm in terms of imperceptibility and robustness has been shown to be satisfactory by many researchers (Yeo & Kim, 2003a; Cvejic & Seppänen, 2003; Cvejic & Tujkovic, 2004). Recently, Das et al. (2005) have performed cryptanalysis against the GPA for image watermarking based on the curve-fitting attack (Das & Maitra, 2004) under the assumption that the information of how to divide the image is known. Note that this cryptanalysis against audio watermarking has not been performed yet. Thus, robustness of the patchwork algorithm against the curve-fitting attack as a cryptanalysis for an audio and blind multiple embedding attack is presented, respectively. Our experimental testing shows that GPA image watermarking and MPA audio watermarking have a satisfactory security against the cryptanalytic methods (Das & Maitra, 2004; Das et al., 2005) since the information of how to divide the image or segment the audio that may be considered as a secret key in such a way that it is difficult to be broken by the attacker. As a result, the security of patchwork schemes is reliable against the fitting-curve attacks. In addition, robustness against jitter attack is also studied. As a conclusion, the patchwork algorithm is sufficiently robust and imperceptible. And, it is also security against curve-fitting attack by saving the size of image block or the length of audio segment for watermarking as a secret key. However, there are some open issues to be discussed in patchwork schemes. One of them is how to fight the desynchronization attacks presented by Xiang et al. (2006), such as jitter attacks discussed in this chapter.

FURTHER RESEARCH DIRECTION

Patchwork is one of the first generation watermarking schemes. It embeds the watermark in the spatial (image)/time (audio) domain or the frequency domain by using linear transformations, such as discrete Fourier transform (DFT), discrete cosine transform (DCT), and discrete wavelet transform (DWT). Spatial or time domain watermarking schemes directly tinker with the pixel or sample amplitude to embed the watermark. In order to reduce the sensitivity to additive noise corruption, the watermark is being suggested to embed in the frequency domain by modifying the transformed coefficients. In such a way, the distortion due to the watermark can be shared by multiple samples and the imperceptibility of the watermark can be improved largely.

Patchwork-based watermarking scheme has been verified very effective to those common signal processing operations, such as image/audio compression, low-pass filtering, and so forth. Recently, one attack called "curve-fitting attack" has been mounted successfully for patchwork watermarking scheme. In addition, patchwork watermarking scheme is also sensitive to various synchronization attacks since the watermark is extracted in some predefined positions. Once the location of image pixels or audio samples is changed, the detector will be incompetent.

In future researches, how to combine other techniques into patchwork watermarking scheme to combat the synchronization attacks is an important issue. Additionally, since "Curve-Fitting Attack" is running under an assumption that how to divide audio or image to embed the watermark is known, there is also room to enhance the security of patchwork watermarking scheme by keeping those to-be marked patchworks as a secret.

ACKNOWLEDGMENT

This work was in part supported by the ITRC, Ministry of Information and Communication, Korea, in part supported by the Second Brain Korea 21 Project.

REFERENCES

Arnold, M. (2000). Audio watermarking: Features, applications and algorithms. In *Proceedings of the IEEE International Conference on Multimedia and Expo* (Vol. 2, pp. 1013-1016).

Bender, W., Gruhl, D., Morimoto, N., & Lu, A. (1996). Techniques for data hiding. *IBM Systems Journal, 35*, 313-336.

Cox, I. J., Kilian, J., Leighton, T., & Shamoon, T. (1997). Secure spread spectrum watermarking for multimedia. *IEEE Transaction on Image Processing, 6*(12), 1673-1687.

Cvejic, N., & Seppänen, T. (2003). Robust audio watermarking in wavelet domain using frequency hopping and patchwork method. In *Proceedings of the 3rd International Symposium on Image and Signal Processing and Analysis* (pp. 251-255).

Cvejic, N., & Tujkovic, I. (2004). Increasing robustness of patchwork audio watermarking algorithm using attack characterization. In *Proceedings of the IEEE International Symposium on Consumer Electronics* (pp. 3-6).

Das, T. K., & Maitra, S. (2004). Cryptanalysis of correlation based watermarking schemes using single watermarked copy. *IEEE Signal Processing Letters, 11*, 446-449.

Das, T. K., Kim, H. J., & Maitra, S. (2005). Security evaluation of generalized patchwork algorithm from cryptanalytic viewpoint. *Lecture Notes in Artificial Intelligence, 3681*, 1240-1247.

Pitas, I., & Kaskalis, T. H. (1995). Applying signatures on digital images. In *Proceedings of the IEEE Workshops on Nonlinear Image and Signal Processing* (pp. 460-463).

Wang, Y. (2004). Estimation-based patchwork image watermarking technique. *Journal of Digital Information Management, 1*, 154-161.

Xiang, S. J., Huang, J. W., & Yang, R. (2006). Time-scale invariant audio watermarking based on the statistical features in time domain. In *Proceedings of the 8th Information Hiding Workshop.*

Yeo, I. K., & Kim, H. J. (2003a). Modified patchwork algorithm: The novel audio watermarking scheme. *IEEE Transactions on Speech and Audio Processing, 11*, 381-386.

Yeo, I. K., & Kim, H. J. (2003b). Generalized patchwork algorithm for image watermarking. *ACM Multimedia Systems, 9*, 261-265.

ADDITIONAL READING

Cox, I.J., Kilian, J., Leighton, T., Leighton, T., & Shamoon, T. (1995). *Secure spread spectrum watermarking for multimedia* (Tech. Rep.) NEC Research Institute.

Bender, W., Gruhl, D., & Morimoto, N. (1996). Techniques for data hiding. *IBM System Journal, 3*(3), 313-336.

Chen, B., & Wornell, G.W. (2001). Quantization index modulation: A class of provably good methods for digital watermarking and information embedding. *IEEE Transactions on Information Theory, 47*(4), 1423-1443.

Moulin, P., & Koetter, R. (2005). Data-hiding codes. *Proceedings of the IEEE, 93*(12), 2083-2127.

Costa, M.H.M. (1983). Writing on dirty paper. *IEEE Transactions on Information Theory, 29*(3), 439-441.

Fridrich, M., Goljan, P., Lisonek, & Soukal, D. (2005). Writing on wet paper. *IEEE Transactions on Signal Processing, 53*, 3923-3935.

Fridrich, J., Goljan, M., & Soukal, D. (2006). Wet paper codes with improved embedding efficiency. *IEEE Transactions on Information Security and Forensics, 1*(1), 102-110.

Swanson, M.D., Zhu, B., Tewfik, A. H. et al. (1998). Robust audio watermarking using perceptual masking. *Signal Processing, 66*(3), 337-355.

Cayre, F., Fontaine, C., & Furon, T. (2005). Watermarking security part I: Theory. In *Proceedings of that SPIE, Security, Steganography and Watermarking of Multimedia Contents VII* (p. 5681).

P'erez-Freire, L., Comesa˜na, P., & P'erez-Gonz'alez, F. (2005). Information-theoretic analysis of security in side-informed data hiding. In *Proceedings of the 7th Information Hiding Workshop.*

Cayre, F., Fontaine, C., & Furon, T. (2005). Watermarking security part II: Practice. In *Proceedings of SPIE, Security, Steganography and Watermarking of Multimedia Contents VII* (p. 5681).

Maes, M. (1998). Two peaks: The histogram attack to fixed depth image watermarks. In *Proceedings of Information Hiding Workshop.*

Malkin, M. (2006). Cryptographic methods in multimedia identification and authentication. Unpublished doctoral dissertation, Stanford University.

Li, W., & Xue, X. (2006). Content based localized robust audio watermarking robust against time scale modification. *IEEE Transactions on Multimedia, 8*(1), 60-69.

Wu, S., Huang, J., Huang, D., & Shi, Y. (2005). Efficiently self-synchronized audio watermarking for assured audio data Transmission. *IEEE Transactions on Broadcasting, 51*(1), 69-76.

Xiang, S., & Kim, H.J. (2007). Invariant audio watermarking in DWT domain. In *Proceedinsg of International Conference on Ubiquitous Information Technology and Applications* (pp. 13-22).

Ruanaidh, J., & Pun, T. (1998). Rotation, scale and translation invariant spread spectrum digital image watermarking. *Signal Process, 66*(3), 303-317.

Lin, C.Y., Wu, M., Bloom, J., Miller, M., Cox, I., & Lui, Y.M. (2001). Rotation, scale, and translation resilient public watermarking for images. *IEEE Transactions on Image Processing, 10*(5), 767-782.

Licks, V., & Jordan, R. (2005). Geometric attacks on image watermarking systems. *IEEE Multimedia, 12*(3), 68-78.

Seo, J.S., & Yoo, C.D. (2006). Image watermarking based on invariant regions of scale-space representation. *IEEE Transactions on Signal Processing, 54*(4), 1537-1549.

Fabien, A., Petitcolas, P., Anderson, R.J., & Kuhn, M.G. (1998). Attacks on copyright marking systems. In *Proceeding of Information Hiding* (pp. 219-239).

Petitcolas, F.A.P. (2000). Watermarking schemes evaluation. *IEEE Signal Processing, 17*(5), 58-64.

Pereira, S., Voloshynovskiy, S., Madueño, M., Marchand-Maillet, S., & Pun, T. (2001). Second generation benchmarking and application oriented evaluation. In *Proceedings of the Information Hiding Workshop.*

Herley, C. (2002). Why watermarking is nonsense. *IEEE Signal Processing Magazine, 19*(5), 10-11.

Moulin, P. (2003). Comments on 'why watermarking is nonsense.' *IEEE SignalProcessing Magazine, 20*(6), 57-59.

Chapter IX
Time-Frequency Analysis of Digital Audio Watermarking

Sridhar Krishnan
Ryerson University, Canada

Behnaz Ghoraani
Ryerson University, Canada

Serhat Erkucuk
University of British Columbia, Vancouver, Canada

ABSTRACT

In this chapter, we present an overview of our time frequency (TF) based audio watermarking methods. First, a motivation on the necessity of data authentication, and an introduction in digital rights management (DRM) to protect digital multimedia contents are presented. TF techniques provide flexible means to analyze nonstationary audio signals. We have explained the joint TF domain for watermark representation, and have employed pattern recognition schemes for watermark detection. In this chapter, we introduce two watermarking methods; embedding nonlinear and linear TF signatures as watermarking signatures. Robustness of the proposed methods against common signal manipulations is also studied in this chapter.

INTRODUCTION

The electronic distribution of multimedia (EDM) through the Internet offers many advantages to content sellers as well as consumers. Due to the digital representation, sellers face reduced cost of manufacturing, transportation, storage, and display. They can reach a larger number of users when compared to distributing by compact disc (CD) or digital versatile disk (DVD). Consumers, apart from the reduced cost, enjoy numerous benefits. They can have access to large collection of multimedia files, and can purchase and enjoy multimedia content instantly. Obtaining music on-line also enables consumers to have more control over what they listen to. They can buy individual singles rather than the whole album. By ordering these songs together, consumers can create their own listening experiences and bypass the context in which artists envisaged their work that would be listened to when purchased.

Despite these potential advantages, both the music and the movie industries are reluctant to distribute multimedia content through Internet. One of the reasons is that these industries are afraid of change. But these industries will eventually accept that EDM will be a significant distribution channel in the future. So the main obstacle in the implementation of EDM is piracy. While there are many advantages associated with digital media and digital media distribution, clear disadvantages are present. Prior to digital technologies, content was created, displayed, and stored in analog means. The advent of personal video recorder in the 1980s presented an opportunity to view video at home; but also provided an opportunity to make an illegal copy. However, when a copy is made from a recorded content, the new copy is inferior in quality to the original one. Any further copies made from that copy are very much reduced in quality to be of any commercial use. This discouraged people from copying and prevented piracy efforts from reaching alarming proportions. The new digital technologies represent multimedia content in digital format (1s and 0s). These bits can be efficiently stored in an optical or magnetic media. Since digital recording is a process where by each bit in the source stream is read and copied to the new medium, an exact replica of the content is obtained. The digital copies can be created with low cost equipment such as a CD recorder.

Though the digital representation helped to make identical copies easily, the full resolution multimedia files comprise large amounts of data. Transferring or storing them took a large amount of bandwidth. In the beginning of the 1990s, several compression technologies were developed and new standards such as motion picture experts group (MPEG) for video and MPEG 1 Layer - 3 (MP3) for audio reduced the size of the multimedia files by an order of magnitude. This coupled with the reduction in the cost of storage media allowed computer users to have thousands of songs or movies stored in their computers. Towards the end of the 1990s, the increasing availability of high-speed Internet provided an easy and cheap way of distributing movies and songs. The development of peer-to-peer (P2P) networks (BitTorrent, 2006; eDonkey, 2006; Kazaa, 2006) to exchange files helped Internet users to have an easy way to search and exchange multimedia files effortlessly on the Internet. The illegal sharing of multimedia content incurred severe losses to the copyright holders. According to the Recording Industry Association of America (RIAA), the volume of sold audio CDs dropped by 5% in 2001 (RIAA, 2001) and by 11% in the first half of 2002 (Valenti, 2002). Motion Picture Association of America (MPAA) estimates that the movie industry annually looses US$3 billion through physical piracy (Valenti, 2002). This figure does not include Internet piracy. Hence developing digital rights management (DRM) to protect digital multimedia content is a crucial problem for which immediate solutions are needed.

DIGITAL RIGHTS MANAGEMENT

DRM is a collection of commercial, legal, and technical measures that enable technically enforced licensing of digital information. DRM makes it possible for content distributors to distribute valuable content electronically, without destroying the copyright holders' revenue stream. Though DRM can be designed to protect any digital information, in this chapter, the focus is restricted only on the technologies that are designed to protect multimedia content. DRM ensures that access to protected content (such as video or audio) is possible only under the conditions specified by the content owner. Any unauthorized access must be prevented because such access is an opportunity for an unprotected version of the content to be obtained. If unprotected content is obtained, then it can be distributed and used in any manner, bypassing DRM. DRM prevents the creation of unauthorized copies (copy protection) and provides a mechanism by which copies can be detected and traced (content tracking). The following section describes the various components of DRM.

Components of DRM

A typical DRM for multimedia has the following components, as shown in Figure 1. The figure shows only the technical aspects of DRM. Other than the technical aspects, DRM includes legislative and regulatory solutions for copy protection and management of digital rights:

1. **Management of digital rights:** The management of digital rights includes expression of rights and distribution of rights.
   - **Expression of rights:** The content producer expresses the conditions under which the content can be accessed using rights expression languages (REL) such as the open digital rights language (ORDL) and the eXtensible rights markup language (XrML). In DRM, the usage rules can be adapted to the business models. For example, access can be restricted to selected users, a limited time, or a limited number of accesses. Initial access to the data may even be free (e.g., the first playback

Figure 1. Components of DRM

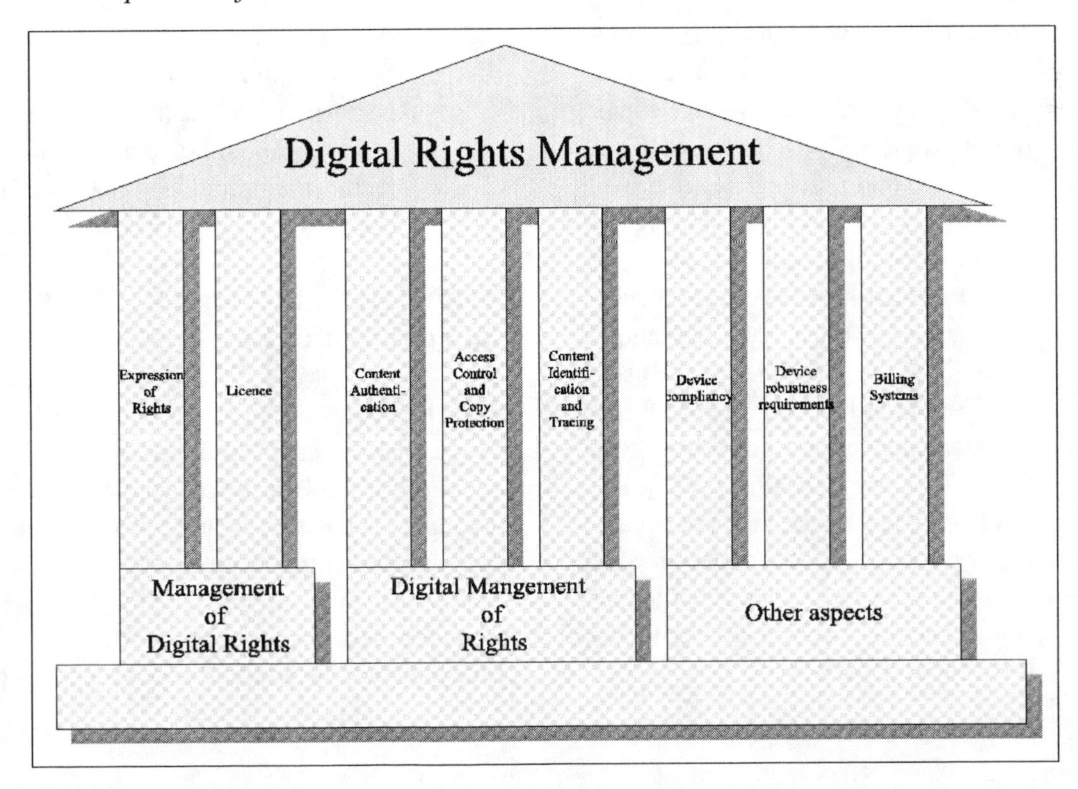

of an audio track), while subsequent access has to be paid for.
- **Licenses:** Licenses are used as a mechanism to distribute rights that are expressed in the rights language. By bringing a license and a digital item together in a device, the device can inspect the right to see what it may do with the digital item.

2. **Digital management of rights:** The digital management of rights includes:
   - **Content authentication:** DRM should protect the authenticity and integrity of the content. Authenticity is securing the content what it claims to be. Integrity means securing the content from any alteration during distribution to the consumer.
   - **Access control and copy protection:** DRM controls who has access to the content and how the content has been used. DRM should prevent the illegal copying of the content once it has been decrypted. Depending on the usage rules, no/one/several/unlimited copies of the multimedia data are allowed, with or without the right to produce copies of the copies. DRM enforces those copy restrictions using sophisticated technology such as watermarking.
   - **Identification and tracing:** In DRM, the authorized users have access to play the content. The playback medium is analog and it is possible for the users to make copies from the analog output. Thus, analog copies in general can hardly be prevented. But it is possible to identify and trace back analog and digital copies of distributed media. This can be done by individual digital watermarking (traitor-tracing) of the distributed data.
3. **Other aspects:**
   - **Device compliancy:** DRM rely on device compliancy to function properly. Device compliancy requires that devices that implement (part of) DRM functionality function according to the rule imposed by DRM. This means that devices do not access digital items in case of absence of a license and also that they do not carry out operations on digital items that are not allowed by the associated rights (e.g., copying or sending content in the clear over unprotected links).
   - **Device robustness requirements:** Since devices manipulate licenses, rights, and keys, the manipulation and storage of these items needs to take place in a secure environment. As a result, DRM imposes hardware and software tamper resistance requirements on devices.
   - **Billing systems:** The business models for media distribution usually involve monetary transactions. Therefore, DRM should contain mechanisms to perform those transactions. Billing systems should be able to handle different pricing models such as pay per use, monthly subscription.

Technologies for Digital Management of Rights

The technologies used for the digital management of rights include encryption, fingerprinting, and watermarking and are described in the following paragraphs. The technologies used in a multimedia DRM are shown in Figure 2.

In DRM, encryption can be used to package the content securely and force all accesses to the protected content through the rules enforced by

DRM. If the content is not packaged securely, the content could be easily copied. Encryption scrambles the content and renders the content unintelligible unless a decryption key is known. Once the content has been encrypted, the encrypted content may be transmitted over a distribution network or recorded onto a media. The encrypted content cannot be decoded or displayed without the decryption key. When the content is to be decoded and displayed, the decryption key is provided to the decoder only after DRM has verified that the conditions for accessing to the content are satisfied. In this way, DRM can secure the distribution of content and ensure that access to the content is consistent with the usage rights. Other examples of using encryption in DRM include device authentication and the secure exchange of keys and authorization information. Encryption packages the content securely and provides restricted access to the content. However, once an authorized user has decrypted the content, it does not provide any protection to the decrypted content. Encryption does not prevent an authorized user from making and distributing illegal copies. Watermarking and fingerprinting are two technologies that can provide protection to the data after it has been decrypted.

A watermark is a signal that is embedded in the content to produce a watermarked content. The watermark may contain information about the owner of the content and the access conditions of the content. When a watermark is added to the content, it introduces distortion. But the watermark is added in such a way that the watermarked content is perceptually similar to the original content. The embedded watermark may be extracted using a watermark detector. Since the watermark contains information that protects the content, the watermarking technique should be *robust*, that is, the watermark signal should be difficult to remove without causing significant distortion to the content. Watermarking can be used in DRM in many ways. Some of the applications include the following (Lin, Eskicioglu, Lagendijk, & Delp, 2005):

- **Copyright or owner identification:** The embedded watermark identifies the owner of the multimedia content. The watermark

Figure 2. DRM technologies

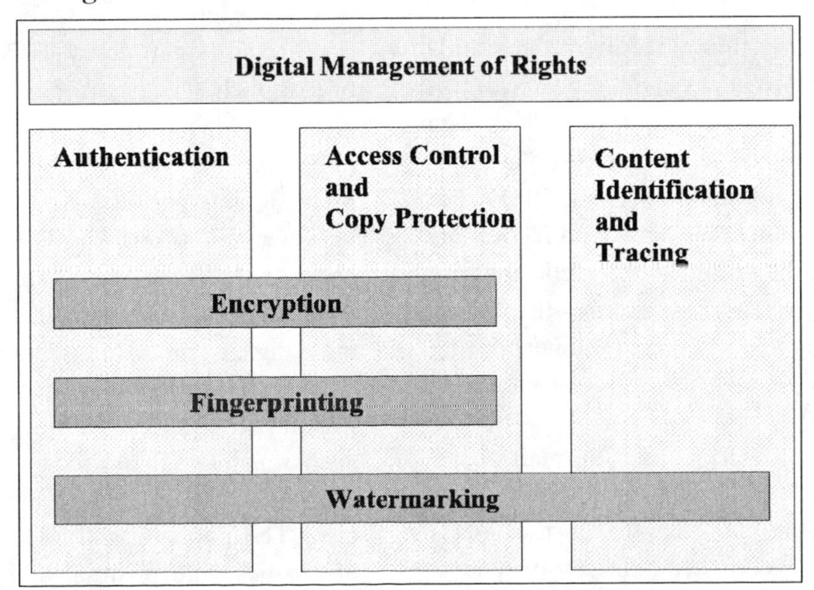

provides a proof of ownership if the copyright notice has been altered or removed.

- **Copy protection:** The watermark encodes the number of times the multimedia content may be (legally) copied. A compliant device checks the watermark and determines whether creating an additional copy is allowed. Each time a copy is made, the watermarked content is modified to decrement the count of allowable copies.
- **Access control:** The watermark encodes the usage and access rights that are granted by the content owner. Compliant devices detect the watermark and comply with the encoded usage restrictions.
- **Content tracking:** In many applications the copyrighted content may be distributed to many users. Some of the users may copy the content and distribute. To identify the particular user who pirates content, the watermark encodes the identification of the user or recipient of the video. This implies that each user obtains a unique or personalized copy of the video. If a copy of the video is found in a suspicious location (such as being shared by a peer-to-peer program), the embedded watermark can identify the source of the suspected copies.

In watermarking, the embedding process adds a watermark before the content is released. But watermarking cannot be used if the content has been already released. According to Venkatachalam, Cazzanti, Dhillon, and Wells (2004), there are about 0.5 trillion copies of sound recordings in existence and 20 billion sound recordings are added every year. This underscores the importance of securing legacy content. Fingerprinting is a technology to identify and protect legacy content.

In multimedia fingerprinting[1], the main objective is to establish the perceptual equality of two multimedia objects, not by comparing the objects themselves, but by comparing the associated fingerprints. The fingerprints of a large number of multimedia objects, along with their associated metadata (e.g., name of artist, title, and album, copyright) are stored in a database. This database is usually maintained online and can be accessed by recording devices.

The introduction briefly described the technologies used for the DRM, namely encryption, fingerprinting, and watermarking. These are essentially content protection technologies. While encryption protects the content during transmission from the producer to the consumer, fingerprinting and watermarking protects the content once the content reaches the consumer. The remainder of the chapter gives a description of the time-frequency (TF) analysis, and its application in watermarking.

TIME-FREQUENCY SIGNAL PROCESSING

Joint time frequency analysis of signals such as radar, sonar, communications, and biomedical signals is necessary to understand and analyze their true nonstationary behavior. One of the popular ways for describing the notion of TF is to understand musical notation. Each musical note corresponds to a specific instant in time (localization in time) and frequency localization or pitch. One of the most commonly known methods of spectral analysis developed by Fourier known as the Fourier transform is quite useful in analyzing periodic and stationary signals. However this transform does not allow for the concept of frequency evolving over time, therefore rendering instantaneous frequency (IF) is meaningless. Since many practical signals have frequency information which changes over time, joint TF analysis is required. The resolution of such transforms is limited by their time duration and bandwidth product in the uncertainty principle. This notion was first examined in quantum mechanics with position and momentum used instead of time and frequency. This work was first noted by Heisenberg (1925), later by Weyl (1927), and then Gabor (1947) in his application to signal theory. This principle stated that the energy spread of a function and its corresponding Fourier transform cannot both be simultaneously small. Furthermore, the concept of instantaneous frequency as first explored by Gabor and Ville involved the use of the Hilbert transform to compose an analytical signal from which the IF could be derived. This approach was faced with a limitation for multicomponent signals requiring a two dimensional distribution such as the sliding Fourier transform to analyze them.

With the advancement in multimedia systems and audio coders, the concept of music analysis remains the same. To analyze a music signal such as in an audio coder, it is necessary to understand and mimic the characteristics and limitations of the human auditory systems (HAS). Here, several characteristics are important. For instance, the human ear is able to perceive the frequencies that create a sound localized in time. Therefore, the model used needs to use joint TF analysis to process the music signal. Since most of the previous work in watermarking area examine audio in either the time or frequency domain, it is assumed that the signals are either wide sense stationary (WSS) or that they have constant frequency components within the discrete Fourier transform window. In reality however, audio signals are nonstationary and multicomponent signals which consist of a series of sinusoids with harmonically related frequencies. Many advanced TF distributions based on Cohen's class of TF representations, such as the Wigner-Ville distribution and Choi-Williams distribution, have been proposed over the years. The STFT provides good frequency resolution in all bands, where Wavelet does a poor job at higher frequency bands.

Using TF analysis, we examine audio watermarking applications. In watermarking audio signals, an imperceptible and statistically undetectable digital signature consisting of a sequence of bits is embedded within the music segment. These bits could be used to prove ownership of intellectual property, track pirated copies of multimedia, and prevent illegal copying among other applications. These embedded bits are referred to as a watermark. In audio watermarking schemes, the watermark has to be statistically undetectable to prevent its removal by unauthorized parties. This can be obtained using spread spectrum techniques which have been used since the 1930s for military applications due to their robustness to jamming attacks as they spread the message across all frequency bands to make the message statisically undetectable to others.

Jammer of detection (a narrowband interference) is possible by using linear frequency modulated (chirp with linear IF) detection algorithms. By taking advantage of this concept, we have developed an audio watermarking scheme in which the chirp is intentionally embedded in the audio content after spreading. Chirp detectors are used in the receiver to enhance the robustness of the watermarking scheme. In addition to linear frequency modulated signals, we proposed another watermarking scheme based on the use of the instantaneous mean frequency (IMF) of the input signal detected by content-based analysis. The watermark is embedded in this perceptually significant region so that it can resist attacks. The details of the watermarking schemes are provided in the subsequent section.

DIGITAL AUDIO WATERMARKING

In recent years, the digital format has become the standard for the representation of multimedia content. Today's technology allows the copying and redistribution of multimedia content over the Internet at a very low or no cost. This has become a serious threat for multimedia content owners. Therefore, there is significant interest to protect copyright ownership of multimedia content (audio, image, and video). Watermarking is the process of embedding additional data into the host signal for identifying the copyright ownership. The embedded data characterizes the owner of the data and should be extracted to prove ownership. Besides copyright protection, watermarking may be used for data monitoring, fingerprinting, and observing content manipulations. All watermarking techniques should satisfy a set of requirements (Arnold, 2000). In particular, the embedded watermark should be:

- Imperceptible
- Undetectable to prevent unauthorized removal
- Resistant to all signal manipulations
- Extractable to prove ownership

Before the proposed technique is made public, all the above requirements should be met.

Watermarking schemes can be classified into blind and nonblind schemes according to the watermark detection procedure used at the receiver (Katzenbeisser & Petitcolas, 2000). The difference between these two schemes is that nonblind schemes detect the watermark using the original signal whereas blind schemes do not use the original signal for watermark detection. Although nonblind schemes are more robust in detecting watermarks, the multimedia industry appears to prefer the blind schemes due to their practicality.

The watermarking literature also describes a second classification made according to the amount of data that a watermark carries. In particular, there are two classes considered. The first class includes the *one-bit watermarking* techniques (Cox, Miller, & Bloom, 2002), which detect only the presence of the watermark. Lee and Ho (2000), propose an algorithm that embeds a narrowband sequence into cepstral coefficients. Cox, Killian, Leighton, and Shamoon, (1997), propose a spread spectrum image watermarking algorithm. Bassia, Pitas, and Nikolaidis (2001) embed a chaotic sequence in the time domain. Kirovski and Malvar (2001) embed a spread spectrum sequence in the modulated complex lapped transform domain. These algorithms can be extended to detect and extract multiple-bit watermarks; however their robustness decreases with increasing number of watermark messages.

The second class of techniques not only detects but also extracts the embedded watermark message. Swanson, Zhu, and Tewfik (1999) propose that watermark embedding algorithm projects the audio signal's frequency subbands onto a secret key. Lie and Chang (2001) present an algorithm which embeds watermarks in time-domain using the energy of consecutive audio blocks. An algorithm embedding the watermark as noise into Fourier coefficients is proposed by Seok and Hong (2001). Gang, Akansu, and Ramkumar (2001) propose an algorithm considering the effects of MP3 compression. All these algorithms embed and extract multiple watermark bits. As a result of signal manipulations, some message bits extracted by the detector may be in error, potentially resulting in the detection of the wrong watermark message.

In order to propose watermarking algorithms that are robust to signal manipulations, we introduced two TF signatures for audio watermarking: instantaneous mean frequency of the signal, and fixed amplitude linear and quadratic phase signal (chirp). The following sections present an overview of the two proposed methods, and their performances.

IMF-Based Watermarking

In our "Audio Watermarking Time-Frequency Characteristics" paper (Esmaili, Krishnan & Raahemifar, 2003), we proposed a watermarking scheme using the estimated IMF of the audio signal. Our motivation for this work is to address two important features of security and imperceptibility and this can be achieved using spread spectrum and instantaneous mean frequency. In fact, the estimated IMF of the signal is examined as an optimal point of insertion of the watermark in order to maximize its energy while achieving imperceptibility.

1. **Watermarking algorithm:** Figure 4 demonstrates the watermark embedding and extracting procedure.

Figure 4 demonstrates the watermark embedding and extracting procedure. In this figure, *Si* is a nonoverlapping block of the windowed signal. Based on Gabor's work on IF (Qian & Chen, 1996), Ville devised the Wigner-Ville distribution (WVD), which showed the distribution of a signal over time and frequency. The IMF of a signal was then calculated as the first moment of the WVD with respect to frequency. Therefore, the IMF of a signal could be expressed as (Krishnan, 2001):

$$f_i(n) = \frac{\sum_{f=0}^{F_m} fTFD(n,f)}{\sum_{f=0}^{F_m} TFD(n,f)} \quad (1)$$

This IMF is computed over each time window of the STFT, and *TFD* (*n; f*) refers to the energy of the signal at a given time and frequency. Note that in equation (1), F_m refers to the maximum frequency of the signal, *n* is the time index, and

Figure 4. Watermark embedding and recovery using IMF

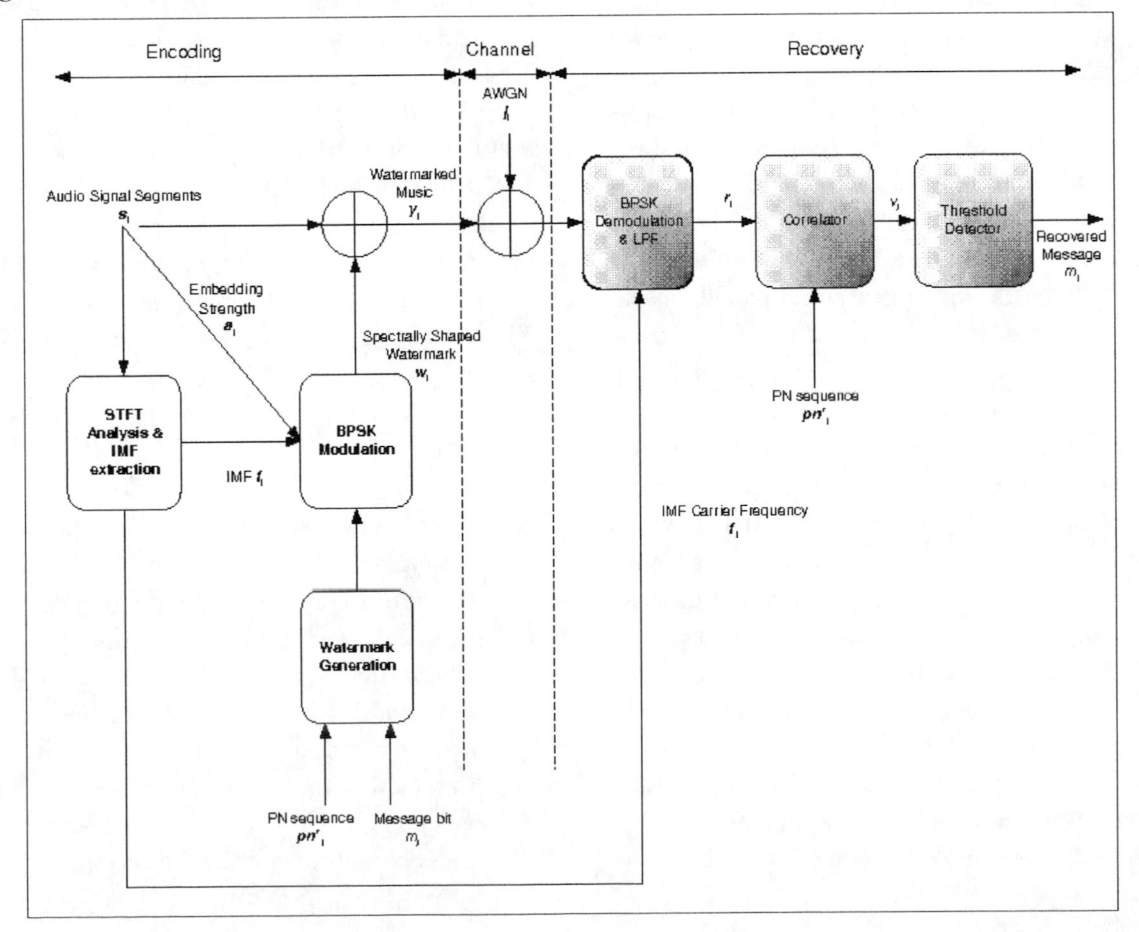

f is the frequency index. From this we can derive an estimate of the IMF of a nonstationary signal assuming that the IMF is constant throughout the window.

The watermark message is defined as a sequence of randomly generated bits and that each bit is spreaded using a narrowband PN sequence, then shaped using BPSK modulation and an embedding strength. The modulated watermarked signal can now be defined by:

$$w_i = m_i hpna_i \,|\cos(2\pi f_i)| \quad (2)$$

where m_i refers to the watermark or hidden message bit before spreading, and pn is the spreading code or the PN sequence which is low-passed filtered by filter h. The FIR low-pass filter should be chosen according to frequency characteristics of the audio signal; the cutoff frequency of the filter was chosen empirically to be 1.5 KHz. f_i refers to the time-varying carrier frequency which represents the IMF of the audio signal. The power of the carrier signal is determined by a_i, and is adjusted according to the frequency masking properties of the HAS.

In order to understand the simultaneous masking phenomenon of the HAS, we will examine two different scenarios of simultaneous masking. First, in the case where a narrowband noise masks a simultaneously occurring tone within the same critical band, the signal-to-mask ratio is about 5 dB. Second, in the case of tone-masking noise, the noise needs to be about 24 dB below the masker excitation level; meaning that it is generally easier for a broadband noise to mask a tonal sound than for the tonal sound to mask a broadband noise. Note that in both cases, the noise and tonal sounds need to occur within the same critical band for simultaneously masking to occur. In our case, the tone- or noise-like characteristic is determined for each window of the spectrogram and not for each component in the frequency domain. We found the entropy of the signal useful in determining whether the window can best be classified as tone-like or noise-like. The entropy can be expressed as:

$$H(n) = \sum_{f=0}^{F_m} P_f(TFD(n,f)) \log_2 P_f(TFD(n,f)), \quad (3)$$

where

$$P_f(TFD(n,f)) = \frac{TFD(n,f)}{\sum_{f=0}^{f=F_m} TFD(n,f)}, \quad (4)$$

Since the maximum entropy can be written as:

$$H_{\max}(n) = \log_2 F_m. \quad (5)$$

We assume that if the entropy calculated is greater than half the maximum entropy, the window can be considered noise-like; otherwise it is tone-like. Based on these values, the watermark energy is then scaled by the coefficients a_i such that the watermark energy will be either 24 dB or 5 dB below that of the audio signal.

In order to recover the watermark and thus the hidden message, the user needs to know the PN sequence and the IMF of the original signal. Figure 4 illustrates the message recovery operation. The decoding stage consists of a demodulation step using the IMF frequencies, and a dispreading step using the PN sequence.

2. **Algorithm performance:** The proposed watermarking algorithm was applied to several different music files ranging between classical, pop, rock, and country music. These files were sampled at a rate of 44.1 kHz, and 25 bits were embedded into a 5 sec sample of the audio signal. Figure 5 gives an overview of the watermark procedure for a voiced pop segment. As can be seen from these plots, the watermark envelope follows

Figure 5. Overview of watermarking procedure for POP voiced segment ("viorg.wav")

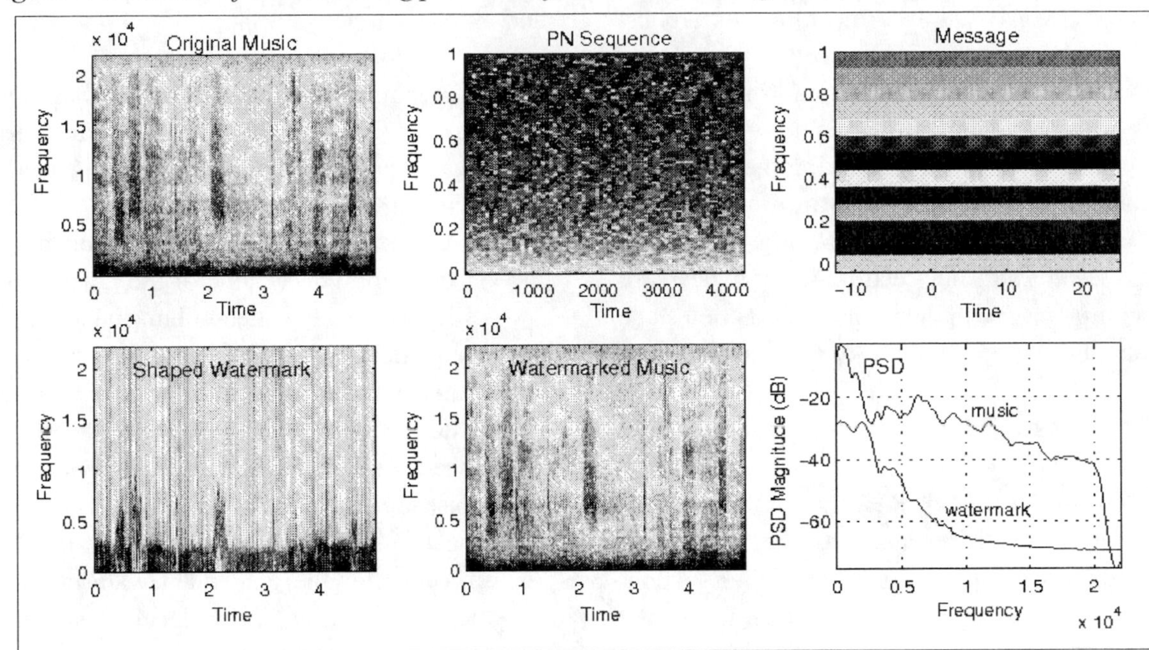

Table 1. Performance of the IMF-based algorithm after various attacks

Attacks	Average BER	Affected Algorithms in StirMark
1. None	0.00	N/A
2. HPF (100 Hz)	0.05	A, D
3. LPF (4 kHz)	0.06	A, C, D
4. Resampling factor (0.5)	0.04	C, D
5. Amplitude change (+/-10dB)	0.08	N/A
6. Parametric equalizer (bass boost)	0.13	A, B, C, D
7. Noise reduction (hiss removal)	0.02	C,D
8. MP3 compression	0.08	N/A

the shape of the music signal. As a result, the strength of the watermark increases as the amplitude of the audio signal increases.

Several robustness tests based on StirMark Benchmark (Petitcolas, 2001) attacks were performed on the five different audio files to examine the reliability of our algorithm against signal manipulations. The summary of these results can be seen in Table 1. A, B, C, and D are four watermarking algorithms; for each algorithm, six audio segments were watermarked, and it was noted whether the watermark was completely destroyed or somewhat changed by the attacks. As can be seen from the above tests, our technique offers several improvements over existing algorithms.

As it was demonstrated in this section, the proposed IMF-based watermarking was a robust watermarking method. In the following section, the proposed chirp-based watermarking technique is introduced. The motivation of using linear chirps as a TF signature is taking the advantage of using a chirp detector in the final stage of watermark decoding to improve the robustness of the watermarking technique, and decrease the complexity of the watermark detection stage compared to the IMF-based watermarking.

Chirp-Based Watermarking

We proposed (Erkucuk, Krishnan, & Zeytinoglu, 2006) chirp-based watermarking; the highlights of the paper are mentioned in the following section. In this watermarking scheme, a linear frequency modulated signal, known as a chirp, is embedded as the watermark message. Our motivation in chirp-based watermarking is utilizing a chirp detection tool in the post-processing stage to compensate bit errors that occur in embedding and extracting the watermark signal.

1. **Watermark algorithm:** Figure 6 provides an overview of the chirp-based watermarking scheme for a spread spectrum watermarking algorithm.
   The watermark message is a 1-bit quantized amplitude version of the normalized chirp b on a TF plane, with initial and final frequencies $f_{\{0b\}}$ and $f_{\{1b\}}$, respectively. Each watermark bit is spread with a secret-key generated binary PN sequence p. The spread spectrum signal $w_{\{k\}}$ appears as wideband noise and occupies the entire frequency spectrum spanned by the audio signal x. In order for the embedded watermark to be imperceptible, the watermark signal is perceptually shaped by a scale factor α, and a low-pass filter. The cut off frequency of the low-pass filter is $0.05 f_{sx}$, where f_{sx} is the sampling frequency of the audio signal. The

Figure 6. Watermark embedding and detecting scheme

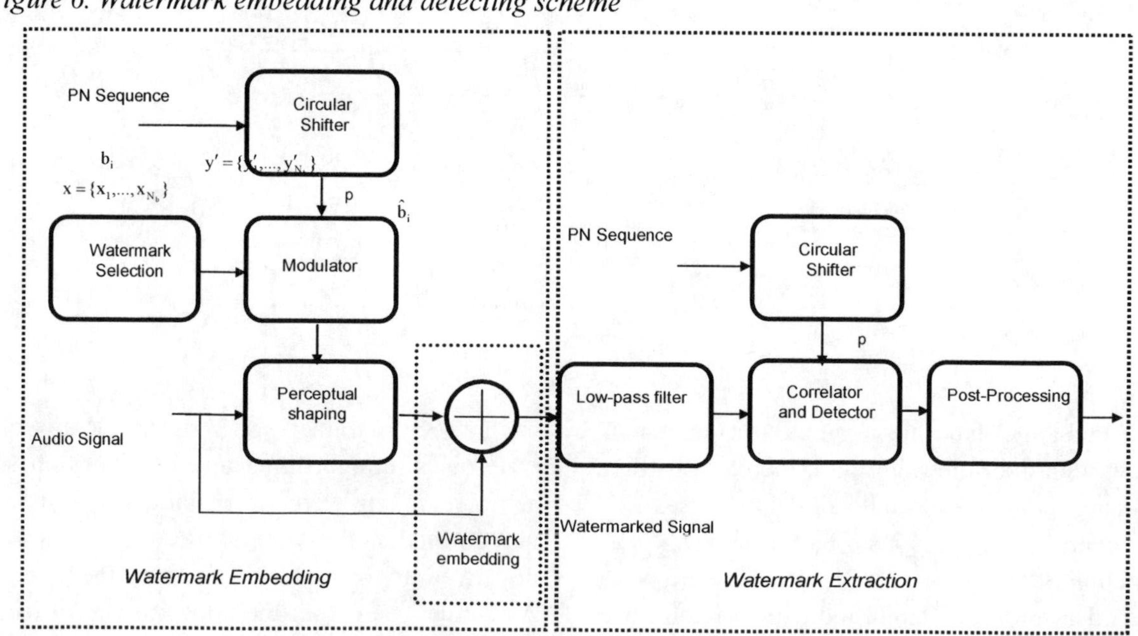

low-pass filtering step allows us to increase the value of α to a value while maintaining imperceptibility. We used the empirically determined value of 0.3 for the embedding strength parameter α.
Since the watermark bit is embedded in the low frequency bands of the transmitted signal, we extractthe watermark bit by processing the low frequency bands of the received signal, and despread the signal using the same PN sequence used in watermark embedding. We repeat the bit estimation process outlined above for each input block, until we have an estimate of all the transmitted watermark bits. While it is possible to combine the estimated bits sequence, we can improve the performance of the watermark extraction algorithm by postprocessing the estimated bits. Here, as we know that the embedded watermark has a chirp structure, by using a chirp detector, the original watermark message can be estimated.

2. **Post-processing of the estimated bits for watermark message extraction:** After all watermark bits are extracted, we first construct the TFD of the extracted watermark. The TF representation resulting from the TFD of the estimated bits can be considered as an image in TF plane. Once we generate the image of the TF plane, a parametric line detection algorithm based on the Hough-Radon transform (HRT) operates searches for the presence of the straight line and estimates its parameters. The HRT is a parametric tool to detect the pixels that belong to a parametric constraint of either a line or curve in a gray level image (Rangayyan & Krishnan, 2001). HRT divides the Hough-Radon parameter space into cells, and then calculates the accumulator value for each cell in the parameter space. The cell with the highest accumulator value represents the parameter of the HRT constraint. Since we are looking for the embedded chirp as straight lines in the TF plane in the application of post-processing of chirp-based watermarking, we can apply the HRT method to detect the embedded chirp. First, the extracted watermark bits are transformed to the TF plane; then the HRT detects the line representing the chirp in TFD. In order to achieve a good detection performance, Wigner-Ville transform (WV) is used as the TFD representation of the signal as it provides fine TF resolution.
3. **Technique evaluation:** We implemented the time-domain spread spectrum watermarking algorithm to embed and extract watermark. The sampling frequency $f_{\{sb\}}$ = 1 kHz to generate the watermark signals. Therefore, the initial and final frequencies, $f_{\{0b\}}$ and $f_{\{1b\}}$ of the linear chirps representing all watermark messages are constrained to [0–500] Hz. As host signals, we used five different audio files with $f_{\{sx\}}$ = 44.1 kHz and 16 bits/sample quantization. These sample audio files represent rock, classical, harp, piano, and pop music, respectively. We embedded watermark messages into audio signals of 40 second duration for a chip length of 10,000 samples per watermark bit (corresponding to an embedding rate of 4.41 bps), and into audio signals of 20 second duration for a chip length of 5,000 samples per watermark bit (corresponding to an embedding rate of 8.82 bps). In both cases, these values result in 176-bit long chirp sequences.
4. To measure the robustness of the watermarking algorithm, we performed 8 signal manipulation tests, which represent commonly used signal processing techniques. Table 2 shows the bit error rate (BER) results expressed as a percentage of the total number of watermark bits for the two chip lengths and for each signal manipulation operation.

Table 2. Bit error rate (in percentage) for 5 different music signals under different signal manipulations

Audio Samples					
Robustness Test	**S1**	**S2**	**S3**	**S4**	**S5**
No signal manipulation	1.14	0.57	0.00	0.57	0.00
MP3 128 kbps	1.14	0.57	0.00	1.70	0.00
MP3 80 kbps	1.14	0.57	0.00	1.70	0.00
4 kHz low-pass filtering	3.42	3.42	1.14	5.68	1.70
Resampling at 22.05 kHz	3.98	3.42	2.27	3.98	1.14
Amplitude scaling	1.14	0.57	0.00	0.57	0.00
Inversion	1.14	0.57	0.00	0.57	0.00
Addition of delayed signal	1.14	0.57	0.00	1.14	0.57
Additive noise	2.27	2.84	1.70	2.27	1.14
Embedding multiple (two) watermarks	2.27	2.84	1.70	2.27	1.14

In all the robustness tests performed, the HRT was able to extract the watermark message parameters correctly even in the worst case scenario. The experiments showed that the HRT-based post processing is able to estimate the correct watermark message up to a BER of 20%, where the maximum BER reported in Table 2 was about 6%.

The proposed chirp-based watermarking using HRT as post-processing step offers a robust watermark extraction performance; however, calculation of WVD and taking HRT on the resulted WVD has a high complexity of maximum:

$$O(N^2 \log_2(N)) + O(N^3) \quad (6)$$

where, N is the length of the chirp.

In order to decrease the complexity of the post processing stage, we could use discrete polynomial phase transform (DPPT) (Le, Krishnan, & Ghoraani, 2006) as a faster chirp estimator to estimate the watermark message. DPPT is a parametric signal analysis approach for estimating the phase parameters of constant-amplitude polynomial phase signals. The DPPT operates directly on the signal in time domain and is a computationally efficient method comparing to HRT. Complexity of DPPT is:

$$O(N \log_2(N)) \quad (7)$$

The proposed chirp-based watermark representation is fundamentally generic and inherently flexible for embedding and extraction purposes such that it can be embedded and extracted in any domain. Accordingly, we can embed the chirp sequence into the audio or image signals using any of the methods proposed by Lie and Chang (2001) and Swanson et al. (1999). For example, if we were to use the algorithm developed by Seok and Hong (2001) we would embed the chirp sequence into the Fourier coefficients. At the receiver, we extract the chirp sequence which is likely to have some bits in error. We then input the extracted chirp sequence to the HRT or DPPT based post-processing stage to detect the slope of the chirp.

Table 3 presents the result of the chirp-based watermarking using DPPT for Images in discrete cosine transform domain (DCT) (Ghoraani & Krishnan, in press). As it is observed in this table, the robustness of the watermarking scheme is satisfactory.

Although the proposed chirp-based watermarking representation is not a classical forward error correction (FEC) code, an analogy can be made between FEC codes and this new representation as they both introduce performance improvements at the expense of code redundancy. FEC codes have been commonly used in watermarking to reduce the bit error rate in order to achieve the desired BER performance. Most commonly used FEC codes for audio watermarking are Bose-Chaudhuri-Hocquenghem (BCH) codes and repetition codes. Table 3 compares the performance of the chirp-based watermarking using DPPT chirp detector, repetition coding, and BCH coding; all codes have a redundancy value of about 11/12. The chirp-based watermarking offers higher amount of BER correction than the pepetition and BCH coding.

In this chapter, we brought highlights of our proposed watermarking schemes by introducing two TF signatures. We used IMF estimation of the signal, nonlinear TF signature, as the watermark signal. Due to complexity of the watermark estimation, then we proposed chirp-based watermarking in which we embedded the linear phase signals as TF signatures. HRT is used as chirp detector in the postprocessing stage to compensate the BERs in the estimated watermark signal. The method could correct the error up to BER of 20%, and the robustness result was satisfactory. Since the HRT had high complexity, and the postprocessing stage was time consuming, we used DPPT instead of HRT in post processing. The DPPT-based post processing was applied on chirp-based image watermarking. Due to error correction property of the chirp-based watermarking, we also compared it with two well-known FEC schemes; it was shown that the chirp-based watermarking offered higher BER correction than the repetition and BCH coding.

Table 3. Performance comparison of the FEC-based post-processing schemes and DPPT-based technique under checkmark benchmark attacks (Pereira, Voloshynovskiy, & Pun, 2001) for 10 images

Attacks	Error correction methods		
	DPPT	REP	BCH(7,63)
Remodulation(4)	95	58	65
MAP(6)	100	97	100
Copy(1)	100	90	100
Wavelet(10)	98	90	92
JPEG(12)	100	100	100
ML(7)	79	57	67
Filtering(3)	100	100	100
Resampling(1)	100	100	100
Color Reduce(2)	75	65	70
Total Detection (%)	95	85	89

FUTURE RESEARCH DIRECTIONS

We used chirps as instantaneous features (time-varying features). We have shown the success of the method for audio signals and images. The watermarking technique that has been applied in our research work was the spread spectrum technique. The utilization of a blind watermarking technique that is robust to geometrical attack could be a promising direction.

It would be also worthwhile to embed nonlinear chirps. There are optimal detectors available to detect nonlinear chirps such as parabolic chirp and sinusoidal FM waveforms, and embedding of these waveforms could provide additional degree of flexibility in terms of robustness and capacity.

Also by exploring other instantaneous features we could improve the watermarking algorithm. Rather than embedding linear chirps, time-varying spectral measures such as the instantaneous mean frequency, instantaneous median frequency, and instantaneous bandwidth information could be used to represent watermarks. For extraction of such instantaneous measures, true time frequency representation (TFR) is needed. The TFRs could be constructed in many ways, and the one that would be appealing would be to construct positive TFRs that satisfy the marginal constraints.

The proposed technique using linear chirps has shown error control properties, and comparison with some common error control coding techniques showed equal or better performances of the chirp detectors used. It would be worthwhile to continue this study from a fully information theoretic perspective, which might open doors for the possible implementation of chirp-based spread spectrum schemes for various other digital communication applications.

ACKNOWLEDGMENT

The authors greatly appreciate the input, suggestions, and contributions of Shahrzad Esmaili, Arunan Ramalingam, Mehmet Zeytinoglu, and Libo Zhang. The funding contributions of NSERC and Micronet are also acknowledged.

REFERENCES

Arnold, M. (2000). *Audio watermarking: Features, applications and algorithms*. Paper presented at the IEEE International Conference on Multimedia and Expo (Vol. 2, pp. 1013-1016).

Bassia, P., Pitas, I., & Nikolaidis, N. (2001). Robust audio watermarking in the time domain. *IEEE Transactions on Multimedia, 3*(2), 232-241.

BitTorrent. (2006). Retrieved March 26, 2007, from http://www.bittorrent.com

Cox, I.J., Killian, J., Leighton, F.T., & Shamoon, T. (1997) Secure spread-spectrum watermarking for multimedia. *IEEE Transactions on Image Processing, 6*(12), 1673-1687.

Cox, I.J., Miller, M.L., & Bloom, J.A. (2002). *Digital watermarking*. San Diego: Academic Press.

edonkey. (2006). Retrieved March 26, 2007, from http://www.edonkey2000.com

Erkucuk, S., Krishnan, S., & Zeytinoglu, M. (2006). A robust audio watermark representation based on linear chirps. *IEEE Transactions on Multimedia, 8*(5), 925-936.

Esmaili, S., Krishnan, S., & Raahemifar, K. (2003). Audio watermarking time-frequency characteristics. *Canadian Journal of Electrical and Computer Engineering, 2*(28), 57-61.

Gang, L., Akansu, A.N., & Ramkumar, M. (2001). *MP3 resistant oblivious steganography*. Paper presented at the IEEE International Conference on Acoustics, Speech and Signal Processing (Vol. 3, pp. 1365-1368).

Ghoraani, B., & Krishnan, S. (in press). Chirp-based image watermarking schemes. *IEEE Transactions on Information Forensics and Security.*

IFPI. (2006). World sales 2001. Retrieved March 26, 2007, from http://www.ifpi.org/site-content/statistics/worldsales.html

Katzenbeisser, S., & Petitcolas, F.A.P. (2000). *Information hiding techniques for steganography and digital watermarking*. Norwood, MA: Artech House.

Kazaa. (2006). Retrieved March 26, 2007, from http://www.kazaa.com

Kirovski, D., & Malvar, H. (2001). *Spread-spectrum audio watermarking: Requirements, applications, and limitations*. Paper presented at the IEEE Workshop on Multimedia Signal Processing (pp. 219-224).

Krishnan, S. (2001). *Instantaneous mean frequency estimation using adaptive time-frequency distributions.* Paper presented at the IEEE Canadian Conference on Electrical and Computer Engineering (pp. 141-146).

Le, L., Krishnan, S., & Ghoraani, B. (2006.). *Discrete polynomial transform for digital image watermarking application.* Paper presented at the IEEE International Conference on Multimedia and Expo (pp. 1569-1572).

Lee, S., & Ho, Y. (2000). Digital audio watermarking in the cepstrum domain. *IEEE Transactions on Consumer Electronics, 46*(3), 744-750.

Lie, W.N., & Chang, L.C. (2001). *Robust high quality time-domain audio watermarking subject to psychoacoustic masking.* Paper presented at the IEEE International Symposium on Circuits and Systems (Vol. 2, pp. 45-48).

Lin, E.T., Eskicioglu, A.M., Lagendijk, R.L., & Delp, E.J. (2005). Advances in digital video content protection. *Proceedings of the IEEE, 93*(1), 171-183.

Pereira, M.S., Voloshynovskiy, S., & Pun, T. (2001). *Second generation benchmarking and application oriented evaluation.* Paper presented at the Information Hiding Workshop.

Petitcolas, A.P., et al. (2001). *StirMark benchmark: Audio watermarking attacks.* Paper presented at the International Conference on Information Technology: Coding and Computing (pp. 49-55).

Qian, S., & Chen, D. (1996). *Joint time-frequency analysis: Method and application.* New York: Prentice Hall.

Rangayyan, R., & Krishnan, S. (2001). Feature identification in the time-frequency plane by using the Hough-Radon transform. *IEEE Transactions on Pattern Recognition, 34*, 1147-1158.

RIAA. (2006). Market data pages of the web site. Retrieved March 26, 2007, from http://www.riaa.org

Seok, J.W., & Hong, J.W. (2001). Audio watermarking for copyright protection of digital audio data. *Electronics Letters, 37*(1), 60-61.

Swanson, M.D., Zhu, B., & Tewfik, A.H. (1999). *Current state of the art, challenges and future directions for audio watermarking.* Paper presented at the IEEE International Conference on Multimedia Computing and Systems (Vol. 1, pp. 19-24).

Valenti, J. (2002, February). *Piracy threatens to destroy movie industry and U.S. economy.* Testimony before the US Senate foreign relations committee.

Venkatachalam, V., Cazzanti, L., Dhillon, N., & Wells, M. (2004). Automatic identification of sound recordings. *IEEE Signal Processing Magazine, 2*(2), 92-99.

ADDITIONAL READING

Kirovski, D., Malvar, H., & Yacobi, Y. (2004, July-September). A dual watermark-fingerprint system. *IEEE on Multimedia*, 59-73

Trappe, W., Wu Min, Wang, Z.J., & Liu, K.J.R.. (2003, April). Anti-collusion fingerprinting for multimedia. *IEEE Transactions on Signal Processing*, 1069-1087.

Zhao, H.V., & Liu, K.J. R. (2006, January). Fingerprint multicast in secure video streaming. *IEEE Transactions on Image Processing*, 12-29.

Dugelay, J.L, Roche, S., Rey, C., & Doerr, G. (2006, September). Still-image watermarking robust to local geometric distortions. *IEEE Transactions on Image Processing*, 2831-2842.

Voloshynovskiy, S., Pereira, S., Pun, T., Eggers, J. J., & Su, J.K. (2001, August). Attacks on digital watermarks: Classification, estimation based attacks, and benchmarks. *IEEE Communications Magazine,* 118-126.

Chih-Wei, T., & Hsueh-Ming, H. (2003, April). A feature-based robust digital image watermark scheme. *IEEE Transactions on Signal Processing*, 950-959.

Rughooputh, H.C.S., & Bangaleea, R. (2002, October). *Effect of channel coding on the performance of spatial watermarking for copyright protection*. Paper presented at the 2002 IEEE African Conference in Africa (Vol. 1, pp. 149-153).

Yimin, J., & Feng-Wen, S. (2001, November). *'Watermarking' for convolutionally/turbo coded systems and its applications*. Paper presented at the 2001 IEEE Global Telecommunications Conference (Vol. 2, pp. 1421-1425).

Loo, P., & Kingsbury, N. (2003, April). Watermark detection based on the properties of error control codes. *IEEE Proceedings on Vision, Image and Signal Processing, 150*(2), 115-121.

Cohen, L. et al. (1985). Positive time-frequency distribution functions. *IEEE Transactions on Acoustics, Speech, and Signal Processing, 33*, 31-38.

Loughlin, P. et al. (1994). Construction of positive time-frequency distributions. *IEEE Transactions on Signal Processing, 42*, 2697-2705.

ENDNOTE

1 The term "fingerprinting" has also been employed as a special case of watermarking in which each legal copy of a multimedia content is watermarked uniquely. However, the same term has been used to name techniques that associate a multimedia signal to a much shorter numeric sequence (the "fingerprint") and use this sequence to identify the content. In this chapter, the term fingerprinting refers to the latter meaning.

Chapter X
Digital Audio Watermarking Techniques for MP3 Audio Files

Dimitrios Koukopoulos
University of Ioannina, Greece

Yannis C. Stamatiou
University of Ioannina, Greece

ABSTRACT

In this chapter, we will present a brief overview of audio watermarking with the focus on a novel audio watermarking scheme which is based on watermarks with a 'semantic' meaning and offers a method for MPEG Audio Layer 3 files which does not need the original watermark for proving copyright ownership of an audio file. This scheme operates directly in the compressed data domain, while manipulating the time and subband/channel domains and was implemented with the mind on efficiency, both in time and space. The main feature of this scheme is that it offers users the enhanced capability of proving their ownership of the audio file not by simply detecting the bit pattern that comprises the watermark itself, but by showing that the legal owner knows a hard to compute property of the watermark bit sequence ('semantic' meaning). Our discussion on the scheme is accompanied by experimental results.

INTRODUCTION

Nowadays, a vast amount of information concerning all aspects of human activity is digitised with the goal of profit. One of the most prominent examples is the music industry that now markets audio data mainly in the form of MPEG Audio Layer 3 files rather than on CDs or DVDs. The MP3 audio files are distributed as digital files through various communication means and can easily be copied and retransmitted. In such an open environment, the issue of security and copyright protection of MP3 audio files is critical. Therefore, there is a necessity for the development of reliable, fast, and robust schemes for protecting MP3 audio files, especially of those sent over the Internet because they are vulnerable to unauthorised copying and manipulation by malicious users. However, any protection intervention should be minimal since data should not carry with them significant overhead date that cause download delays. Today, there are some audio watermarking techniques that give an answer to these concerns providing mechanisms that allow the transparent embedding of copyright protection information into the actual audio data. The aim of our chapter is to present a new watermarking method based on complexity theory and cryptography and using some recent developments in *zero knowledge interactive proofs* and *threshold phenomena* in computationally intractable combinatorial problems. We will show how theoretical advances in these areas can help us build protocols that provide ways to prove one's ownership of a digital item based on one's *knowledge* of a piece of information that is computationally difficult to extract if not known in advance, and, further, how the demonstration of this knowledge can be accomplished *without* revealing it. Our goal is also to show how theory can be used as a vehicle to build the protocols and, most importantly, build up the confidence required of people in order to trust the security of the protocols that they will be eventually required to accept and use within the context of protection of digital works of art.

SIGNIFICANCE AND RESEARCH QUESTIONS

In this work we will describe a watermarking paradigm that views a watermark/signature or, generally, any bit-sequence that is used as a "trademark," as an instance of a hard computational problem. The main idea is that a watermark is a specially chosen instance of a computationally hard problem, such as 3-SAT or 3-COLORING (see next sections for definitions). The instance is constructed to have a solution known only to the individual (the individual's secret). Then a *zero knowledge interactive proof* (ZKIP) protocol can be used in order to prove knowledge of the solution *without* revealing it and, thus, prove ownership of the copyright of the item bearing the watermark. We will also show how instances can be chosen so that they have increased solution difficulty (based on the theory of threshold phenomena in computationally intractable problems), and we will demonstrate some protocols that can be used for the identification of an individual's identity during a transaction. By viewing signatures as instances of hard problems, we can exploit a variety of properties of the problems in order to produce signatures with desirable characteristics, such as increased solution difficulty, a thing that cannot be easily accomplished in more conventional signature schemes. Our work essentially links the watermark verification process (of a person, a work of art, etc.). Thus, even if the signature is stolen and subsequently used by a forger, the signature cannot be used without knowledge of the property, which is only in the mind of its creator. One of the main open issues, however, is the development of hardness criteria, that is, sets of properties that can be employed in order to construct signatures with difficult to compute properties. The theory of threshold phenomena of intractable problems gives only a partial answer. Characterising hard instances remains a very interesting problem in the field of computational complexity and we believe that it can be successfully applied to many other disciplines, one of which is audio watermarking.

MPEG AUDIO LAYER III COMPRESSION STANDARD

The MPEG audio compression algorithm is a lossy algorithm that was developed by the Motion Picture Experts Group as an ISO standard for the high fidelity compression of digital audio. The algorithm works by exploiting the perceptual properties of the human auditory system. MPEG audio offers a choice of three independent layers of compression. From the three layers, MPEG Layer III offers the best audio quality, particularly for bit rates around 64 kbits/s per channel. This layer is also well suited for audio transmission over ISDN.

The MPEG compression algorithm of audio sequences operates as follows. Input audio samples are fed into the encoder. The input audio stream passes through a filter bank that divides the input into multiple subbands of frequency. The input audio stream simultaneously passes through a psychoacoustic model that determines the ratio of the signal energy to the masking threshold for each subband. Thus, the psychoacoustic model creates a set of data to control the quantiser and coding. The quantiser and coding block use the signaltomask ratios to decide how to apportion the total number of code bits available for the quantisation of the subband signals to minimise the audibility of the quantisation noise. Its output is a set of coding symbols from the mapped input samples. Finally, the block frame packing takes the representation of the quantised subband samples and formats this data and side information into a coded bit stream. Ancillary data not necessarily related to the audio stream can be inserted within the coded bit stream. The decoder's operation is the reverse. First, it deciphers the coded bit stream. Then, it restores the quantised subband values, and reconstructs the audio signal from the coded subband values.

The MPEG audio stream consists of audio frames. Each frame contains audio data, a 4byte header, an error detection code (CRC), and ancillary data. Frame header contains information about the bit stream such as syncword, layer, bit rate, sampling frequency, and mode. The payload of any MPEG Layer III frame contains information about MDCT coefficients that are produced by processing the filter bank outputs with a modified discrete cosine transform (MDCT). Also, it contains scale factor bands that cover several MDCT coefficients and have approximately critical band widths. Additionally, it uses variablelength Huffman codes to encode the quantised samples to get better data compression. Layer 3 processes the audio data in frames of 1152 samples. The audio data representing these samples do not necessarily fit into a fixed length frame in the code bit stream. The encoder can donate bits to a reservoir when it needs less than the average number of bits to code a frame. Later, when the encoder needs more than the average number of bits to code a frame, it can borrow bits from the reservoir. The encoder can only borrow bits donated from past frames; it cannot borrow from future frames. The Layer 3 bit stream includes a 9bit pointer, 'main_data_begin,' with each frame's side information that points to the location of the starting byte of the audio data for that frame. Although the main_data_begin limits the maximum variation of the audio data to 29 bytes (header and side information are not counted because for a given mode they are of fixed length and occur at regular intervals in the bit stream), the actual maximum allowed variation would often be much less.

REVIEW OF LITERATURE

Audio Watermarking Techniques

As it has been mentioned by previous works (Arnold & Kanka, 1999; Basia, Pitas, & Nikolaidis, 2001; Bender, Gruhl, Morimoto, & Lu, 1996; Boney, Tewfic, & Hamdy, 1996; Dittmann, Steinebach, & Steinmetz, 1999; Seok, Hong, & Kim, 2002; Qiao & Nahrstedt, 1998), there is a number of properties an audio watermarking algorithm should have in order to be efficient:

- *Inaudibility*, the watermark embedding should not be accompanied by loss of audio quality
- *Statistical invisibility*, the algorithm should prevent unauthorised watermark detection/removal or alteration
- *Similar compression characteristics with the original signal*
- *Robustness*, the algorithm should be robust against various attacks for malicious users
- *Embedded directly in the data*
- *Support multiple watermarks*
- *Low redundancy*
- *Self-clocking*

There are three different approaches for the classification of watermarking schemes. The first approach differentiates watermarking schemes to the schemes that embed watermarks to PCM-data (raw audio data) (Arnold & Kanka, 1999; Basia et al., 2001; Bender et al., 1996; Boney et al., 1996) and the ones that embed watermarks in the compressed-domain (Arttameeyanant, Kumhom, & Chamnongthai, 2002; Dittmann et al., 1999; Koukopoulos & Stamatiou, 2001; Li, Zhang & Zhang, 2004; Qiao & Nahrstedt, 1998; Yeo & Kim, 2003). The schemes that embed watermarks to PCM-data operate with all audio formats because they presuppose that the audio file is uncompressed in order to embed the watermark and then recompress again. However, this approach creates two significant problems. First, it is not sure that the watermark will survive the coding/decoding procedure, and second, it is time consuming, which is a luxury not available in real-time applications. In order to be efficient for online applications, audio data are commonly stored and transmitted in compressed format, such as MPEG audio. Thus, the watermarking schemes would be preferred to target the compressed data domain when they are going to be used to online applications. There is the problem of watermark technique adaptability with this approach because these techniques are oriented to specific audio formats and they should be transformed in order to be able to operate with other audio formats.

Several popular techniques for audio watermarking in the compressed-domain are based on audio scale factors. Additional techniques are the Sandford et al. (1997) and MP3Stego (Petitcolas, 1998) schemes that have low robustness. In Sandford et al., the auxiliary information is embedded as a watermark into the host signal created by a lossy compression technique. On the other hand, MP3Stego hides information in MP3 files during the compression process. The watermark data is first compressed, encrypted, and then hidden in the MP3 bit stream. The hiding process takes place at the heart of the Layer III encoding process namely in the inner loop. Any attempt to remove or distort the watermark, including re-encoding the audio content, will lead to perceptible distortion of the original audio content. The watermark embedding and detection can be done very fast. The detected watermark information can provide proofs of copyright and distribution sources. On the other hand, for some audio streams, there are only a few scale factors for each frame. Thus, our frame-based watermarking scheme does not have much data to watermark.

The second approach classifies schemes depending on whether they operate in the frequency-domain (Arnold & Kanka, 1999; Boney et al, 1996) or time-domain (Basia et al., 2001; Dittmann et al., 1999; Koukopoulos & Stamatiou, 2001; Qiao & Nahrstedt, 1998). Usually, watermarking in the time domain handles time distortions well, while special care should be taken against spectral distortions, while in the frequency domain the opposite happens. Some of the methods in the frequency domain exploit the frequency characteristics of the audio signal in order to embed the watermark, minimising audible distortions even for high amplitude watermarks. However, the techniques that are applied only in the time domain or the frequency domain suffer from robustness problems. For their overcoming, the watermarking schemes use combinations of manipulations in the time and frequency domain.

The third approach separates watermarking schemes based on the use (Boney et al., 1996; Qiao & Nahrstedt, 1998) or not (Arnold & Kanka, 1999; Basia et al., 2001; Dittmann et al., 1999; Koukopoulos & Stamatiou, 2001) of the original signal for watermark detection. It is desirable not to use the original signal because it results in storage waste and it adds the danger of its usage by malicious users. But if the original signal is not needed, then the watermarked signal is not immunised against inversion attacks by third parties that could prove ownership by subtracting their watermark from the original marked by the true owner.

Complexity Theory and the Difference between Discovering and Verifying

Let us fix a set of Boolean variables, that is, variables that may assume only one of two possible values: 0 ("false") and 1 ("true"). Call this set V. The *satisfiability problem,* or *SAT,* consists in deciding whether a Boolean formula given as a conjunction (that is logical AND) of disjunctions (that is logical OR) of variables in V or their negation (literals) evaluates to true for some value assignment to the variables. For example consider the set of variables $V = \{x_1, x_2, x_3\}$ and the following formula (AND = $\wedge$, OR =$\vee$):

$$f = \{x_1 \vee x_2 \vee x_3\} \wedge \{\neg x_1 \vee \neg x_2 \vee \neg x_3\}$$

This formula is written in *conjunctive normal form* (CNF), as we say, and in order to be satisfiable all its constituents, also called clauses, must be satisfiable. It is easy to see that by assigning $x_1 = 1, x_2 = x_3 = 0$, the formula is made satisfiable. But let us now consider what happens if we add some more clauses:

$$f' = \{x_1 \vee x_2 \vee x_3\} \wedge \{\neg x_1 \vee \neg x_2 \vee \neg x_3\} \wedge \{x_1 \vee \neg x_2\} \wedge \{x_2 \vee \neg x_3\} \{x_3 \vee \neg x_1\}$$

A little thought will reveal to us that this formula is not satisfiable. The task of deciding whether the formula is satisfiable seems to be a really formidable one. If you consider the small example we gave in the previous section, trying to find a truth assignment to the variables that satisfy some subset of the clauses causes problems in other clauses of the formula. And there seems to be no easy way of extracting some clever structural information about the clauses of the formula so that locating a satisfying truth assignment becomes easier than simply testing whether each of the possible 2^n such assignments satisfy all the clauses.

But what if we were to assert that the formula is satisfiable? Of course you would ask to have a piece of evidence, or certificate, for our assertion. And what's more, this certificate should be checkable fast. At least faster than it would take you to check the validity of my assertion by trying all the 2^n truth assignments, a huge number by any standards (n is a free parameter and can be unboundedly large). And what would be a better and faster checking certificate than giving you a truth assignment on which the formula evaluates to true? Then you would just evaluate the formula and, indeed very quickly (just check whether each clause contains a literal that is true under the truth assignment that we gave you), you would be convinced of my assertion. The problems that have the property that for each of the instances (e.g., formulas) that have the property sought by the problem (e.g., is the formula satisfiable?) there exists an easily checkable certificate for this: constitute the important computational complexity class NP (nondeterministically polynomially fast solvable problems). We know that there is some obscure point here: how did we manage to find the truth assignment that we gave you as a certificate in the first place? This is of no concern for the class NP; if one gives you a certificate (satisfying truth assignment), then you can check it easily and this is what matters. Moreover, if it happens for a problem to also exist a way to discover easily (fast) the certificate, then we say that the problem belongs to the class P (polynomially fast solvable problem). The reason for the word nondeterministically lies in the point of view that giving the certificate to someone is equivalent to him/her guessing it, since we are only interested in the existence of such a certificate and not of actually locating it which may be hard to do fast. Of course, for problems in P not only such a certificate exists, but it can also be found fast. It should be borne in mind that for computational complexity, fast means always in a number of steps which is bounded from above by a polynomial in the size of the problem instance. The field of Computational Complexity has its roots in the seminal papers of Cook (1971), Karp (1972), and Levin (1973) and has as its milestone the celebrated Cook-Levin theorem: there is no fast algorithm that can determine whether a Boolean formula is satisfiable unless there is also an algorithm that solves fast all the problems whose "yes" instances possess short certificates. Problems with this property are called *NP-complete*. It should be stressed that for this problem class, we only consider worst case complexity. That is, every algorithm that runs in polynomial time and decides correctly these problems must take at least superpolynomial time on some instances. It may well be the case that the algorithm decides fast the majority of the instances. However, for some of them, the algorithm must fail to finish quickly (unless, of course, P = NP). (An excellent introduction to NP-completeness can be found in Garey and Johnson (1979))

Different views of complexity were considered by Kolmogorov (1965), Solomonoff (1964), and Chaitin (1966) who, independently, developed the concept of complexity of *finite objects* (now called *Kolmogorov complexity*). Moreover, Levin (1986) introduced the concept of completeness in the sense of *average case*.

Interactive Proofs: How Randomness Expands Verification Capabilities

In this section, we will focus on the basic ideas underlying the concept of an interactive proof. This concept was introduced in 1985 by the seminal papers of Goldwasser, Micali, and Rackoff (1985) and Babai (1985). We will demonstrate this concept using the *graph nonisomorphism problem*.

Two graphs $G_1 = (V, E_1)$ and $G_2 = (V, E_2)$ are isomorphic if there exists a permutation π (i.e., 1-1 and onto function) of the set of vertices $\{1, \ldots, n\}$ to itself such that a pair of vertices i, j is adjacent in G_1 if and only if the pair $\pi(i), \pi(j)$ is adjacent in G_2. This permutation is called *adjacency preserving permutation*. The *graph Isomorphism problem*, or GI, asks whether two given graphs G_1 and G_2 are isomorphic, that is, whether there *exists* an adjacency preserving permutation of the set of vertices (see the books of Papadimitriou (1994) and Garey and Johnson (1979) for more formal definitions and information on this problem). This permutation is, of course, a short and easily verifiable certificate or proof that can convince anyone that the two graphs are isomorphic. Thus, we can easily see that the problem GI belongs to the class NP.

The *complementary* problem, called *graph nonisomorphism* or GNI, asks whether two given graphs G_1 and G_2 are *nonisomorphic*. Using, again, the definition of isomorphism, two graphs G_1 and G_2 are isomorphic are nonisomorphic if there is *no* adjacency preserving permutation of the set of vertices. Generally, problems complementary to problems that belong to NP constitute another important complexity class called *co-NP*. For example, the problem of deciding whether a given formula *is not* satisfiable or UNSAT is in co-NP since its complementary problem of deciding satisfiability is in NP. Does GNI belong to the class NP? What kind of short certificate, like the adjacency preserving permutation used in GI, would convince one fully that two given graphs are nonisomorphic? It seems that one has to check *all* possible permutations and verify that they are really not adjacency preserving in order to be convinced of the fact that two graphs are nonisomorphic. Unfortunately, for a set of n vertices, there are $n! = 1 \cdot 2 \ldots n$ possible permutations to check and their number grows really huge in the order of n^n as n increases. This is in sharp contrast with the short certificate used to prove isomorphism (the same asymmetry between satisfiability and nonsatisfiability) and it is widely believed that GNI does not belong to NP, giving rise to the conjecture that co-NP $\neq$ NP.

According to the discussion above, it seems that there exists a huge gap between verifying properties in NP (e.g., isomorphism between pairs of graphs and satisfiability of formulas) and properties in co-NP (e.g., isomorphism between pairs of graphs and satisfiability of formulas), and moreover, it seems that proofs of properties in co-NP are larger and more time consuming to check. Let us, however, have a look at the following protocol (see, e.g., Goldreich (1995) and Motwani and Raghavan (1995) for accessible presentations), where by P we denote the powerful *prover* and by V the *verifier* who is confined to use only randomised computations of polynomial time.

- V chooses at random a number $i \in \{1,2\}$ as well as a random permutation σ of the set $\{1, \ldots, n\}$. In addition, it constructs the graph H that results from the application of σ on the vertices of $G_i : H = \sigma(G,i)$. Finally, V sends H to P and asks of P to return a number $i \in \{1,2\}$ so that G_j is isomorphic to the graph H.
- P returns the required number j.
- Now that V has received j it checks whether $i = j$. If yes, then it accepts as a fact that G_1 and G_2 are isomorphic. If not, V rejects.

Let us first assume that G_1 and G_2 are indeed nonisomorphic and V chooses $i = 1$. Since G_1 and G_2 are nonisomorphic, the graph H constructed by V during its first step above is isomorphic to G_1 (by construction) but *not* to G_2 (otherwise G_1 and G_2 would be isomorphic). Then the powerful P can decide that H is indeed isomorphic to G_1 and not to G_2 and respond with $j = 1$, leading V to accept the fact that G_1 and G_2 are nonisomorphic. The same argument holds if V chooses $i = 2$. What, now, if G_1 and G_2 are isomorphic? Then P is unable to discern which was the graph, G_1 or G_2, that was used by V in order to produce H. Thus P can at best respond at random and, thus, its response j can be correct with probability only 1/2. Repeating the protocol for t times, we obtain that the probability the V gets fooled by P into believing that two graphs are not isomorphic when they are equal to $(1/2)^t$, a quantity that vanishes to 0 exponentially fast. On the other hand, if the graphs are not isomorphic, then V always verifies this fact with certainty.

Concluding, it seems that randomness can bridge the gap between the efficiency of verifying properties defined using an existential quantifier (e.g., *there exists* a permutation) against properties defined using the universal quantifier (e.g., *no* permutation exists); or in others words, to substantially shorten deterministic proofs of facts at the cost of abandoning absolute certainty (but who needs absolute certainty when one can reach certainty levels arbitrarily close to certainty at little additional cost, i.e., protocol rounds?).

Zero Knowledge Interactive Proofs: Verification without Revelation

In this section we will describe a well known ZKIP protocol for the *3-COLORING* problem. This problem asks whether a given graph $G = (V, E)$ is 3-colorable if its vertices can be colored using at most three colors so that no two adjacent vertices are assigned the same color. A color assignment that respects this constraint is called a 3-coloring of G. The 3-COLORING problem is *computationally intractable* (*NP-complete*) (see, e.g., Papadimitriou, 1994, and Garey & Johnson, 1979).

A ZKIP (introduced also by Goldwasser et al. (1985)) has the desirable property that one may provide convincing evidence of knowledge (e.g., of a 3-coloring of a graph) without revealing it. Let P be the person who knows a 3-coloring of the vertices of a graph $G = (V, E)$ and V be the person who wants to be convinced of P's knowledge of the coloring. Let also 00, 11, and 01 be a bit representation of the three colors. For some vertex $v \in V$, its color will be denoted by $C(v) = b_1 b_0$, where b_1, b_0 are each 0 or 1. The ZKIP protocol, proposed by Goldreich, Micali, and Widgerson (1991), is comprised of a number of rounds. At each round, the following steps are executed (see, also, Papadimitriou, 1994)):

- P produces a random permutation of the three colors. Thus, the color of each vertex is changed according to the permutation.
- Then P generates $|V|$ cryptosystems (based, for instance, on RSA), one for each vertex v.
- For each vertex v, P *probabilistically* encrypts (i.e., adding a random number to its code) the new color $C'(v)$ of v. Suppose that $C(v) = b_1' b_0'$, where b_1', b_0' are the two bits representing the new color.
- Finally, P reveals to V all the encrypted colors of each vertex v_i.
- V chooses at random an edge connecting two vertices v_i, v_j. Then P reveals the keys used for the encryption of the bits of the colors of these vertices and V decrypts their colors, that is, $C'(v_i)$ and $C'(v_j)$, using these keys.
- The above steps are repeated $k|E|$ times where k represents the *reliability* of the protocol.

Now if P did not really possess a 3-coloring, there would be a nonzero probability that P would get caught at some step. When $k|E|$ *independent* rounds have been executed, this probability is at least $1 - e^{-k}$. This expression approaches 1 exponentially fast with k, meaning that almost surely P will be caught lying after a number of repetitions of the ZKIP protocol. The important characteristic of the ZKIP protocol above is that not only does it provide strong evidence that the signature is ours, but by permuting and encrypting the colors, it effectively erases P's knowledge about our coloring as well. At each round, V is only offered the obvious fact: that in a 3-coloring of a graph, the vertices of an edge are colored differently (but V cannot tell which are the colors of the vertices). This offers the possibility to P of reusing the signature since no information is revealed about it.

Threshold Phenomena: Randomness Meets Complexity

In this section, we will survey a number of experiments and results related to threshold phenomena in two computationally intractable problems: 3-SAT (SAT restricted to clauses of length 3) and 3-COLORING. We will see how these phenomena can be used as an aid for discovering hard instances of these problems to be used in person identification protocols. In contrast with ZKIP schemes based on number theoretical problems where there is no notion of threshold behaviour, ZKIP schemes using intractable combinatorial problems such as the 3-COLORING can have the ability of involving instances with 'tunable' difficulty.

Locating 'Hard' Formulas: Experiments with 3-SAT

Following observations about concentration of hard instances of the NP-complete problem 3-COLORING (the work in Cheeseman, 1991; see next section) around some well definable area, a series of further experimental observations was reported by Mitchell, Selman, and Levesque (1992) for the SAT problem (e.g., Kirkpatrick & Selman, 1994; Selman, 1995). In order to conduct experiments that study Boolean formulas, one has to choose a method by which it is possible to check a formula for satisfiability as fast as possible. Of course, 3-SAT is a hard problem so we cannot have such an algorithm, but we can at least use some systematic technique that does significantly better than trying blindly all possible truth assignments. One commonly used algorithm that can be used for deciding satisfiability of any Boolean formula is the well known Davis-Putnam (or DP for short) procedure. The experiments were conducted as follows: for each value of n (number of variables), a set of formulas was randomly produced for various values of m (number of clauses), and their satisfiability was tested using the DP procedure. For each of them, the number of DP calls was recorded. This number is representative of the running time of the DP procedure because it is these calls that are the most time consuming. The results of the various runs are depicted in Figure 1.

The remarkable observation is that the DP procedure appears to have difficulties with formulas whose *m/n* ratio clusters around a value between 4 and 4.5. For larger or smaller ratios, DP decides satisfiability considerably faster. The plot in Figure 2 shows the percentage of satisfiable formulas for various values of the clauses to variables ratio *m/n*.

With the help of these figures, it can be seen that the *m/n* ratio around which the difficult formulas are clustered, is also the point where the generated formulas suddenly change from being almost all satisfiable to being almost all unsatisfiable. We say that random formulas undergo a phase transition, where the two phases of a formula are the 'satisfiable' and the 'unsatisfiable.' We should, however, keep in mind that it is not generally true that phase transitions are a defining characteristic of NP-complete problems. It has been rigorously proved by the independent works of Chvátal and

Figure 1. Hard instance location for various values of m/n

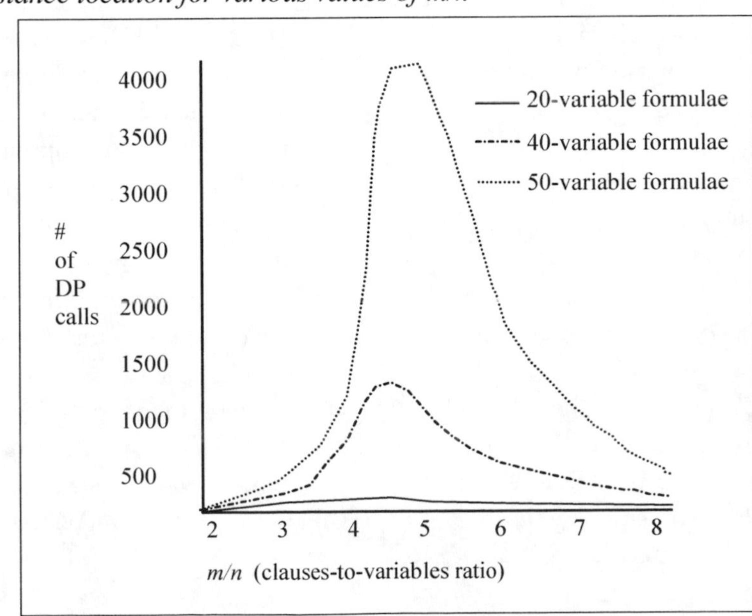

Figure 2. Percentage of satisfiable formulas for various values of m/n

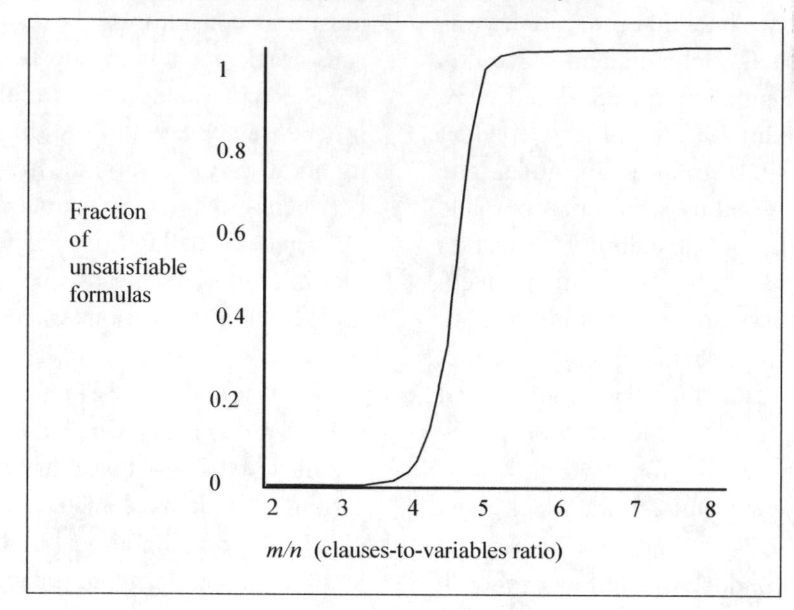

Reed (1992) and Goerdt (1996) that 2-SAT random formulas undergo a phase transition at $m/n = 1$ even though 2-SAT is in P. Also, for some NP-complete problems, like the traveling salesman problem, hard instances do not cluster around some parameter like the SAT m/n ratio.

Producing Random Hard Instances of the 3-Coloring Problem. As we have already discussed, the problem of coloring a given graph using at most three colors is computationally intractable. However, it is easy to construct a graph, randomly, having a prescribed 3-coloring on its vertices. An algorithm that achieves this is the following:

- Let p_1, p_2, and p_3 be real numbers between 0 and 1 with sum equal to 1.
- For $j = 1, \ldots, n$, vertex v_j is assigned to color class C_k with probability $p_k, k = 1,2,3$.
- For each pair u,v of vertices not belonging to the same color class, an edge (u,v) is introduced with probability p.

Given p, the above algorithm produces a *solved* instance of the 3-COLORING problem with expected color class sizes $|C_k| = p_k n, k = 1,2,3$. The algorithm provides both a 3-colorable graph along with a coloring of its vertices. However this does not suffice. What is actually needed is a 3-colorable graph for which it is also hard to find a 3-coloring. Although the current theory does not offer an effective way to generate directly or, at least, recognise a hard instance of a computationally intractable problem, there are a number of heuristic approaches that one may follow in order to produce an instance that has some increased probability of being hard. An interesting possibility for creating such instances comes from the area of threshold phenomena in combinatorial problems.

Let G be a random graph with m edges and n vertices and r the ratio m/n. In 1991, Cheeseman, Kanefsky, and Taylor demonstrated experimentally that for values of r that cluster around the value 2.3, randomly generated graphs with rn

edges were either almost all 3-colorable or almost none 3-colorable depending on whether $r < 2.3$ or $r > 2.3$ respectively. This suggested that there is some value r_0 for r around which an abrupt transition can be observed from almost certain 3-colorability to almost certain non 3-colorability of the random graphs. However, from a complexity-theoretic perspective, their crucial observation was that graphs with edges to vertices ratio around r_0 caused the greatest difficulty to the most efficient graph coloring algorithms.

To return to the algorithm that produces solved instances of 3-colorability, it seems that the edge probability p should be selected so that the *expected* ratio $E[r] = E[m]/n$ is around r_0. Setting $p_k = 1/3, k = 1,2,3,$ for having balanced color partitions and solving for p, we obtain $p = 3r_0 / n$ and we sample from the 'hard-instances' region. On the negative side, *solved* instances on the hard instance region may possess a very large number of 3-colorings and, thus, enable a fast discovery of one of them (a similar concern for 3-SAT was expressed in Achlioptas et al., 2000). A heuristic approach that can be adopted to protect the instances from this deficiency is to ensure that, except for the solution which the instance is forced to have, no other solution exists. Such an approach was followed by Motoki and Uehara (1999) for the 3-SAT problem, where it is shown that if sufficiently many clauses are chosen at random from the clauses satisfying a given truth assignment, then the formula that results possesses a unique solution with probability that tends to 1 with the number of available variables. However, due to the increased number of possible values per variable for 3-COLORING (3 values vs. 2 for 3-SAT), the technique of Motoki and Uehara (1999) cannot be directly generalised in the presented case. However, it can still be claimed that a number of special colorings are excluded from being legal colorings for this instance after it is constructed (with the algorithm given before); these are the colorings that result from the prescribed coloring with the change of a single color to a 'higher numbered' color. Therefore, assuming that colors are numbered as 0, 1, and 2, the colorings that are excluded (see below) are the ones resulting from the initial coloring by changing the colors of single vertices to a 'higher' color. More formally, a coloring P is a partition of the vertex set V into three vertex sets V_0, V_1 and V_2 so that no edge connects two vertices belonging to the same partition. A vertex u of color i is *unmovable* if every change to a higher indexed color j invalidates the coloring P. Thus, u of color i is unmovable if it is adjacent with at least one vertex of color class V_j, such that $j > i$. A coloring P is *rigid* if all its vertices are unmovable. A good thought would, then, be to start with initial coloring that has high probability of being rigid. In Kaporis, Kirousis, and Stamatiou (2000), the probability of rigidity is computed for a given coloring P given (1) the fractions x, y, z of the chosen edges connecting vertices from V_0V_1, V_0V_2 and V_1V_2 respectively and (2) the fractions α, β, γ of the vertices assigned color 0, 1, and 2 respectively. As a heuristic, we can use the probabilities estimates so that the values of $x, y, z, \alpha, \beta, \gamma$ can be adjusted to maximise the probability of rigidity. As α, β, γ enter in the probability estimate, the form of the initial coloring has an effect on how probable it is for other 'neighboring' colourings to be illegal.

Another, more rigorous and interesting approach is to construct a *certain* random class of solved 3-COLORING instances with the following property: if an algorithm can be found capable of solving an instance of the class with positive (no matter how small) probability, then there is also an algorithm solving one of the NP-complete problems *in the worst case*, which would be most surprising. In Ajtai (1996), this goal was accomplished for a *lattice problem*.

KOUKOPOULOS-STAMATIOU AUDIO WATERMARKING SCHEME FOR MP3 AUDIO FILES

General Scheme

Our algorithm (Koukopoulo & Stamatiou, 2005) extends the one that was proposed for MPEG Audio Layer 2 files by Dittmann et al. (1999). The main differences are: (1) the application of a mechanism that produces unique crypto-keys that are robust in malicious attacks because they can be detected correctly even if a big percentage of them has been damaged; (2) the use of a control mechanism that ensures that the embedding of the key in the MP3 audio file scale factors is done suitably (no audible distortions and preservation of scale factors bits number) and; (3) the specification of similar patterns with the one that is wanted to be embedded extremely fast as an one step procedure, improving extremely the time complexity of the algorithm.

As input the algorithm takes an MP3 audio file, a unique key that is produced with the use of the watermarking key creation algorithm or simple text and a set of three patterns that represent the binary digits '0' and '1' and a synchronisation bit, which is used between the bits or between prespecified group of bits and at the end of the embedding key for self-clocking and robustness against cropping.

The next step is the transformation of the binary string that represents the watermarking key into a sequence of the given patterns and the extraction of the scale factors from the frames of the MP3 audio file. The scale factors in MP3 files have variant number of bits that depends from 0 to 4 bits, while in MPEG Audio Layer 1 and 2 formats, they consist of a fixed number of

Figure 3. General scheme

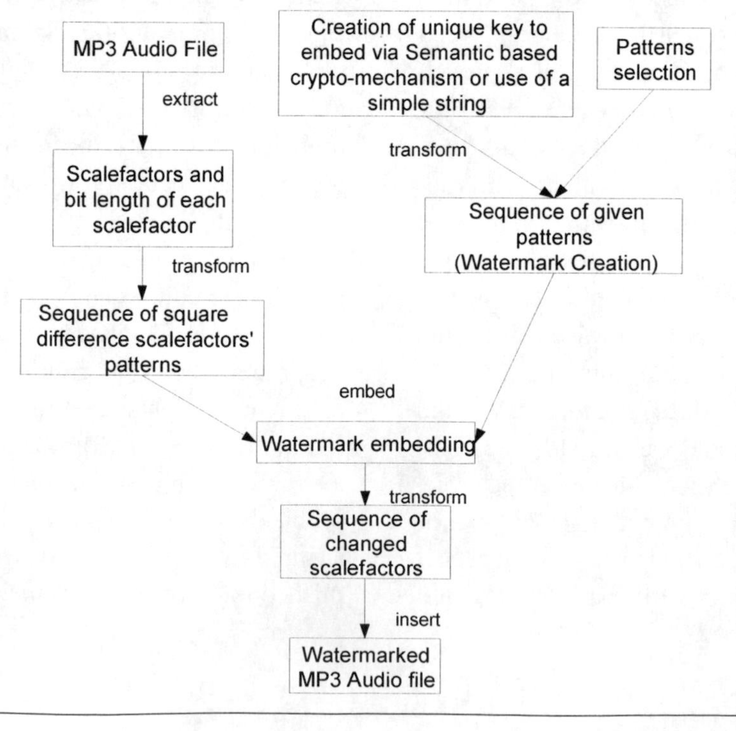

bits equal to 6. Because of that, along with the scale factors, another stream of information is extracted by an MP3 audio file, which specifies the number of bits used in the bit representation of each scale factor. Based on the scale factors, difference patterns are calculated. These patterns correspond to the difference between the first scale factor that can be considered as a starting point and the following scale factors in the scale factor stream. Assuming, as an example, the list of scale factors {6, 8, 5, 3}, the first scale factor is considered as a starting point and the pattern is {2, -1, -3}.

In the following step, the sequence of patterns produced by the watermarking key is embedded into the sequence of difference patterns produced by the scale factors of the MP3 audio file. The embedding is achieved by a mechanism that changes the scale factors difference patterns until a sufficient number of them matches the desired sequence of patterns. The mechanism guarantees that it has not introduced an audible distortion in the watermarked audio file and that each modified pattern does not violate the number of bits the underlying scale factors consist of. The second goal of the applied mechanism is achieved by applying a control submechanism that takes as input the stream that carries the information about the number of bits each scale factor consists of. Finally, the produced sequence of patterns from the previous step is transformed to a sequence of scale factors that are embedded in the source audio file, creating a watermarked MP3 audio file.

Extraction Mechanism

Unlike MPEG Layers 1 and 2, the audio coded data in MP3 audio files do not necessarily fit into a fixed length frame in the code bitstream. The encoder can donate bits to a reservoir when it needs less than the average number of bits to code a frame. Later, when the encoder needs more than the average number of bits to code a frame, it can borrow bits from the reservoir. The encoder can only borrow bits donated from past frames; it cannot borrow from future frames. The Layer 3 bitstream includes a 9bit pointer, 'main_data_begin,' within each frame's side-information that points to the location of the starting byte of the audio data for that frame. For practical considerations, the standard stipulates that the audio data variation cannot exceed what would be possible for an encoder with a code buffer limited to 7,680 bits.

The first step of the extraction mechanism is the decoding of the information that is contained in the side information field of the frame that has constant length of 17 bytes for single mode and 32 bytes for the other modes and appears in regular intervals of the bit-stream. Then, the 'main_data_begin' pointer that is contained in the side information is used for the specification of the location of the starting byte and the location of the ending byte. After that, the audio data are loaded in a buffer without the headers, and side information fields that can be cut are loaded inside the audio data. The audio data are divided in two granules each of which are divided to two channels; if the mode is not the single one or if the mode is single, its granule contains a channel. The channels consist of the scale factors region, the Huffman code bits region. Finally, in order to extract the scale factors of each channel/granule pair, we use the information we take from the decoding of side information field.

Embedding Mechanism

The embedding mechanism takes as input the pattern we want to embed, the unwanted patterns, and the scale factors of frames where we have decided to embed our pattern. The embedding mechanism also uses as parameters the minimal number of patterns in the specified area of frames that must be equal to the pattern of the information bit we want to embed in order to consider that the wanted pattern has been embedded; the maximal tolerance that determines the changes that the patterns can suffer (as concerns the audibility) to match the wanted pattern and the stream of bit lengths of the scale factors, which patterns examined.

The embedding mechanism does not consist of one loop based on the parameter of tolerance as in [5]. The first step is the investigation of the scale factors pattern stream for unwanted patterns and their removal by changing them slightly. The next step is looking for patterns similar to the wanted one and their counting. If the wanted pattern appears more times than the parameter that specifies the minimal number of its appearances, then the mechanism is satisfied and it extracts the pattern as a result. Otherwise, the mechanism tries to find similar patterns to the wanted one.

The specification of the similar patterns is achieved using a control mechanism that uses as parameters the maximal tolerance concerning audibility and the number of bits each scale factor consists of. First, the sums of squares of the difference of the patterns of all the given scale factors with the wanted pattern are estimated and sorted in increasing order. Then, the pattern, which sum is at the beginning of the sorted list, is selected as candidate for changing. This pattern is checked by the control mechanism to be specified if the tolerance parameter is satisfied and the bit size of the underlying scale factor is not violated. If the check is successful, it is replaced by the wanted pattern. Then, it is checked if the minimal number of patterns parameter is satisfied. If not, the corresponding sum in the sorted list is replaced by a big value and the list of sums is sorted again and the same procedure is repeated.

This procedure can be considered as a one-step procedure, compared to the procedure in [5] where the searching for similar patterns consists

Figure 4. Watermark embedding

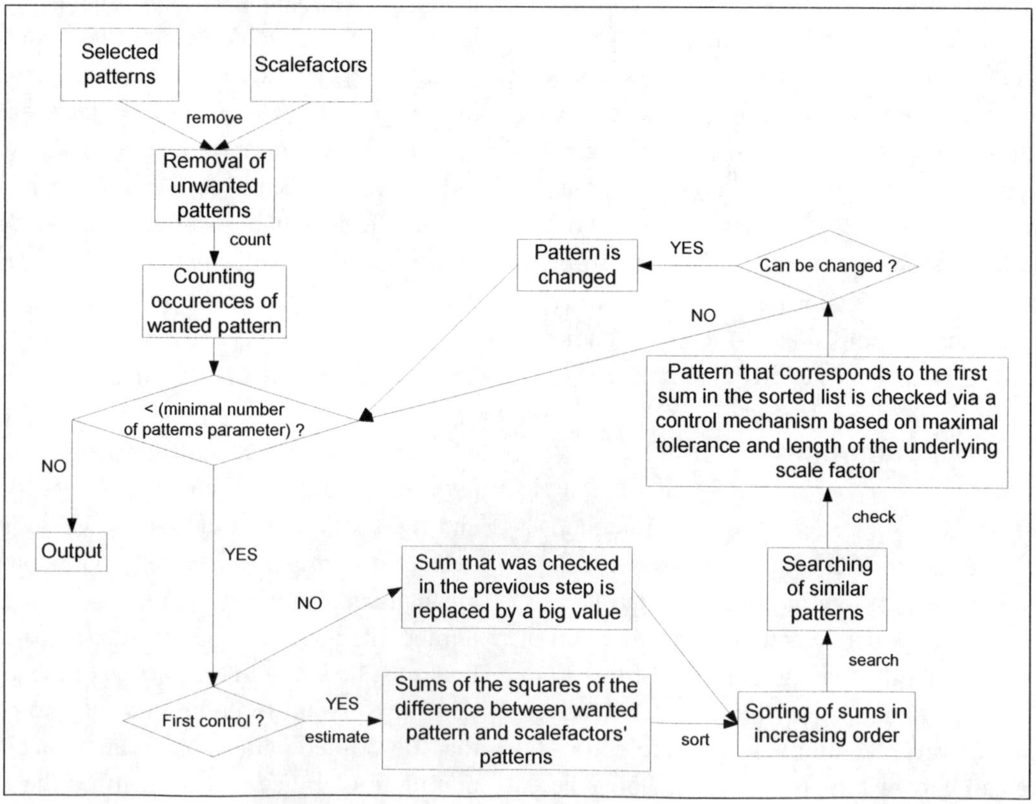

of a loop that is based on the one-by-one increase of the tolerance parameter until the maximal tolerance is reached. Thus, most of the changes will affect patterns that differ as little as possible from the wanted pattern, minimising the distortion in audio quality, whereas the number of bits of the underlying scale factors is not violated.

We should always take in care the variant feature of the number of bits that are used for the scale factors because otherwise the watermarked audio file can be damaged due to overwriting different data from scale factors or overlapping between scale factors. This may affect the audibility of the audio file because it can lead to pattern changes that are not at the lowest levels of tolerance. However, in practice with suitable selection of the set of three patterns, there is only a slight deterioration of the quality of the watermarked audio file.

Detection Mechanism

The detection mechanism depends basically on the method we choose to embed the watermarking key and the crypto mechanism with the help of which the semantically unique crypto-key is created. Thus, if a fixed number of frames has been used for the embedding of each information bit, the given fixed region of each embedding pattern, which is given as input to the detection mechanism, is searched and the presence of the three patterns is counted. The pattern with the most hits is selected. This method is sufficient when there is no trimming. If trimming exists, then the synchronisation bits should be used for the resynchronisation of the algorithm at the beginning of each watermark. Our algorithm handles this problem efficiently because the watermarking key is long enough to spread all over the scale factors.

If the number of frames/bit used for watermark embedding is not given, but we know the minimal number of equivalent consecutive patterns and the tolerance, then the detection procedure is more complicated because it demands the algorithm to search for the dominant of the three patterns in the frames with respect to the parameter of the minimal number of patterns. When the dominant pattern changes, then the previous dominant pattern is considered as a found one. In this method, the synchronisation bit-pattern is used for the separation of information bit-patterns along with weighting of the frames and filtering of very short dominance phases in order for our algorithm to be robust against noise and false detection. Also, the crypto key plays a big role towards the minimisation of noise and false detection artifacts because it is embedded all over the scale factors and it is robust against significant damages of it. If the file containing the graph (adjacency matrix) that is used for the creation of the watermarking key is attacked, then we expect that at least half of the embedded (in the scale factors) adjacency matrix will be left intact (assuming that each of its bits is flipped with probability ½). But even then, we can still show a 3-colouring (subset of the 3-colouring of the original graph) surviving, in this way, the attack.

IMPLEMENTATION SOLUTIONS OF OUR WATERMARKING SCHEME

The use of algorithms in a real-time system requires several modifications to have their needs to adjust to the limited resources of a computer system. Two crucial factors that should be taken into account are time and space. The delays, from which the used algorithms suffer, should be decreased and the memory requirements should be as small as possible. In this section, we will present the engineering solutions we gave to these problems in order to make our algorithm efficient for real-time application. Also, we present several issues that improve the audio quality.

In order to achieve the first goal, delays reduction, we should search for the components of the algorithm that are time consuming and then to see if they can be improved or not. It can be noticed that the software module, which is responsible for scale factors extraction from MPEG audio files, contributes a significant portion to the total delay, which depends on the file size. However, this delay cannot be decreased, as it is proportional to the real music time of an audio file. The module, which is responsible for the transformation of the watermarking key to a sequence of given patterns, is proportional to the key length, but it is very fast. Therefore, it contributes a minor portion to the total delay that is less than a second.

Another important module is the one that embeds the watermark to the scale factors of the audio file. The initial algorithm for watermark embedding in MPEG Layers 1 and 2 audio files that was proposed by Dittman et al. [5] consists of a loop that is used to find out similar patterns with the pattern that is wanted to be embedded in the scale factors patterns stream of the original audio file. In each round of this loop, the patterns of the scale factors are processed one by one for the estimation of the sums of squares difference with the wanted pattern. In the first round of the loop only patterns, which sum of squares difference to the wanted pattern is one, will be changed; in the next round, the used tolerance is increased until the given maximal tolerance or the minimal number of patterns is reached. This loop was found that delays the watermarking procedure unexpectedly because it cannot find the appropriate patterns fast enough. In order to decrease the delay, we tried to find a way to break the loop. The chosen solution improved the time duration of the whole procedure dramatically. The solution we gave is based on the idea that the patterns with the smallest sum of the squares difference with the wanted pattern can be found if we estimate the sum of squares differences of the patterns of all the given scale factors with the wanted pattern and sort them in increasing order. Then, the wanted patterns are the patterns at the beginning of the sorted list that can be picked if the tolerance parameter is not violated. This solution can be applied after the counting of wanted pattern occurrences in scale factors' patterns sequence for the location of similar patterns.

In order fore the watermark embedding module not to consume much time, we only change the scale factors of the frames that are enough to embed the watermark once. So, we save time not only at this module, but the software module that embeds the changed scale factors in the original audio file, too. The changed scale factors embedding procedure in the source audio file is limited only to the embedding of the changed scale factors frames. Considering the watermark detection part of our system, it does not need the original audio file to extract the embedding watermark and it is very fast as the only consuming part of it, the scale factors extraction mechanism, is applied.

Another factor that affects the delay of the system is the selection of patterns for the representation of binary digits '0' and '1' and the synchronisation bit. The selected patterns should appear rare in the audio stream. Especially, care should be taken when MP3 audio files are watermarked where scale factors have variant sizes. An improper selection of patterns in that case could result in a big delay for finding the appropriate scale factors for modification in order for the watermark to be embedded. For example, if the chosen pattern has in its triplet a number bigger than four and the scale factors have sizes one or two bits in many consecutive frames, it will have as a result a big delay. In order to handle this problem in the MP3 case, the patterns that have been selected are triplets that consist of numbers close to zero.

Moreover, the predefined number of frames for the embedding of a bit's pattern plays a big role in system delay. If we embed each bit in a small number of frames, the system will be fast but the audio quality will be low because the changed factors will be concentrated in a small region. If we choose a big number of frames to embed our bit then we will face bigger delays because we will have many more scale factors to process in each step. So, an appropriate trade-off should be found. After exhaustive testing, we decided that a good trade-off for fast system execution and acceptable audio quality in the resulted watermarked audio file is a selection of 5 to 15 frames for the embedding of each bit. Also, we found out that in the MP3 case, we can use less number of frames to embed each bit than in the other cases.

Concerning the space requirements, we introduce a mechanism in the scale factors extraction procedure and in the changed scale factors embedding procedure that permits the process of audio data in small portions of 9800 bytes in each step. Thus, we do not need to preserve a big array for storing audio data and other data that are required for the needed manipulations. With this way, we make our system independent of the memory resources of the system, in which it is executed, as it needs to use small amounts of memory.

Another important issue is the processing of audio data in MP3 audio files for watermark embedding and detection. The problem starts from the fact that audio data of a frame are not located in the frame as happens in MPEG Layers 1 and 2. They can be in the audio data field of previous frames. The solution we selected is the loading of audio data in a special buffer and their processing as a whole. Also, we use a special mechanism for the specification of the starting and ending byte of each scale factors region of each frame because the scale factors of a frame can be spread in many frames and the audio data field that contains scale factors has a special organisation in granules and channels where Huffman bits intervene between scale factors areas.

Finally, it should be mentioned that the watermarking algorithm for MP3 audio files that is proposed here is the first compressed-domain watermarking algorithm proposed for MP3 audio files. In its design and implementation, we faced a lot of problem mainly because of the different encoding it uses for audio data and scale factors respectively. Furthermore, another basic problem is owed to the variant nature of scale factors size. Scale factors have 0 to 4 bits size. This fact differentiates mainly the scale factors extraction algorithm and watermark-embedding algorithm. In the scale factors extraction phase, not only the scale factors should be extracted but their sizes should be extracted, too. Also, in the watermark embedding phase, when we search for similar patterns in scale factors pattern stream, we should check not only the sum of squares difference to be as small as possible for audio quality preservation but, also, we should control the size of scale factors that are proposed to be changed not to be violated. This is done using a special control mechanism.

EVALUATION RESULTS OF OUR AUDIO WATERMARKING SCHEME

Evaluation

In the context of this algorithm a lot of issues arise. The first of them is how somebody should choose the appropriate group of patterns to represent the information bits and the synchronisation bit in order to enhance the security of the watermark and the inaudibility of watermark embedding. Basically, there are two rules for this selection that became evident after experimental evaluation. The first rule states that patterns, which occur often in the context of scale factors, should not be selected, and the second one that patterns with larger steps than two among the components of their triple should be preferred. Also, when multiple keys are inserted inside the scale factor stream, then the used patterns in different keys should differ as much as possible. Our algorithm is more secure than others with similar characteristics because it uses a key with semantic meaning. Thus, even if somebody guesses correctly the used patterns, he will not be able to *deduce* the semantics of the key.

Another important issue is that of robustness. Our algorithm operates at the compressed domain and not at the time domain. That is, we embed the watermark bits in the scale factors regions of the frames of an MP3 audio file that are special fields inside the audio data stream and not in the original raw audio data. Therefore, our algorithm is robust against attacks in audio data. Also, it is robust against local attacks or even random attacks to the whole region of scale factors because the watermarking key patterns are distributed over the whole range of subbands and the key allows its detection even if a percentage of it has been damaged. Such attacks are not really a danger because they lead to audible distortions and damage the audio quality. We should mention that, as far as time-domain attacks that are applied to raw audio data, they require the watermarked MP3 audio file to be first decompressed. But, our algorithm cannot handle the decoding of the MP3 audio file and the recoding of it. However, the malicious users do not prefer this kind of attack, because it leads to a serious loss of quality. This occurs because in a watermarked MP3 audio file, we have modified some of the scale factors, which are used for the production of the real audio data at the decompression phase. This modification precludes decompression without affecting the sound quality. So in this sense, we can claim that our algorithm is robust against such attacks. Another kind of attack is the creation of a mono channel from the two stereo channels. This attack is handled efficiently because the patterns in the time axis survive this attack even if the patterns over the subband axis are damaged. Finally, in the context of robustness, we should mention that our algorithm could be used both for authentication and copyright protection. This is because of the nature of our watermarking key, which is a graph with a known 3-colouring, that can survive against inversion attacks (a malicious user subtracts the watermark from the original marked by the true owner) because it is spread all over the scale factors range and the colouring of even a small part of it can be used by the legal owner for proving ownership of the file that contains it.

Our algorithm has been tested in laboratory conditions where has been proved that the watermark embedding inserts a slight distortion but not a strong one. It should be mentioned that the distortion is more obvious in the case of spoken poems when there is no background sound. Our algorithm can be used for online distribution of audio files and in CD-ROMS and DVD from music industry for authentication and copy-protection purposes because it needs only small transfer rates for online use and the used calculations have low complexity because they are just additions or subtractions on integers or bytes. Also, the key's nature makes it essential for copyright protection.

The audio files that have been used in our experiments are:

- **Youthinasia:** Disco music (44.1 kHz, 128 kbps)
- **Dream:** Male ethnic singing with native instruments (44.1 kHz, 128 kbps)
- **Ipomoni:** Greek folk music (44.1 kHz, 128 kbps)
- **Blue:** Greek pop music (44.1 kHz, 96 kbps)

The following experiments (time tests) have been conducted in a Pentium 4 at 3.4 GHz. Repeating these time tests in PCs with different processor speeds from 733 MHz to 3.1 GHz, it was observed that the estimated times (embedding-detection time) are proportional to the raw processor speed. All the times that have been presented in the following experiments are in the same range as the real music time on the described platform. Furthermore, we should mention that these times refer to the users of our algorithm that want to make watermark embedding and detection, and not to the audience of watermarked audio files who do not face any artifacts or delay hearing the watermarked audio files.

Table 1 shows that the perceived quality of watermarked files is approximately the same as the quality loss produced by MPEG-compression. Most results are in the range of two, which means a difference like between two stereo-sets.

Table 2 gives some test results of the time needed in our system to embed a watermark in MP3 audio files. Again the scale factors extraction component of the system is the most time consuming. We observe that less bit rate leads to more frames, which results in the encoding of more audio data. That's way the Blue audio file has more real music time than Ipomoni while it has less size.

Another important part of our algorithm is the watermark detection mechanism that suffers from the significant time needed to extract the scale factors from the original file. Table 3 gives some time results of the detection procedure along with the percentage of successful watermark detection.

Except from the above basic time tests, we used the Dream audio file to study the impact of the use of different watermarking keys in the same audio file on watermark embedding and detection time (Table 4).

In Table 4 we can observe that using different one-character keys (9-bit watermarks-8 bits the character + 1 the synchronisation bit) there is a slight difference in watermark detection time; while using a two-characters key (17-bit

Table 1. Averaged test results

Example	MPEG	1 wm
Youthinasia	1,5	1,6
Dream	1,5	1,6
Ipomoni	1,5	1,6
Blue	1,5	1,6

Table 2. Watermark embedding time tests

File Name-Size	Embedding Time (sec)	Real Music Time (sec)
Youthinasia-103Kb	6	6
Dream-1,3Mb	55	78
Ipomoni-2,9Mb	128	178
Blue-2,2Mb	92	180

Table 3. Watermark detection time tests

File Name-Size	Detection Time (sec)	Detection (%)
Youthinasia-103Kb	4,5	100
Dream-1,3Mb	54	100
Ipomoni-2,9Mb	108	100
Blue-2,2Mb	90	100

Table 4. Time tests with different watermarking keys

Dream	Embedding Time (sec)	Detection Time (sec)
One-character watermark (w)	57	54
One-character watermark (p)	56	54
Two-characters watermark (re)	53	54

watermarks), the detection time becomes bigger than the embedding time. These results are logical if we have in mind the embedding algorithm that searches for similar patterns to use them for watermark embedding. Therefore, it is logical for different keys to have different probabilities of finding similar patterns in the audio stream. The second result, in which the use of a two-characters key results in more detection time than embedding time, comes to strengthen the previous conjecture. This implies that the selection of the watermarking key is more crucial for watermark embedding/detection time than the size of the key.

CONCLUSION

The watermarking scheme that was presented in this chapter adopts and integrates in the audio watermarking domain concepts whose security and efficiency is well-known in cryptography. This gives the users the option of proving ownership of the audio file by showing that they know a property of the watermarking string. Also, the scheme is not computation or space demanding and, thus, it offers the capability of watermark embedding and detection in real time, with no detectable distortion of the sound.

By now, the interplay among all the concepts outlined in this chapter, towards the goal proving copyright ownership, should be apparent. First, each person (*P*) first acquires (from a trusted source such as the Certification Authority of a Public Key Infrastructure) a sufficiently large graph, which becomes *P*'s public key, that is, the entity that represents *P* in all transactions. At the same time, *P* acquires (from the trusted source again) a 3-coloring of the graph, that can be thought of as an analogous to a secret key in a public key cryptosystem and should not be publicised. The graph assigned to each person *P* should be constructed so as to have some 'hardness properties' that will prevent a forgerer from discovering easily a 3-coloring, impersonating *P*. Second, each time *P* needs to prove ownership of an audio file, *P* becomes the prover in the protocol described earlier in the chapter and the other person becomes the verifier *V*.

Of course, the above brief description is too general and avoids many important technological issues. For instance, where is the 3-coloring stored? USB token, smart card, diskette, and so forth. How is the 3-coloring extracted from the place it is stored in order to be used in the ZKIP? How can the ZKIP avoid the creation of a huge number of encryptions, one for each graph vertex? How large a graph can be stored in an audio file of high quality? And the list is endless. We believe that these issues, although important in practice, can always be settled given the advances in hardware as well as control software. On the other hand, these questions provide a set of future research issues to tackle in order to provide the watermarking domain with practical robust watermarking signals with *provable* cryptographic properties.

Apart from purely technological issues mentioned above, there is another, more important one that is also an open question in Computational Complexity Theory: the *characterisation* of hard instances. Until now, all such characterisations are either of a heuristic nature or too theoretical to be used in practice and, thus, it is not possible to guarantee 'hardness' using an algorithmic method.

REFERENCES

Achlioptas, D., Gomez, C., Kautz, H., & Selman, B. (2000). Generating satisfiable problem instances. In *AAAI*.

Ajtai, M. (1996). *Generating hard instances of lattice problems* (ECCC Report TR96-007). Electronic Colloquium on Computational Complexity.

Antonopoulou, H. (2002). A user authentication protocol based on the intractability of the 3-COLORING problem. *Journal of Discrete mathematical Sciences & Cryptography, 5*(1), 17-21.

Armeni, S., Christodoulakis, D., Kostopoulos, I., Kountrias, P. D., Stamatiou, Y. C., & Xenos, M. (2003). An information hiding method based on computational intractable problems. In Y. Manolopoulos & S. Evripidou (Eds.), *8th Panhellenic Conference on Informatics Advances in Informatics* (Vol. 2563, pp. 262-278). Springer-Verlag.

Armeni, S., Christodoulakis, D., Kostopoulos, I., Kountrias, P. D., Stamatiou, Y. C., & Xenos, M. (2003). Secure information hiding based on computationally intractable problems. *Journal of Discrete Mathematical Sciences & Cryptography, 6*(1), 21-33.

Arnold, M., & Kanka, S. (1999). *MP3 robust audio watermarking.* Paper presented at the DFG VIIDII Watermarking Workshop, Erlangen, Germany.

Arttameeyanant, P., Kumhom, P., & Chamnongthai, K. (2002). *Audio watermarking for Internet.* Paper presented at the IEEE ICIT'02 (Vol. 2, pp. 976-979).

Babai, L. (1985). *Trading group theory for randomness.* Paper presented at the 17th Annual ACM Symposium on Theory of Computing (pp. 421-420).

Basia, V., Pitas, I., & Nikolaidis, N. (2001). Robust audio watermarking in the time-domain. *IEEE Transactions on Multimedia, 3*(2), 232-241.

Bender, W., Gruhl, D., Morimoto, N., & Lu, A. (1996). Techniques for data hiding. *IBM Systems Journal, 35*(3/4), 313-336.

Boney, L., Tewfic, A., & Hamdy, K. (1996). *Digital watermarks for audio signals.* Paper presented at the IEEE International Conference on Multimedia Computing and Systems (pp. 473-480).

Brandenburg, K. (1999). *MP3 and AAC explained.* Paper presented at the AES 17th International Conference on High Quality Audio Coding.

Chaitin, G.H. (1966). On the length of programs for computing finite binary sequences. *Journal of the ACM, 13*, 547-570.

Cheeseman, P., Kanefsky, B., & Taylor, W.M. (1991). *Where the really hard problems are.* Paper presented at the International Joint Conference on Artificial Intelligence (pp. 331-337).

Chvátal, V., & Reed, B. (1992). *Mick gets some (the odds are on his side).* Paper presented at the 33rd IEEE Annual Symposium on Foundations of Computer Science (pp. 620-627).

Cook, S. (1971). *The complexity of theorem-proving procedures.* Paper presented at the 3rd Annual ACM Symposium on Theory of Computing (pp. 151-158).

Dittmann, J., Steinebach, M., & Steinmetz, R. (2001). *Digital watermarking for MPEG audio layer 2.* Paper presented at the Multimedia and Security Workshop at ACM Multimedia.

Garey, M.R., & Johnson, D.S. (1979). *Computers and Intractability, a guide to the theory of NP-completeness.* W.H. Freeman and Company.

Goerdt, A. (1996). A threshold for unsatisfiability. *Journal of Computer and System Sciences, 53*, 469-486.

Goldreich, O. (1995). Randomness, interactive proofs and zero-knowledge: A survey. In R. Herken (Ed.), *The universal turing machine: A half-century survey.* Springer-Verlag.

Goldreich, O., Micali, S., & Widgerson, A. (1991). Proofs that yield nothing but their validity or all languages in NP have Zero-Knowledge proof systems. *Journal of the ACM, 38*(1), 691-729.

Goldwasser, S., Micali, S., & Rackoff, C. (1985). *The knowledge complexity of interactive proof-*

systems. Paper presented at the 17th Annual ACM Symposium on Theory of Computing (pp. 291-304).

Herken, R. (1995). *The universal turing machine: A half-century survey.* Springer-Verlag.

ISO/IEC 11172-3. Coding of moving pictures and associated audio for digital storage media at up to about 1.5 Mbit/s-Part 3.

Kaporis, A.C., Kirousis, L.M., & Stamatiou, Y.C. (2000). A note on the non-colorability threshold of a random graph. *Electronic Journal of Combinatorics, 7*(R29).

Karp, R.M. (1972). Reducibility among combinatorial problems. In Miller & Thatcher (Eds.), *Complexity of computer computations* (pp. 85-103). Plenum Press.

Kirkpatrick, S., & Selman, B. (1994). Critical behavior in the satisfiability of random Boolean expressions. *Science, 264*, 1297-1301.

Kolmogorov, A. (1965). Three approaches to the concept of the amount of information. *Probl. of Inform. Transm, 1*(1).

Koukopoulos, D., & Stamatiou, Y.C. (2001). *A compressed domain watermarking algorithm for MPEG Layer 3.* Paper presented at the Multimedia and Security Workshop at ACM Multimedia (pp. 7-10). ACM Press.

Koukopoulos, D., & Stamatiou, Y. C. (2005). A watermarking scheme for MP3 audio files. *International Journal of Signal Processing, 2*(3), 206-213.

Levin, L. (1973). Universal'nyĭe perebornyĭe zadachi [Universal search problems]. *Problemy Peredachi Informatsii, 9*(3), 265-266.

Levin, L. A. (1986). Average case complete problems. *SIAM Journal on Computing, 15*, 285-286.

Li, X., Zhang, M., & Zhang, R. (2004). *A new adaptive audio watermarking algorithm.* Paper presented at the Fifth World Congress, Intelligent Control and Automation (Vol. 5, pp. 4357-4361).

Mitchell, D., Selman, B., & Levesque, H. (1992). *Hard and easy distributions of SAT problems.* Paper presented at the Tenth National Conference on Artificial Intelligence (pp. 459-465).

Motoki, M., & Uehara, R. (1999). Unique solution instance generation for the 3-satisfiability (3SAT) problem (Tech. Rep. No. C-129). Japan, Sciences Tokyo Institute of Technology, Dept. of Math. and Comp.

Motwani, R., & Raghavan, P. (1995). *Randomized algorithms.* Cambridge University Press.

Papadimitriou, C.H. (1994). *Computational complexity.* Addison-Wesley.

Petitcolas, F. (1998). *MP3Stego.* Computer Laboratory, Cambridge.

Qiao, L., & Nahrstedt, K. (1998). *Non-invertible watermarking methods for MPEG video and audio.* Paper presented at the Multimedia and Security Workshop at ACM Multimedia (pp. 93-98).

Sandford, S. et.al. (1997). *Compression Embedding.* [US Patent 5,778,102].

Selman, B. (1995). *Stochastic search and phase transitions.* Paper presented at the International Joint Conference on Artificial Intelligence (Vol. 2, 998-1002).

Seok, J., Hong, J., & Kim, J. (2002). A novel audio watermarking algorithm for copyright protection of digital audio. *ETRI Journal, 24*(3), 181-189.

Solomonoff, R.J (1964). A formal theory of inductive inference. *Information and Control, 7*(1), 1-22.

Stamatiou, Y.C. (2003). Threshold phenomena: The computer scientist's viewpoint. *European Association of Theoretical Computer Science Bulletin (EATCS), 80*, 199-234.

Yeo, I.K., & Kim, H. J. (2003). Modified patchwork algorithm: A novel audio watermarking scheme. *IEEE Transactions, Speech and Audio Processing, 11,* 381-386.

Additional Reading

Asarin, E.A. (1998). On some properties of finite objects random in the algorithmic sense. *Soviet Math. Dokl., 36,* 109-112.

Balcázar, J.L., Diaz, J., & Gabarró, J. (1988). *Structural complexity.* Springer-Verlag.

Biddle, P., England, P., Peinado, M., & Willman, B. (2003). The darknet and the future of content protection. *Digital rights management: Technological, economic, legal and political aspects* (pp. 344-365). Springer.

Buchheit, M., & Kügler, R. (2004). *Secure music content standard: Content protection with codeMeter.* Paper presented at the Virtual Goods 2004 - International Workshop for Technology, Economy, Social and Legal Aspects of Virtual Goods, Germany.

Calude, C. (1988). *Theories of computational complexity* (Chapter 4). North Holland.

Chaitin,G. J. (1987). *Algorithmic information theory.* Cambridge University Press.

Cheng, S., Yu, H., & Xiong, Z. (2002). *Enhanced spread spectrum watermarking of MPEG-2 AAC audio.* Paper presented at the IEEE International Conference on Acoustics, Speech, and Signal Processing (pp. 3728-3731).

Cox, Miller, & Bloom. (2002). *Digital watermarking.* Academic Press.

Daley, R.P. (1980). Qualitative and quantitative information in computation. *Inform. Contr., 45*, 236-244.

Egidi, L., & Furini, M. (2005, May). Bringing multimedia contents into MP3 files. *IEEE Communications Magazine.*

Erdös, P., & Spencer, J. (1974). *Probabilistic methods in combinatorics.* Academic Press.

Gács, P. (1983). On the relation between descriptional complexity and algorithmic probability. *Theoretical Computer Science, 22*, 71-93.

Gács, P., & Chaitin, G.J. (1989). Algorithmic information theory. *Journal Symbolic Logic, 54*, 624-627.

Gang, L., Akansu, A., Ramkumar, M., & Xie, X. (2001). *Online music protection and MP3 compression.* Paper presented at the International Symposium on Intelligent Multimedia, Video and Speech Processing (pp. 13-16).

ISO/IEC 13818. Coding of moving pictures and associated audio (MPEG-2).

Li, M., & Vitányi, P. (1993). *An introduction to Kolmogorov complexity and its applications.* Springer Verlag.

Lookabaugh, T., Vedula, I., & Sicker, D. (2003). *Security analysis of selectively encrypted MPEG-2 streams.* Paper presented at SPIE – Multimedia Systems and Applications VI.

Loveland, D.W. (1969). A variant of the Kolmogorov complexity concept of complexity. *Inform. Contr., 15*, 510-526.

Martin-Löf, P. (1966). The definition of random sequences. *Inform. Contr., 9*, 602-619.

Martin-Löf, P. (1966). On the concept of a random sequence. *Theory Probability Appl., 11*, 177-179.

Moghadam, N., & Sadeghi, H. (2005). Genetic content-based MP3 audio watermarking in MDCT domain. Paper presented at the *IEC* (pp. 348-351), Prague.

Noll, P. (1997). MPEG digital audio coding. *IEEE Signal Processing Magazine*, 59-81.

Page, T. (1998). Digital watermarking as a form of copyright rrotection. *Computer Law & Security Report.* Elsevier.

Pan, D. (1995). A tutorial on MPEG audio compression. *IEEE Multimedia, 2*(2), 60-74.

Podilchuk, C., & Delp, E. (2001). Digital watermarking, algorithms and applications. *IEEE Signal Processing Magazine*, 33-46.

Simitopoulos, D., Tsaftaris, S., Boulgouris, N., Briassouli, A., & Strintzis, M. (2004). Fast watermarking of MPEG-1/2 streams using compressed-domain perceptual embedding and a generalized correlator detector. *EURASIP Journal on Applied Signal Processing.*

Sipser, M. (1983). *A complexity theoretic approach to randomness.* Paper presented at the *15th* ACM Symposium Theory of Computing (pp. 330-335).

Steinebach, M., & Dittmann, J. (2003). *Capacity-optimized mp2audio watermarking.* Paper presented at the SPIE- Security and watermarking of Multimedia Contents V (pp. 44-54).

Thorwirth, N., Horvatic, P., Zhao, J., & Weis, R. (2000). *Security methods for MP3 music delivery.* Paper presented at the Asilomar Conference on Signals, Systems and Computers (pp. 1831-1835).

Van Lambalgen, M. (1989). Algorithmic information theory. *Journal Symbolic Logic, 54*, 1389-1400.

Vovk, V.G. (1987). On a randomness criterion. *Soviet Math. Dokl., 35*, 656-660.

Wang, C., Chen, T., & Chao, W. (2004). *A new audio watermarking based on modified descrete cosine transform of MPEG audio layer III.* Paper presented at the 2004 IEEE International Conference on Networking, Sensing and Control.

Watanabe, O. (1992). *Kolmogorov complexity and computational complexity.* Springer-Verlag.

Xu, C., & Feng, D. (2000). *Robust and efficient content-based digital audio watermarking.* Paper presented at the International Conference of Multimedia Systems.

Zurek, W.H. (1989). Algorithmic randomness and physical entropy. *Physical Review: Series A, 40*(8), 4731-4751.

Zurek, W.H. (Ed.). (1991). *Complexity, entropy and the physics of information.* Addison-Wesley.

Zvonkin, A.K., & Levin, L.A. (1970). The complexity of finite objects and the development of the concepts of information and randomness by means of the theory of algorithms. *Russian Math. Surveys, 25*(6), 83-124.

Chapter XI
Digital Watermarking of Speech Signals

Aparna Gurijala
Michigan State University, USA

ABSTRACT

The main objective of the chapter is to provide an overview of existing speech watermarking technology and to demonstrate the importance of speech processing concepts for the design and evaluation of watermarking algorithms. This chapter describes the factors to be considered while designing speech watermarking algorithms, including the choice of the domain and speech features for watermarking, watermarked signal fidelity, watermark robustness, data payload, security, and watermarking applications. The chapter presents several state-of-the-art robust and fragile speech watermarking algorithms and discusses their advantages and disadvantages.

INTRODUCTION

Digital watermarking is the process of embedding data (comprising the *watermark*), ideally imperceptibly, into a host signal (the *coversignal*) to create a *stegosignal*. Although the term "coversignal" is commonly used in watermarking literature (Cox, Miller, & Bloom, 2002) to denote the host signal (data to be protected), the name used for the watermarked result, the "stegosignal," is borrowed from steganography (Johnson, Duric, & Jajodia, 2000). In the last decade many algorithms have been proposed for multimedia watermarking. Early work emphasized watermarking algorithms that could be universally applied to a wide spectrum of multimedia content, including images, video, and audio. This versatility was deemed conducive to the implementation of multimedia watermarking on common hardware (Cox, Kilian, Leighton, & Shamoon, 1997). However, many watermarking applications, such as copyright protection for digital speech libraries (Ruiz & Deller, 2000), embedding patient information in medical records (Anand & Niranjan, 1998; Miaou,

Hsu, Tsai, & Chao, 2000), or television broadcast monitoring (Kalker, Depovere, Haitsma, & Maes, 1999), involve embedding information into a single medium. Also, the attacks and inherent processing distortions vary depending on the nature of the multimedia content. For example, an attack on watermarked images may involve rotation and translation operations to disable watermark detection. However, such an attack is not applicable to audio data. Watermarking algorithms that are specifically designed for particular multimedia content can exploit well-understood properties of that content to better satisfy the robustness, fidelity, and data-payload constraints. Unlike general audio, speech is characterized by intermittent periods of voiced (periodic) and unvoiced (noise-like) sounds. Speech signals are characterized by a relatively narrow bandwidth, with most information present below 4 kHz. Broadband audio watermarking algorithms involving human auditory system (HAS) based perceptual models (Boney, Tewfik, & Hamdy, 1996) may not be effective for speech. Also, well-established analytical models for speech production exist (Deller, Hansen, & Proakis, 2000) and can be effectively exploited in the watermarking process.

The main objective of the present chapter is to provide a comprehensive overview of the factors to be considered while designing a speech watermarking system and the typical approaches employed by existing watermarking algorithms. The various speech watermarking applications and the algorithmic requirements will also be described in detail. The speech watermarking techniques presented in this chapter are presently deployed in a wide range of applications including copyright protection, copy control, broadcast monitoring, authentication, and air traffic control. The algorithms presented in this chapter are classified into robust watermarking and fragile watermarking categories according to the intended application. Furthermore, the chapter will describe the (commonly employed) signal processing, geometric, and protocol attacks on speech watermarking techniques. The chapter also discusses existing methods for objectively evaluating and controlling the quality/fidelity of watermarked speech.

Figure 1. A general watermarking system

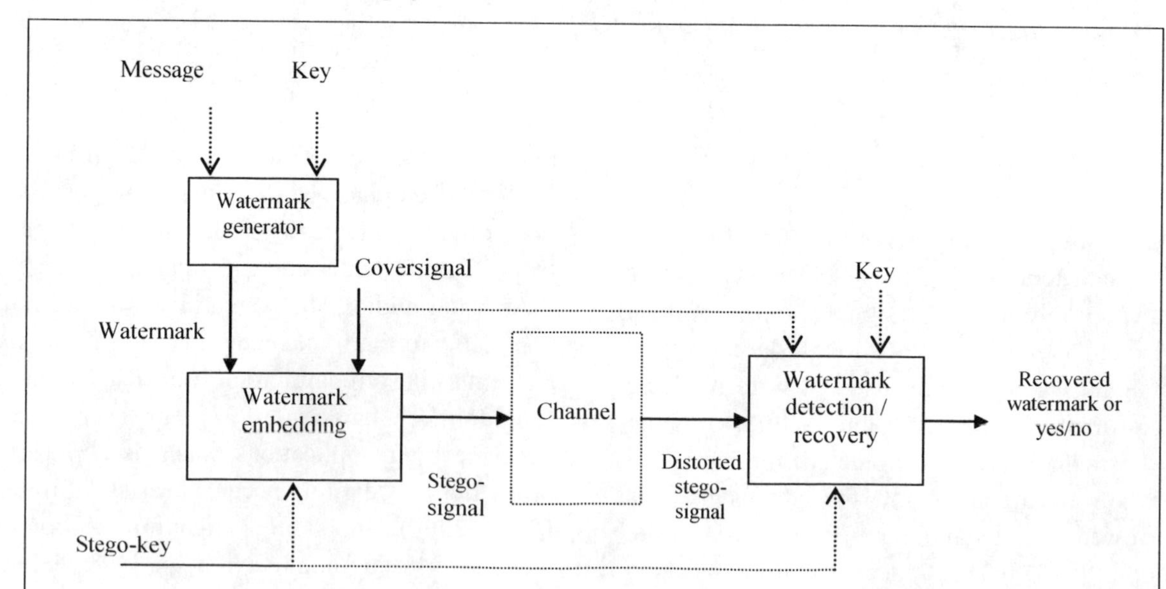

BACKGROUND

A general speech watermarking system consists of a watermark generator, a watermark embedding component, and a watermark detector as depicted in Figure 1. The transmission channel represents possible attacks or distortion on the stegosignal from the time the watermark is embedded until there is a need for watermark detection or recovery. Certain inputs are indicated by dotted lines in Figure 1, meaning that they may not be present in all techniques. There are two kinds of keys that may be used in a watermarking system. The watermark key (key in Figure 1) is used to encrypt the watermark so that the message contained in the watermark is not obvious. In Figure 1, the stego-key determines the locations in the coversignal where the watermark is to be embedded and thus introduces additional protection by making the watermark's location unknown. A watermark may take different forms: an encrypted or modulated speech sequence or an image, a pseudo-random sequence, a sequence of symbols mapped from a message, and so on. Consequently, the inputs to watermark generators are highly diverse.

Speech watermarking techniques may be additive, multiplicative, or quantization-based, and the watermarks may be embedded in the time domain, or in a transform domain. Each technical variation tends to be more robust to some forms of attack than to others, and for this and other application-specific reasons, particular strategies may be better suited to certain tasks. Watermark detection (or recovery) is classified as being *blind* or *informed.* In informed detection, a copy of the unmarked host signal is required. Some informed watermark detectors require only partial information about the host signal. Detectors that do not require a representation of the host signal for watermark detection are called blind detectors. The additional information available during watermark detection and recovery in informed detection can be used to better satisfy watermarking requirements. For example, the coversignal information available at the detector can serve as a registration pattern to undo any temporal or geometric distortions of the host signal as it appears in the stegosignal (Cox et al., 2002). In case of a "cropping attack," wherein speech samples are randomly deleted, a dynamic programming algorithm can be used for watermark recovery from the desynchronized stegosignal (Gurijala & Deller, 2001). On the other hand, blind watermarking can be deployed for a wider range of applications. For applications in which watermark detection must be performed by the public, not just by the content owner or distributors, informed watermarking is inapplicable.

A common approach to additive watermark detection employs classic binary decision theory (Poor, 1994). The hypotheses are $H_0 : Z_R = X$ and $H_1 : Z_R = X + W$ where Z_R is the received speech, X is the original speech, and W is the watermark signal (Barni, Bartolini, Rosa, & Piva, 2003; Hernandez, Amado, & Perez-Gonzalez, 2000). A Bayesian or Neyman-Pearson paradigm is followed in deriving the detection thresholds. Many of these approaches do not consider the effect of noise while designing the detector. Several watermark detectors are based on correlation detection in the time or in transform domain (Linnartz, Kalker, & Depovere, 1998; Miller & Bloom, 1999). That is, the correlation between the original and recovered watermarks, or the correlation between the original watermark and the possibly distorted stegosignal (the *recovered speech*), is compared against a threshold. Correlation detectors are optimal when the watermark and noise are jointly Gaussian, or, in case of blind detectors, when the watermarked signal and noise are jointly Gaussian. This is true for watermark patterns that are spectrally white.

The design of a watermarking system involves balancing several conflicting requirements. Watermark robustness, stegosignal fidelity, watermark data payload, and security are important requirements of any watermarking system. First, embedded watermarks must be imperceptible, that is, the stegosignal must be of high *fidelity*. Second, watermarks must be *robust*, that is, they must be able to survive *attacks* (Voloshynovskiy, Pereira, Pun, Su, & Eggers, 2001); those deliberately designed to destroy or remove them, as well as distortions inadvertently imposed by technical processes (e.g., compression) or by systemic processes (e.g., channel noise). These fidelity and robustness criteria are generally competing, as greater robustness requires more watermark energy and more manipulation of the coversignal, which, in turn, lead to noticeable distortion of the original content. Related measures of a watermark's efficacy include *data payload* (the number of watermark bits per unit of time) (Cox et al., 2002), and watermark *security* (the inherent protection against unauthorized removal, embedding or detection) (Cox et al., 2002). In fact, when stegosignal fidelity parameter is fixed, robustness and data payload cannot be increased simultaneously (Cvejic & Seppänen, 2003). A trade-off also exists between watermark security and stegosignal fidelity as evidenced by the fact that audible watermarks tend to be less secure. Audible watermarks are easily identifiable and this facilitates their removal from the speech signal. The computational complexity of watermarking algorithms is also a factor for consideration, especially for real-time applications.

Speech Watermarking Applications

Speech watermarking applications may be based on embedding either robust or fragile watermarks. Speech watermarking can be used for conventional watermarking applications such as content authentication, broadcast monitoring, content management, copy control, copyright protection, and transaction tracking. Speech watermarking is currently used for air traffic control (Hagmüller, Horst, Kröpfl, & Kubin, 2004) and to provide time-stamp information for speaker verification (Faundez-Zanuy, Hagmuller, Kubin, & Kleijn, 2005). Additionally, watermarking requirements vary depending on the application(s) in which they are employed. For example, in many authentication applications, watermark robustness to compression is desired, while robustness to high-level information tampering is not required.

- **Content authentication:** Watermarking is used for establishing the authenticity of speech content. Traditional cryptographic authentication schemes involve hash functions and are used to ensure that every bit in the data stream is unmodified. In case of speech signals, minor changes can often be made to their content without affecting the perceptual quality. Moreover, multimedia content is typically subjected to nonmalicious, yet distorting, signal-processing operations and channel noise, which result in minor perceptual changes to the content. For speech authentication, fragile watermarking provides a flexible alternative to techniques based on cryptographic hash functions or digital signatures. Authentication is performed by determining whether certain defining characteristics and rules present in the original signal exist in the test signal (Wu & Liu, 2004). Two important issues to be addressed while designing an authentication system include (a) choice of features to embed, and (b) security considerations to prevent forgery or manipulation of embedded data (Wu & Liu, 2004).
  In authentication watermarking systems, another important issue is the ability of the algorithm to pinpoint the areas of the signal where tampering has occurred.
- **Broadcast monitoring:** Digital watermarking is used for tracking the broadcast and distribution of information over television, radio, and the Internet. In broadcast monitoring applications, the content is embedded with a unique identification tag and optional information about the content owner or

distributor. Such a scheme can be used for automatically monitoring the broadcast of advertisements by a particular television station, for example. A watermark decoder determines the number of times an advertisement was broadcast and whether the advertisement was broadcast at the contracted time slot. This information is conveyed to the content owner or distributor. Digital watermarking for broadcast monitoring helps ensure broadcast compliance and detects unauthorized use of content.

- **Content management:** Digital watermarks can be used as tags containing identifying or supporting information for content management. In this case, the embedded digital watermarks may link to a central server or remote system containing metadata, keywords, rights and permissions, and other pertinent information. This application of watermarking will greatly facilitate content management in environments such as digital libraries where large databases are to be handled.
- **Copy control:** Watermarks can be used for preventing unauthorized copying and subsequent distribution of digital content. Attempts to integrate digital watermarking technology into DVD and CD writers have been variously successful. To prevent unauthorized DVD copying, watermarks containing copy-control instructions are embedded into the content. As an example, the watermark may include instructions indicating whether or not a single copy of the content may be made, or if not further copying is permitted. This application of watermarking is of great interest to the movie and music industries for the prevention of piracy.
- **Copyright protection:** Traditional copyright protection mechanisms for analog content are generally inapplicable to digital content. Hence, watermarking of digital content for the communication of copyright and ownership information has been proposed as a solution. Watermarking facilitates detection of unauthorized use of the content, and as witnessed by the recent Napster case, there is evidence that the courts will consider watermarks as a legitimate of copyright protection (Seadle, Deller, & Gurijala, 2002). Additionally, watermarking is also deployed for owner identification (Cox et al., 2002). Application of digital watermarking for copyright protection has always been a controversial issue. This is mainly because such an application entails robustness to a wide range of attacks that no existing (perhaps future) watermarking algorithm exhibits.
- **Transaction tracking:** Digital watermarks are also used for transaction and traitor tracking. Every copy of the digital content is securely embedded with unique identification information before distribution. If an unauthorized copy of the content is found, the extracted watermarks will track the last authorized recipient.
- **Air traffic control:** Speech watermarking has also been developed for air traffic control (Hagmüller et al., 2004). In an air traffic control environment, there are several aircrafts communicating with the controller in a single very high frequency (VHF) channel. The pilot of an aircraft starts the communication by indicating the aircraft call sign. Generally, the aircraft registration number serves as the call sign. There is potential for confusion if two flights on the same VHF channel at any time have similar sounding call signs. By hiding unique information about an aircraft in the voice message, any ambiguity over aircraft identification is prevented. Speech watermarking is thus used to provide automatic identification of the aircraft.

- **Biometrics security:** Speech watermarking can be used in conjunction with a speaker verification system for enhanced security. A potential attack against biometric systems is the replay attack wherein digitally stored biometrics data are resubmitted to the feature extractor at a later time. A watermark containing time stamp information is encrypted and embedded in the speech data, input by a legitimate user, to prevent replay attacks (Faundez-Zanuy et al., 2005). The feature extractor has the secret key and a built-in watermark decoder. The speech transmitted to the feature extractor is passed through the watermark decoder. After decoding and decrypting the embedded watermark, the time stamp information is read and it is determined if the speech is from a trusted party.

A related technology is speech *fingerprinting*, a term which is used to denote two slightly different techniques. The term fingerprinting may be used to denote the process of securely embedding unique identification codes in each copy of a multimedia signal for the purpose of traitor tracing. More significantly, digital fingerprinting is used to extract unique features from multimedia content and storing them in association with metadata (identification data) (Cano, Batlle, Kalker, & Haitsma, 2002). At a later time, the stored features may be compared with newly extracted features from the given content file for identification purposes. Applications of fingerprinting include preventing the distribution of copyrighted material, monitoring voice broadcasts, and providing consumers with supplementary information about the digital content.

SPEECH WATERMARKING

This section discusses in detail the existing trends and the factors to be considered while designing speech watermarking algorithms.

Choice of Domain and Speech Features for Watermarking

Speech watermarks may be embedded in either the time or transform domain of the signal. Audio watermarking algorithms based either in time, discrete cosine transform (DCT), fast Fourier transform (FFT), or wavelet domain have been applied to speech signals. A simple way to embed audio watermarks is by modifying the amplitude of speech samples in the time domain. Alternately, speech watermarks can be created by modifying the magnitude or phase coefficients in the DCT, FFT, or wavelet coefficients. Some existing speech watermarking algorithms exploit unique characteristics of speech signals that are not present in other multimedia data. An interesting characteristic of speech signals is the existence of mathematical models of the speech production process. Several robust and fragile speech watermarking algorithms exploit parametric models of speech for embedding watermark information. Embedding information in a parameter domain implies alteration of signal properties that are not linearly related to the signal samples. This renders the embedded information more difficult to separate from the signal samples.

The parameters associated with speech production models include the linear prediction (LP) coefficients, log area ratio (LAR) parameters, inverse sine (IS) coefficients, line spectrum pair (LSP) parameters, and reflection (PARCOR) coefficients (Haykin, 1996). Other related parameters include the autocorrelation coefficients and the cepstral coefficients (Deller et al., 2000). Some robust watermarking algorithms embed information in the model parameters. The advantage of this strategy is that watermark information is concentrated in the few LP coefficients during the watermark-embedding and recovery processes, while it is dispersed in time and spectrally otherwise. These representations can provide a good robustness and fidelity trade-off. For example, localization of watermark content in the frequency domain is more effectively controlled through direct manipulation of LSP coefficients. LSP coefficients can be manipulated to improve watermark robustness for a given fidelity requirement. Also, since LAR coefficients have the highest correlation with subjective quality (Quackenbush, Barnwell, & Clements, 1988), they can be directly altered to preserve stegosignal fidelity. In order to obtain the LSP parameters, the z-domain representation of the Mth order LP inverse filter is decomposed into the following polynomials:

$$P(z) = A(z) + z^{-(M+1)}A(z^{-1})$$
$$Q(z) = A(z) - z^{-(M+1)}A(z^{-1}) \quad (1)$$
$$A(z) = \frac{P(z)+Q(z)}{2}$$

where $A(z) = 1 - \sum_{m=1}^{M} a_m z^{-m}$, the z-domain representation of the inverse filter (Deller et al., 2000). The zeros of the polynomials P and Q constitute the LSP parameters. LSPs are alternate representations of LP parameters and have been used for robust speech watermarking (Hatada, Sakai, Komatsu, & Yamazaki, 2002).

Speech watermarking may also involve embedding information in the autocorrelation coefficients (Gurijala & Deller, 2001). The Levinson-Durbin (L-D) recursion is used for conversion of the modified autocorrelation values to modified LP coefficients. The reflection coefficients (**k**) constitute an alternative representation to LP coefficients. The PARCOR coefficients are obtained as a by-product of the L-D recursion. Watermark information can be added to the reflection coefficients. However, while embedding the watermark, it should be ensured that $|\mathbf{k}_i| \neq 1$ for any m, otherwise finding the reflection coefficients is an ill-conditioned problem. Other sets of speech parametric models for embedding watermark information include the LAR and inverse sine parameters.

Watermarking algorithms have also been developed to operate in the bit-stream domain of compressed or coded speech signals (Yuan & Huss, 2004). The dominant approach to bit-stream watermarking involves integration of watermarking with the speech compression and coding process. A distinguishing aspect of speech signals is the existence of voice coders or vocoders for speech compression, in addition to waveform coders. Vocoders involve the estimation of pitch and parametric models of short-time speech segments and are especially popular for low bit-rate compression (up to 2.4 kbps). The LP, reflection, LSP, LAR, and IS coefficients are commonly employed during the coding process. Speech is synthesized by passing an excitation signal through the all-pole model of the speech production process. For voiced speech, the excitation signal is a periodic pulse train with the period equal to the pitch period of the short-term speech segment. In case of unvoiced sounds, the excitation signal is a white noise sequence. An important class of speech coders is the analysis-by-synthesis (ABS) speech coders (Deller et al., 2000). In ABS coders, three types of excitation signals are available. The excitation signal is dynamically selected based on the perceptual quality of the synthesized speech. An important advantage of integrating watermarking with the coding process is a reduction in computational complexity.

Many fragile watermarking algorithms involve the extraction of speech features which are highly correlated with the semantics. Pitch is an inherent characteristic of a person's voice. An individual's pitch is not constant, but typically varies within a narrow range which is a function of the individual larynx. The pitch range for men is usually in the interval 50-250 Hz while women tend to have pitch in the 120-500 Hz range (Deller et al., 2000). Pitch and duration modification of short-time speech segments are commonly used speech features (Celik, Sharma, & Tekalp, 2005). The variable nature of these features ensures imperceptible data hiding and their robustness to low-bit rate speech compression facilitates robust and semi-fragile watermarking.

Perceptual Aspects of Speech

An important requirement of watermarking is to ensure the inaudibility of the embedded watermarks. Although it is impossible to embed perfectly imperceptible watermarks, it is generally possible to substantially minimize their audibility and their effect on stegosignal fidelity. Fidelity is a measure of perceptual similarity between the coversignal and the stegosignal. It is important to differentiate fidelity from *signal quality* (Cox et al., 2002). Audio signal quality might be low even prior to watermarking. However, the resulting watermarked signal can be of high fidelity. The watermarking process must not affect the fidelity of the audio signal beyond an application-dependent standard.

An understanding of the perceptual aspects of speech signals aids in the design of watermarking algorithms that preserve stegosignal fidelity. Consonants are known to play a major role in speech intelligibility despite representing a smaller fraction of speech energy (Deller et al., 2000). Formants or resonant frequencies characterize the spectrum of phonemes. There are 3-5 formants in the nyquist band after sampling, for any sound. The second formant is known to be more important for sound perception than the first or third. Other factors influencing speech perception include short-term spectrum, formant location, formant bandwidth, formant amplitude, and spectral slope (Deller et al., 2000).

Typically the watermarking process involves a time-, parameter-, or transform-domain scaling parameter for shaping the information to be hidden using the HAS-based psychoacoustic auditory model. If $X = \{x_i\}_{i=1}^{N}$ represents the values of the original audio signal in the time or frequency domain, $W = \{w_i\}_{i=1}^{N}$ is the watermark to be embedded, and α is a scaling parameter, then the resulting watermarked signal $Y = \{y_i\}_{i=1}^{N}$ is obtained as,

$$y_i = x_i + \alpha_i w_i, \; i = 1,2, \ldots, N. \quad (2a)$$

$$y_i = x_i \left(1 + \alpha_i w_i\right) i = 1,2, \ldots, N. \quad (2b)$$

$$y_i = x_i (e^{\alpha_i w_i}) \; i = 1,2, \ldots, N. \quad (2c)$$

The above equations represent a simple and effective way to control the stegosignal fidelity using a scaling parameter α (Cox et al., 1997, 2002). When the signal values x_i are small, $\alpha \prec 1$ ensures that the corresponding watermark values are also low in magnitude and the stegosignal is not noticeably distorted.

Time and frequency domain speech watermarking techniques can exploit the temporal and frequency masking properties of the HAS, because of which louder sounds tend to mask weaker sounds in both the frequency and the time domain. The adult human can typically hear sounds in the 20 Hz to 16 kHz range. The effective frequency range is 300 Hz to 3.4 kHz for telephone quality speech, and wideband speech has an audible frequency range of 50 Hz to 7 kHz (Boney et al., 1996). The effect of frequency on the human ear is not linear, but logarithmic. The loudness of a sound is perceived differently depending on the frequency. Sensitivity of the HAS is determined by measuring the minimum sound intensity level required for perception at a given frequency (Cox et al., 2002). Also, loud sounds tend to mask faint, but barely audible, sounds. Temporal masking can be classified into premasking and postmasking effects. In premasking, a faint signal is rendered inaudible before the occurrence of the louder masker. While in postmasking, a faint signal immediately following a loud masker is rendered inaudible. If two signals occur close together in frequency, then the stronger signal masks the weaker signal resulting in frequency masking. The masking threshold depends on frequency, the intensity level, and the noise-like or tone-like nature of the masker and masked signal. A sinusoidal masker will require greater intensity or loudness to mask a noise-like signal, than conversely.

The LP model has also been effectively used to mask the embedded watermark and minimize its audibility (Cheng & Sorensen, 2001). The LP model can preserve the magnitude spectral information of the speech (Deller et al., 2000). The magnitude spectrum is more significant to speech perception than the phase information. By filtering the frequency domain watermark signal with the LP model, the watermark is shaped-like and masked by the magnitude spectral information of the speech.

Evaluation of Stegosignal Fidelity

Measuring the fidelity of a watermarked signal is a complex problem. Objective evaluation functions or subjective evaluations may be employed. Proper subjective evaluation of fidelity involving a large number of human observers is rare. Many claims of fidelity are made via experiments involving a single listener on a small subset of trials (Cox et al., 2002). Since auditory sensitivity varies slightly across individuals, credible evaluation requires a large number of listeners and a large number of experiments.

Objective evaluation of fidelity is less expensive and more definitive for a given model, without the logistical difficulties and subjective nature of human evaluation. However, it is difficult to find an objective evaluation model that closely approximates human evaluation. Objective evaluation models attempt to quantify the differences between the coversignal and the stegosignal. A commonly-used objective evaluation model in speech watermarking algorithms is the mean squared error (MSE) function,

$$MSE = \frac{1}{N}\sum_{n=1}^{N}(y_n - x_n)^2 \tag{3}$$

The MSE between the stegosignal and the coversignal is used as an indicator of fidelity. MSE does not accurately capture perceptual impact of the changes to the host signal as it weighs all changes equally (Cox et al., 2002).

Another simple and mathematically tractable measure of fidelity is the signal-to-noise ratio (SNR), or, in the watermarking context, *coversignal-to-watermark (power) ratio* (CWR), defined as:

$$CWR = 10\log_{10}\frac{E_X}{E_W} = 10\log_{10}\frac{\sum_{n=1}^{N} x_n^2}{\sum_{n=1}^{N} w_n^2} \tag{4}$$

where $w_n = y_n - x_n$ is the sample-wise difference between the stegosignal and the coversignal at time n. The CWR averages the distortion of the coversignal over time and frequency. However, CWR is a poor measure of speech fidelity for a wide range of distortions. The CWR is not related to any subjective attribute of fidelity, and it weights the time-domain errors equally (Deller et al., 2000).

A better measure of fidelity for speech signals can be obtained if the CWR is measured and averaged over short speech frames. The resulting fidelity measure is known as *segmental* CWR (Deller et al., 2000), defined as:

$$CWR_{seg} = \frac{1}{K}\sum_{j=1}^{K}10\log_{10}\left[\sum_{l=k_j-L+1}^{k_j}\frac{x_l^2}{(y_l - x_l)^2}\right] \tag{5}$$

where $k_1, k_2, \ldots, k_K$ are the end-times for the K frames, each of length L. The segmentation of the CWR assigns equal weight to the loud and soft portions of speech. For computing CWR_{seg}, the duration of speech frames is typically 15 - 25 mS with frames of 15 mS used for the experimental results presented in this chapter. Some of the other objective measures of quality (fidelity, in the present case) of speech signals include the Itakura distance (Deller et al., 2000), the weighted-slope spectral distance, and the cepstral distance. According to Wang, Sekey, and Gersho (1992),

CWR_{seg} is a much better correlate to the auditory experience than the other objective measures mentioned above.

Watermark Data Payload, Robustness, and Security

Embedded watermark messages vary across applications. These messages are typically coded or mapped into symbols or vectors before being embedded into the host signal. The watermark data payload, or *information embedding rate*, is measured in terms of number of bits embedded per unit time of the host signal. If an N bit (binary) watermark is embedded, then the watermark detector must identify which one of the 2^N possible watermarks is present in the speech signal. For many audio watermarking algorithms, the watermark payload increases at least linearly with an increase in the sampling rate. Different speech watermarking applications require different data payloads. Copy control applications may require 4 to 8 bits of watermark information to be embedded in every 10 seconds of music (Cox et al., 2002). Broadcast monitoring may require 24 bits of information per second of the broadcast segment such as a commercial (Cox et al., 2002).

When watermarking is deployed for applications such as content protection, authentication, or copy control, the watermarked signal may be subjected to various deliberate and inadvertent attacks (Petitcolas, Anderson, & Kuhn, 1998; Voloshynovskiy et al., 2001). It is common for multimedia data to be subjected to processing such as compression, format conversion, and so on. Additionally, an attacker might try to deliberately remove the watermark or prevent its detection. Robustness refers to the ability of the watermark to tolerate distortion from any source to the extent that the quality of the coversignal is not affected beyond a set fidelity standard, or that the watermark detection or recovery processes are not hindered. Even watermarks for fragile watermarking applications are required to be selectively robust against certain attacks. Some of the factors affecting the robustness include the length of the audio or speech signal frame to be watermarked, the choice of watermark sequence, the relative energy of the watermark, the spectral content of the watermark, and the temporal locations and duration of the watermarks in the stegosignal. In broader terms, watermark robustness also depends on the watermark embedding, recovery, and detection algorithms. Cox et al. (1997) argue that watermarks must be embedded in perceptually significant components of the host signal to achieve robustness to common signal distortion and malicious attacks (Petitcolas et al., 1998; Voloshynovskiy et al., 2001). The perceptually irrelevant components of an audio signal can be easily removed by lossy compression of other signal manipulations without much loss in signal fidelity. A trade-off also occurs between the quantity of embedded data and the robustness to host signal manipulation. While robustness is an indispensable characteristic of watermarking algorithms, there is no existing watermarking algorithm that is robust to all possible attacks. Some of the important attacks on audio watermarking systems are discussed below:

- **Noise and signal processing operations:** Multimedia content may be subjected to signal processing operations such as lossy compression, digital-to-analog (D/A) conversion, analog-to-digital (A/D) conversion, filtering, quantization, amplification, and noise reduction. High frequency regions are not appropriate for robust data hiding, since lossy compression algorithms remove substantial information from this part of the spectrum. Speech signals may also be subjected to pitch modification. Finally, watermarks must be robust to additive noise for content transmitted across channels.
- **Geometric attacks:** A *geometric attack* distorts the watermark through temporal modifications of the stegosignal. Speech

watermarks may be attacked by geometric distortions such as jitter and cropping. Speech signals may also be subjected to delay and temporal scaling attacks. In a jitter attack or a cropping attack, arbitrary samples of the stegosignal are either duplicated or removed (Petitcolas et al., 1998). Cropping and jitter attacks result in desynchronization of the coversignal and the stegosignal and watermark recovery and detection are hindered.

- **Protocol attacks:** *Protocol attacks* exploit loopholes in the security of watermarking algorithms. Collusion, averaging, brute force key search, oracle, watermark inversion, ambiguity (Craver, Memon, Yeo, & Yeung, 1998), and copy attacks are categorized as protocol attacks (Voloshynovskiy et al., 2001). In a *collusion attack*, an attacker or a group of attackers having access to a number of differently watermarked versions of a work, conspire to obtain a watermark-free copy of the work. In an *inversion* or a *dead-lock attack*, the attacker inserts his or her own watermark into the watermarked signal in such a way that it is also present in the inaccessible and nonpublic host signal. In a *copy attack*, a watermark is copied from one audio signal to another, without any prior knowledge of the watermarking algorithm (Kutter, Voloshynovskiy, & Herrigel, 2000). This can be particularly devastating for authentication applications. Many protocol attacks compromise watermark security.

Security is an important attribute of even fragile watermarking algorithms, though they are required to be only selectively robust. On other hand, any watermarking technique with the desired robustness characteristics will not be of any use for most applications (except content management) if it is not secure. A watermark's security refers to its ability to withstand attacks designed for unauthorized removal, detection, or embedding. For example, if the purpose of watermarking is copyright protection, an adversary may be interested in removing the watermark, or replacing the existing watermark in the content with his or her own copyright information, or in confirming if a watermark is indeed present. A watermarking technique must not rely on the secrecy of the algorithm for its security. A watermarking scheme generally derives its security from secret codes or patterns *(keys)* that are used to embed the watermark. Additionally, cryptographic algorithms may be employed to encrypt the watermark before it is embedded in the content. Only a breach of keying strategies should compromise the security of a watermarking technique; public knowledge of the technical method should not lessen its effectiveness (Cox et al., 2002).

Watermarking technology is also vulnerable to the problem of the *analog hole*. When watermarked digital content is played back in analog form, the watermark information is lost. The analog content can then be reconverted to digital content without any watermark information. Generally there is a loss in audio signal quality due to D/A and A/D conversions. The significance of the analog hole problem is related to the amount of perceptual distortion that can be tolerated on the attacked watermarked signal.

ROBUST SPEECH WATERMARKING

This section will provide a description of some of the state-of-art speech watermarking algorithms. Many of these algorithms are based on developments in key speech technology areas such as coding, enhancement, and recognition. Existing robust speech watermarking algorithms are mostly based on one of three approaches: spread-spectrum signaling, synthesis-based, and parameter-based. In many cases, principles of spread spectrum watermarking (Cox, 1997) and quantization index modulation (QIM) (Chen

& Wornell, 2001) are commonly used for data hiding.

Spread-Spectrum Signaling

In *spread spectrum* (SS) *signaling* (Cheng & Sorensen, 2001) a narrow-band SS watermark signal is embedded in perceptually significant coefficients of the host signal, much like SS watermarking. The watermark message and a PN sequence are first modulated using binary phase-shift keying (BPSK). The center frequency of the carrier is 2025 Hz and the chip rate of the PN sequence is 1775 Hz. The SS watermark signal is lowpass filtered before modulation using a 7th order IIR Butterworth filter with a cut-off frequency of 3400 Hz. Linear predictive coding is used to embed high-energy watermark signals while ensuring imperceptibility. Linear predictive analysis involves the estimation of the all-pole component of speech of the form $A(z) = 1/(a_0 + a_1 z^{-1} + \ldots + a_M z^{-M})$. The LP coefficients of an Mth order all-pole LP model are typically computed using the Levinson-Durbin recursion (Deller et al., 2000). In general, the coversignal is assumed to be generated by the LP model:

$$x_n = \sum_{m=1}^{M} a_m x_{n-m} + \xi_n \tag{6}$$

The sequence $\{\xi_n\}_{n=1}^{N}$ is the prediction residual associated with the estimated model. In SS signaling, a bandwidth expansion operation is performed on the host signal. The poles are moved closer to the center of the unit circle, thereby increasing the bandwidth of their resonances. This is because the all-pole filter tends to have narrow spectral peaks. Frequencies near the peaks are subjected to masking effects and are unlikely to be perceived by an observer. By expanding the bandwidth of these resonances, more watermark frequencies can be effectively masked and this facilitates the embedding of more robust watermarks. A bandwidth parameter $\gamma = (0,1), \gamma \in \mathbb{R}$, is used to scale the LP parameters. For $m = [1, 2, \ldots, M]$: $\dot{a}_i = \gamma a_i$.

The modulated watermark signal is shaped by the LP spectrum of the coversignal. The filtered watermark signal is scaled by an instantaneous gain factor. The latter measure reduces perceptual distortion. In regions of silence in the coversignal, the watermark signal is scaled by a constant gain factor and then added to the coversignal. Watermarking contributes to a small amount of noise, which is not uncommon to any recording. Let E_{ξ_n} be the normalized per sample energy of the prediction residual associated with the all-pole model for one frame and E_{x_n} be the normalized per sample energy of the speech signal for one frame. In regions of speech activity, the watermark gain (g_n) is calculated as a linear combination of the gains for silence (λ_0), normalized per sample energy of the prediction residual associated with the all-pole model (λ_1), and normalized per sample energy of the speech signal (λ):

$$g_n = \lambda_0 + \lambda_1 E_{\xi_n} + \lambda E_{x_n} \tag{7}$$

The possibly distorted stegosignal can be expressed as $z_n = w_n + x_n + I_n = y_n + I_n$, where w_n is the watermark signal and I_n represents the distortion. The LP coefficients of the received signal $\{z_n\}$ are estimated. This is followed by inverse LP filtering of the received signal resulting in the signal $\{\hat{\overline{w}}_n\}$. The inverse-filtering operation converts voiced speech into periodic pulses and unvoiced speech into white noise. The inverse filtering also decorrelates the speech samples. For watermark detection, a correlation detector is constructed. The estimate $\{\hat{\overline{w}}_n\}$ is compared with the synchronized and BPSK-modulated spreading function for the current speech frame, $\{\overline{w}_n\}$. That is:

$$\sum_{n=1}^{N} \overline{w}_n \hat{\overline{w}}_n \underset{H_0}{\overset{H_1}{\gtrless}} 0 \tag{8}$$

The watermark receiver requires perfect synchronization between the whitened stegosignal and the PN spreading sequence. Synchronization is achieved by using the PN sequence for spread spectrum modulation to drive a phase lock loop (Cooper & McGillem, 1986). These techniques have been tested in low-noise environments such as in the presence of additive white Gaussian noise with a 20 dB SNR (Cheng & Sorensen, 2001).

Parametric Watermarking

Parametric watermarking (Gurijala & Deller, 2003; Gurijala, Deller, Seadle, & Hansen, 2002) has both spectrum-spreading and integration-by-synthesis aspects, but is fundamentally different from these approaches. The well known robustness of the LP model to practical anomalies occurring in coding, recognition, and other applications, suggests that some representation of these parameters might provide an effective basis for embedding durable watermarking data. There is a difference in the way LP modeling is applied in this watermarking application relative to its conventional deployment in speech coding and analysis. In coding applications, the goal is to find a set of LP coefficients that optimally model quasi-stationary regions of speech. In parametric watermarking, the LP parameters are derived according to the usual optimization criterion (to minimize the total energy in the residual) (Deller et al., 2000) with the understanding that the aggregate time-varying dynamics will be distributed between the *long-term* parametric code and the residual sequence.

The coversignal or a particular frame of the coversignal $\{x_n\}$ is assumed to be generated by an Mth order LP model:

$$x_n = \sum_{m=1}^{M} a_m x_{n-m} + \xi_n \tag{9}$$

The sequence $\{\xi_n\}$ is the prediction residual associated with the estimated model. The autocorrelation method may be used for deriving the LP coefficients $\{a_M\}_{m=1}^{M}$ (Deller et al., 2000). The prediction residual is obtained by using the LP parameters in an inverse-filter configuration:

$$\xi_n = x_n - \sum_{m=1}^{M} a_m x_{n-m}. \tag{10}$$

Once computed for a frame of speech to be watermarked, the LP parameters are modified in a predetermined way to produce a new set, $\{\tilde{a}_m\}$. The perturbations to the parameters constitute the watermark $\{w_m = \tilde{a}_m - a_m\}$. The stegosignal is constructed by using the modified LP parameters as a suboptimal predictor of the coversignal and by adding the prediction residual $\{\xi_n\}$.

$$y_n = \sum_{m=1}^{M} \tilde{a}_m x_{n-m} + \xi_n \tag{11}$$

The watermark information is concentrated in the few LP coefficients during the watermark-embedding and recovery processes, while it is dispersed in time and spectrally otherwise. The coversignal is required for watermark recovery. The watermark recovery process involves least-square-error (LSE) estimation of modified LP coefficients, and this further contributes to watermark robustness. First the prediction residual associated with the coversignal is subtracted from the possibly distorted stegosignal z_n:

$$\hat{w}_m = \hat{\tilde{a}}_m - a_m \tag{12}$$

The modified LP coefficients are estimated by computing the LSE solution, say, $\{\hat{\tilde{a}}_m\}_{m=1}^{M}$ to the overdetermined system of equations: $d_n \approx \sum_{m=1}^{M} \alpha_m y_{n-m}$. The watermark is recovered by subtracting the original LP coefficients from the estimated LP coefficients:

$$\hat{w}_m = \hat{\tilde{a}}_m - a_m \tag{13}$$

The watermark detection is treated as a binary decision problem in the presence of additive noise in the LP domain (Gurijala & Deller, 2005). A correlation detector is implemented based on the hypotheses:

$$H_0 : \hat{w}_m = v_m$$
$$H_1 : \hat{w}_m = w_m + v_m$$

The null hypothesis is that no watermark is present and only noise is transmitted $\{v_m\}_{m=1}^{M}$, while under H_1, both watermark $\{\hat{w}_m\}_{m=1}^{M}$ and noise samples $\{v_m\}_{m=1}^{M}$ are present in additive combination. The noise in the LP domain v_m is approximated by the distribution $N(\mu,\sigma^2)$, when noise $\{I_n\}_{n=1}^{N}$ is added to the stegosignal in the time domain such that the SNR is,

$$S = 10\log_{10}\frac{\sum_{n=1}^{N}\tilde{y}_n^2}{\sum_{n=1}^{N}I_n^2}.$$

To account for possible deviations from the true LP noise distribution, the detection threshold is selected to be a value several (l) standard deviations away from the mean.

$$\sum_{m=1}^{M}\hat{w}_m w_m \underset{H_0}{\overset{H_1}{\gtrless}} \mu + l\sigma \tag{14}$$

Parametric watermarking provides a good trade-off between robustness, fidelity, and watermark data payload. Parametric watermarking is robust to a wide range of attacks including additive noise, cropping, low pass filtering, and speech coding up to a bit rate of 16 k bits per second.

Speech Watermarking Through Transform Encryption Coding

Another interesting speech watermarking technique is the method based on *transform encryption coding* (TEC) (Ruiz & Deller, 2000). Although this technique was originally developed for speech signals, it is equally applicable to other forms of audio. TEC involves the application of all-pass pre/postfilters to conventional transform coding to improve noise resilience, to increase the coding efficiency, and to facilitate secure transmission of coded images (Kuo, Deller, & Jain, 1996). The all-pass prefilter scrambles the phase component of the signal prior to transform coding. If the all-pass filter is selected to have pseudo-random coefficients, the resulting prefiltered signal is encrypted and is unintelligible. Further, use of such a prefilter results in uncorrelated transform coefficients that are Gaussian distributed and hence independent. The forward TEC operation on a signal frame $X = [x_1, x_2, \ldots, x_N]$ is defined by:

$$X' = \mathrm{T}(X) = FT^{-1}\{|FT(X)|\ e^{j\Theta(X)}e^{j\Phi}\} \tag{15}$$

where Φ is the phase spectrum of a particular all-pass filter and FT represents the Fourier transform. Quasi m-arrays are generally used to obtain the phase coefficients of the all-pass filter (Kuo & Rigas, 1991; Kuo et al., 1996). The two-dimensional quasi m-arrays are arranged into a row vector before the application of TEC. The signal X is uniquely recoverable from the TEC-encrypted signal X' using the following inverse operation.

$$X = T^{-1}(X') = FT^{-1}\{|FT(X')|\ e^{j\Theta(X')}e^{-j\Phi}\} \tag{16}$$

In TEC-based audio watermarking, the watermark (image, speech, or audio signal) is encrypted in accordance with equation (15) by using a set of quasi m-arrays Φ_W. The host audio signal is similarly encrypted using a different set of quasi m-arrays Φ_X. In order to ensure watermark imperceptibility, the encrypted watermark signal is scaled by a gain factor k which is determined by a prespecified *coversignal-to-watermark* (CWR_{dB})

Equation 17.

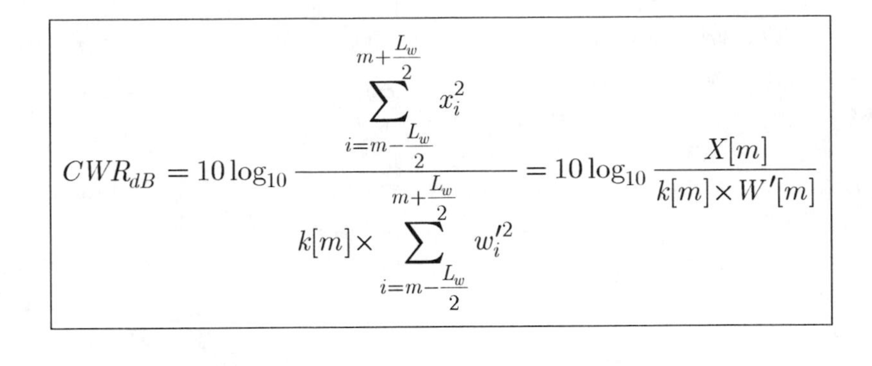

$$CWR_{dB} = 10\log_{10}\frac{\sum_{i=m-\frac{L_w}{2}}^{m+\frac{L_w}{2}} x_i^2}{k[m]\times\sum_{i=m-\frac{L_w}{2}}^{m+\frac{L_w}{2}} w_i'^2} = 10\log_{10}\frac{X[m]}{k[m]\times W'[m]}$$

ratio. The CWR_{dB} is a number that denotes the following relation, (see equation (17)).

In this equation, $X[m]$ and $W'[m]$ represent the short-term energy measures for the host signal and the encrypted watermark, respectively, at time m. The short-term energy measures are calculated over a rectangular window of size L_w and centered at m. The gain factor is determined for every audio frame of size L_w by:

$$k[m] = \frac{X[m]}{W'[m]}10^{-\frac{CWR_{dB}}{10}} \quad (18)$$

The encrypted watermark is then embedded into an encrypted host audio signal resulting in the encrypted stegosignal, Y',

$$Y' = T_X(X) + k\times T_W(W) = X' + k\times W' \quad (19)$$

The stegosignal is reconstructed by decrypting (equation [16]) the encrypted audio and this process encrypts the already-encrypted embedded watermark. Thus, the embedded watermark is twice encrypted,

$$Y = X + T_X^{-1}\{k\times T_W(W)\} = X + k\times W'' \quad (20)$$

If there is a need to recover the watermark from the stegosignal, then the coversignal is subtracted from the received stegosignal to obtain an estimate of the twice-encrypted watermark,

$$\hat{W}'' = Z - X \quad (21)$$

An estimate of the embedded watermark is obtained by applying the inverse TEC operations,

$$\hat{W} = \hat{k}^{-1}\times T_W^{-1}\{T_X\{\hat{W}''\}\} \quad (22)$$

Knowledge of the two sets of quasi m-arrays is necessary for watermark recovery, and this contributes to increased watermark security. There are several other watermarking algorithms based on modifying the phase of audio signals (Ansari, Malik, & Khokhar, 2004; Dong, Bocko, & Ignjatovic, 2004). TEC-based is highly robust to additive noise. Phase-based watermarking techniques are vulnerable to MPEG coding and IIR filtering.

FRAGILE SPEECH WATERMARKING

Fragile watermarking algorithms are used for data authentication. An important issue of fragile watermarking algorithms is their ability to identify areas of the signal where tampering has occurred. In the case of fragile watermarking algorithms, robustness to selective manipulations or attacks is desired. Fragile watermarking can be further classified into fragile, semi-fragile, content-fragile, and invertible categories depending on the required degree of robustness (Steinebach, Lang,

& Dittmann, 2003). Semi-fragile watermarking algorithms are moderately robust to distortion, but cannot distinguish between normal processing operations and deliberate manipulations of the content. On the other hand, content-fragile watermarking schemes can distinguish between the effect of malicious tampering and inadvertent processing of the content. In invertible fragile watermarking, changes to every bit or sample of the stegosignal are recognized by the extracted watermark and the original or coversignal is reconstructed.

Fragile Watermarking Through Pitch and Duration Modification of Speech Segments

An example of a content-fragile speech watermarking algorithm for authentication applications based on pitch and duration modification of quasi-periodic speech segments is presented here (Celik et al., 2005). The significance of these features makes them suitable for watermarking and the variability of these features facilitates imperceptible data embedding. The coversignal is segmented into phonemes. A phoneme is a fundamental unit of speech that conveys linguistic meaning (Deller et al., 2000). Certain classes of phonemes such as vowels, semivowels, diphthongs, and nasals are quasi-periodic in nature. The periodicity is characterized by the fundamental frequency or the pitch period. The *pitch synchronous overlap and add* (PSOLA) *algorithm* is used to parse the coversignal and to modify the pitch and duration of the quasi-periodic phonemes (Molines & Charpentier, 1990). The pitch periods $\{\rho_m\}$ are determined for each segment of the parsed coversignal. The average pitch period is then computed for each segment,

$$\rho_{avg} = \sum_{m=1}^{M} \rho_m / M \tag{23}$$

The average pitch period is modified to embed the pth watermark bit w_p by using dithered QIM (Chen & Wornell, 2001):

$$\rho_{avg}^{w} = Q_w(\rho_{avg} + \eta) - \eta \tag{24}$$

where Q_w is the selected quantizer and η is the pseudo-random dither value. The individual pitch periods are then modified such that:

$$\rho_m^{w} = \rho_m + (\rho_{avg}^{w} - \rho_{avg}) \tag{25}$$

The PSOLA algorithm is used to concatenate the segments and synthesize the stegosignal. The duration of the segments is modified for better reproduction of the stegosignal. As required by authentication applications, watermark detection does not use the original speech. At the detector, the procedure is repeated and the modified average pitch values are determined for each segment. Using the modified average pitch values, the watermark bits are recovered.

The algorithm is robust to low-bit-rate speech coding. This is because it uses features that are preserved by low-bit-rate speech coders such as QCELP, AMR, and GSM-06.10 (Spanias, 1994). Robustness to coding and compression is necessary for authentication applications. On the other hand, the fragile watermarking algorithm is designed to detect malicious operations such as re-embedding and changes to acoustic information (e.g., phonemes).

CONCLUSION

The chapter described the principles of speech watermarking. The main considerations for speech watermarking such as the intended application, choice of domain and speech features, and stegosignal fidelity, watermark robustness, data payload, and security requirements are discussed in detail. Speech watermarking requirements vary depending on the application. The chapter also described some important robust and fragile speech watermarking algorithms.

REFERENCES

Anand, D., & Niranjan, U. C. (1998, October). *Watermarking medical images with patient information.* Paper presented at the IEEE EMBS Conference, Hong Kong, China.

Ansari, R., Malik, H., & Khokhar, A. (2004, May). *Data-hiding in audio using frequency-selective phase alteration.* Paper presented at the IEEE International Conference on Acoustics Speech and Signal Processing, Montreal, Canada.

Barni, M., Bartolini, F., Rosa, A.D., & Piva, A. (2003). Optimal decoding and detection of multiplicative watermarks. *IEEE Transactions on Signal Processing, 51*(4).

Boney, L., Tewfik, A.H., & Hamdy, K.N. (1996, June). *Digital watermarks for audio signals.* Paper presented at the IEEE International Conference on Multimedia Computing and Systems, Hiroshima, Japan.

Cano, P., Batlle, E., Kalker, T., & Haitsma, J. (2002, December). *A review of algorithms for audio fingerprinting.* Paper presented at the International Workshop on Multimedia Signal Processing, U.S. Virgin Islands.

Celik, M., Sharma, G., & Tekalp, A.M. (2005, March). *Pitch and duration modification for speech watermarking.* Paper presented at the IEEE International Conference on Acoustics Speech and Signal Processing, Philadelphia.

Chen, B., & Wornell, G.W. (2001). Quantization index modulation: A class of provably good methods for digital watermarking and information embedding. *IEEE Transactions on Information Theory, 47*(4).

Cheng, Q., & Sorensen, J. (2001, May). *Spread spectrum signaling for speech watermarking.* Paper presented at the IEEE International Conference on Acoustics Speech and Signal Processing, Salt Lake City, Utah.

Cooper, G.R., & McGillem, C.D. (1986). *Modern communications and spread spectrum.* New York: McGraw-Hill Book Company.

Cox, I.J., Kilian, J., Leighton, T., & Shamoon, T. (1997). Secure spread spectrum watermarking for multimedia. *IEEE Transactions on Image Processing, 6*(3), 1673-1687.

Cox, I.J., Miller, M.L., & Bloom, J.A. (2002). *Digital watermarking.* San Diego: Academic Press.

Craver, S.A., Memon, N., Yeo, B.L., & Yeung, M. (1998). Resolving rightful ownerships with invisible watermarking techniques: Limitations, attacks, and implications. *IEEE Journal of Selected Areas in Communications: Special issue on Copyright and Privacy Protection, 16*(4), 573-586.

Cvejic, N., & Seppänen, T. (2003, September). *Robust audio watermarking in wavelet domain using frequency hopping and patchwork method.* Paper presented at the 3rd International Symposium on Image and Signal Processing and Analysis, Rome, Italy.

Deller, J.R., Jr., Hansen, J.H.L., & Proakis, J.G. (2000). *Discrete time processing of speech signals* (2nd ed.). New York: IEEE Press.

Dong, X., Bocko, M., & Ignjatovic, Z. (2004, May). *Data hiding via phase modification of audio signals.* Paper presented at the IEEE International Conference on Acoustics Speech and Signal Processing, Montreal, Canada.

Faundez-Zanuy, M., Hagmuller, M., Kubin, G., & Kleijn, W. B. (2005, April). *The Cost-277 speech database.* Paper presented at the 13th European Signal Processing Conference, Barcelona, Spain.

Gurijala, A., & Deller, J.R., Jr. (2001, May). *Robust algorithm for watermark recovery from cropped speech.* Paper presented at the IEEE International Conference on Acoustics Speech and Signal Processing, Salt Lake City, Utah.

Gurijala, A., & Deller, J.R., Jr. (2003, September). *Speech watermarking by parametric embedding with an l_∞ fidelity criterion.* Paper presented at the Interspeech Eurospeech, Geneva, Switzerland.

Gurijala, A., & Deller, J.R., Jr. (2005, July). *Detector design for parametric speech watermarking.* Paper presented at the IEEE International Conference on Multimedia and Expo, Amsterdam, The Netherlands.

Gurijala, A., Deller, J.R., Jr., Seadle, M.S., & Hansen, J.H.L. (2002, September). *Speech watermarking through parametric modeling.* Paper presented at the International Conference on Spoken Language Processing, Denver, Colorado.

Hagmüller, M., Horst, H., Kröpfl, A., & Kubin, G. (2004, September). *Speech watermarking for air traffic control.* Paper presented at the 12th European Signal Processing Conference, Vienna, Austria.

Hatada, M., Sakai, T., Komatsu, N., & Yamazaki, Y. (2002, July). Digital watermarking based on process of speech production. In *Proceedings of the SPIE: Multimedia Systems and Applications V*, Boston.

Haykin, S. (1996). *Adaptive filter theory* (3rd ed.). NJ: Prentice Hall.

Hernandez, J.J., Amado, M., & Perez-Gonzalez, F. (2000). DCT-domain watermarking techniques for still images: Detector performance analysis and a new structure. *IEEE Transactions on Image Processing, 9*(1).

Johnson, K.F., Duric, Z., & Jajodia, S. (2000). *Information hiding: Steganography and watermarking: Attacks and countermeasures.* MA: Kluwer Academic Publishers.

Kalker, T., Depovere, G., Haitsma, J., & Maes, M. (1999, January). *A video watermarking system for broadcast monitoring.* Paper presented at the IS\&T/SPIE's 11th Annual Symposium on Electronic Imaging '99: Security and Watermarking of Multimedia Contents, San Jose, California.

Kuo, C.J., Deller, J.R, & Jain, A.K. (1996). Pre/post-filter for performance improvement of transform coding. *Signal Processing: Image Communication Journal, 8*(3), 229-239.

Kuo, C.J., & Rigas, H.B. (1991). Quasi m-arrays and gold code arrays. *IEEE Transactions on Information Theory, 37*(2), 385-388.

Kutter, M., Voloshynovskiy, S., & Herrigel, A. (2000, January). *The watermark copy attack.* Paper presented at SPIE Security and Watermarking of Multimedia Contents II, San Jose, California.

Linnartz, J.P.M.G., Kalker, A.C.C., & Depovere, G.F. (1998). Modeling the false-alarm and missed detection rate for electronic watermarks. In L. D. Aucsmith (Ed.), *Notes in computer science* (Vol. 1525, pp. 329-343). Springer-Verlag.

Miaou, S.G., Hsu, C.H., Tsai, Y.S., & Chao, H. M. (2000, July). *A secure data hiding technique with heterogeneous data-combining capability for electronic patient records.* Paper presented at the World Congress on Medical Physics and Biomedical Engineering: Electronic Healthcare Records, Chicago.

Miller, M.L., & Bloom, M.A. (1999, September). *Computing the probability of false watermark detection.* Paper presented at the Third Workshop on Information Hiding, Dresden, Germany.

Molines, E., & Charpentier, F. (1990). Pitch-synchronous waveform processing techniques for text-to-speech synthesis using diphones. *Speech Communication*, 453-467.

Petitcolas, F.A.P. (2000). Watermarking schemes evaluation. *IEEE Signal Processing Magazine, 17.*

Petitcolas, F.A.P., & Anderson, R.J. (1999, June). *Evaluation of copyright marking systems.* Paper presented at the IEEE Multimedia Systems, Florence, Italy.

Petitcolas, F.A.P., Anderson, R.J., & Kuhn, M. G. (1998, April). *Attacks on copyright marking systems.* Paper presented at the Second Workshop on Information Hiding, Portland, Oregon.

Poor, H.V. (1994). An *introduction to signal detection and estimation* (2nd ed.). Springer-Verlag.

Quackenbush, S. R., Barnwell, T. P., & Clements, M. A. (1988). *Objective measures of speech quality.* NJ: Prentice Hall.

Ruiz, F.J., & Deller, J.R., Jr. (2000, June). *Digital watermarking of speech signals for the national gallery of the spoken word.* Paper presented at the International Conference on Acoustics, Speech and Signal Processing, Istanbul, Turkey.

Seadle, M.S., Deller, J.R., Jr., & Gurijala, A. (2002, July). *Why watermark? The copyright need for an engineering solution.* Paper presented at the ACM/IEEE Joint Conference on Digital Libraries, Portland, Oregon.

Spanias, A. (1994). Speech coding: A tutorial review. *Proceedings of the IEEE, 82*(10), 1541-1582.

Steinebach, M., Lang, A., & Dittmann, J. (2002, January). *StirMark benchmark: Audio watermarking attacks based on lossy compression.* Paper presented at the Security and Watermarking of Multimedia Contents IV, Electronic Imaging 2002, Photonics West, San Jose, California.

Voloshynovskiy, S., Pereira, S., Pun, T., Su, J.K., & Eggers, J. J. (2001). Attacks and Benchmarking. *IEEE Communication Magazine, 39*(8).

Wu, M., & Liu, B. (2004). Data hiding in binary image for authentication and annotation. *IEEE Transactions on Multimedia, 6*(4), 528-538.

Wang, S., Sekey, A., & Gersho, A. (1992). An objective measure for predicting subjective quality of speech coders. *IEEE Journal on Selected Areas in Communications, 10*(5), 819-829.

Yuan, S., & Huss, S. (2004, September). *Audio watermarking algorithm for real-time speech integrity and authentication.* Paper presented at the ACM Multimedia and Security Workshop, Magdeburg, Germany.

Chapter XII
Robustness Against DA/AD Conversion: Concepts, Challenges, and Examples

Martin Steinebach
Fraunhofer SIT, Germany

ABSTRACT

This chapter discusses the robustness of digital audio watermarking algorithms against digital-to-analogue (D/A) and analogue-to-digital (A/D) conversions. This is an important challenge in many audio watermarking applications. We provide an overview on distortions caused by converting the signal in various scenarios. This includes environmental influences like background noise or room acoustics when taking microphone recordings, as well as sound sampling effects like quantisation noise. The aim is to show the complexity of influences that need to be taken into account. Additionally, we show test results of our own audio watermarking algorithm with respect to analogue recordings using a microphone which proves that a high robustness in this area can be achieved. To improve even the robustness of algorithms, we briefly introduce strategies against downmixing and playback speed changes of the audio signal.

INTRODUCTION

Due to its name, digital audio watermarking is often seen as a technique only robust to attacks in the digital domain. If this is true for an audio watermarking algorithm, it is vulnerable to an attack called the *analogue hole* or *analogue gap*, making it possible for attackers to circumvent protection mechanisms based on digital watermarking. Here the marked signal is sent to an analogue output and digitised again to get rid of the watermark. Audio watermarking algorithms not robust against DA/AD conversions share a weakness with other DRM mechanisms which are known to be frequently circumvented in this way.

In this chapter we will discuss the digital-to-analogue (D/A) and analogue-to-digital (A/D) robustness of a watermarking algorithm developed by us, introduce a number of challenges and requirements caused by analogue recordings, and provide some strategies for increasing DA/AD robustness.

ANALOGUE RECORDING CATEGORIES

When robustness to analogue conversion is of importance for an audio watermarking algorithm, we need to identify the estimated scenario in which the watermark will be used. This is given by the application type. DRM-protected audio CDs will usually only be copied by simple analogue line-out and line-in chains. Soundtrack protection for cinemas on the other hand may include transfer to analogue recording media, playback speed modification, and microphone recordings with background noise. The challenge for the watermark to survive the two scenarios is very different, as in the second scenario multiple attacks occur.

We distinguish between the following categories of analogue recordings:

1. **Line-out/line-in:** This is the simplest attack. A digital audio signal is converted to an analogue signal by a sound card, sent to an analogue output (often called line-out) and is sampled again from an analogue input (line-in) at the same time, often using the same soundcard for playback and recording. If high quality equipment is used, only a minimum of distortion or noise will be induced to the marked audio signal. This includes effects unavoidable by DA/AD conversion like quantisation noise, clock rate errors, or clock rate differences when using more than one soundcard. The resulting audio file therefore will feature a certain amount of additional noise and the number of samples it consists of may be subject to minimal changes. Still, as the signal leaves the digital domain, all file-specific information is lost (e.g., DRM certificates).
2. **Line-out/microphone-in:** This attack is similar to the previous category but features a microphone for recording instead of a line-in signal. The characteristics of the microphone will have a much stronger influence on the signal than feeding the signal into the recording device via line-in. Additionally, to record the signal by microphone, a speaker and an amplifier are necessary to play back the audio. Both devices may add distortions, and in the case of the speaker, the frequency range may be a relevant issue for some watermarking algorithms. Most microphone recordings are monophonic; therefore a downmix of a multichannel audio signal occurs.
3. **Line-out/analogue recording/line-in:** Adding an analogue recording media like a magnetic tape to Category 1 leads to some significant new challenges for the watermarking algorithm. While the recorded sound is played by the analogue recording device, small changes in playback speed may occur due to device mechanics or tape aging effects. The changes can vary constantly. In addition to this, noise suppression methods may be used by the recording device filtering high frequencies.
4. **Line-out/analogue recording/microphone in with background noise:** When we add the challenges of Categories 2 and 3, we come to the worst case scenario for analogue recordings. We need to take into account both the recording device and the microphone. In addition we assume a noisy environment with background noise like people laughing within a cinema.

The challenge for designing an audio watermarking algorithm increases with the category numbers. But quality reduction of the audio material is also strong in the latter categories; therefore audio watermarking parameters may be adjusted to increase embedding strength but at the cost of transparency in some applications.

RESULTING DISTORTIONS

DA/AD attacks on audio watermarks can be seen as consisting of a set of more simple manipulations. In the following we describe the most important manipulations which can all be part of a DA/AD process. In addition, distortions due to recording equipment are also induced into the audio material. Some of these are discussed in the following section. But when compared to the effects of an environmental microphone recording, the effects caused by quantisation or sampling clock differences are minimal and will not influence the detection rate of a watermarking algorithm robust against microphone recordings. This does not include the microphone used in the process, which can have an effect on the detection results due to its characteristics as mentioned below:

- **Coloured noise:** The addition of noise is always present in DA/AD conversion. It can be caused by the analogue equipment, media, and errors during sampling. White and coloured noise can be observed, making it hard to filter out noise effects effectively during watermark detection.
- **Background noise:** When using microphones, background noises like such as laughter in a cinema may be present in the final recording. Most often these noises are of low volume and therefore do not effect watermark detection quality. But in some cases, background noise may feature a narrow frequency range, like the humming of a refrigerator or a feedback signal. This can become a challenge for some watermarking algorithms not robust against the addition of single frequency sounds.
- **Environmental characteristics:** When recording with a microphone, not only will the desired audio information will be recorded, but always also the acoustic characteristics of the environment. This includes the reverberation of small rooms or echoing in an open environment. In comparison to background noises, these distortions can be seen as passive as they only modify existing audio information but are not sources of new sounds. Depending on the environment (e.g., room size, wall material, or presence of furniture), characteristics will vary. The subsequent section on acoustics, reverberation, and echo discusses these aspects in more detail.
- **Filtering:** The removal of high or low frequencies can occur during playback, sampling, or within the analogue devices. Analogue media, microphones, speakers, and other hardware often have only a limited frequency range. It is often lower than the digital resolution; for example cutting frequencies below 60 Hz or above 16 kHz.
- **Modification of frequency energy:** Within analogue equipment, not all frequencies may be treated equally. For some frequencies, energy may be increased while for others it may be reduced. Only high quality equipment is said to be neutral with respect to frequency transmission. Figure 1 illustrates the effect of a microphone recording on the spectrum of an audio file.
- **Modification of playback speed:** When using analogue media, the speed of the playback may differ slightly from the recording speed. This may be an active decision like in record players with pitch control or cinema projectors. But analogue tape recorders' playback speed may also slightly vary from model to model resulting in speed modifications when moving a tape from one recorder to another.

- **Jittering:** Jittering can be observed in analogue tape decks where playback speed varies due to low quality mechanics or tape material aging effects. This rather random change of playback speed can be challenging for watermark synchronisation. A comparatively minimal jitter can also be caused by digital sampling due to clock inaccuracies.
- **Cropping:** An analogue playback or recording may not consist of the complete original audio recording, resulting in a cropped copy.
- **Downmixing from n channels to mono:** Low quality microphone recordings most often are monaural, mixing both channels of a stereo recording into a single channel. Phase modifications can occur due to a recording position not in the middle between both speakers. In the case of surround recordings, a greater number of channels, for example, a 5+1 signal, may be mixed down to one channel. The watermark therefore needs to be dominant in the sum of all channels to be retrievable after downmixing.

DISTORTIONS CAUSED BY DA/AD CONVERSION HARDWARE

Whenever audio data are converted from digital-to-analogue or analogue-to-digital by a soundcard within a computer, distortions of the audio will occur. The best known source for audio degradation is the quantisation effect during sampling analogue sound to a digital signal. This will cause a certain amount of noise that can be estimated by the resolution of the samples used in the sampling process. In addition, a jitter effect will be added to the sound as the time period for taking an individual sample will vary slightly causing a quality reduction in sampled high frequencies (Zölzer, 1997). The Nyquist–Shannon sampling theorem (Shannon, 1949) of course must be taken into account for audio sampling. A watermark will be attacked during AD conversion if it is embedded in frequencies more than half of the maximum sampling frequency of the audio unit. Another source of noise is the influence of the environment on the sound when sampled or played back. Computers emit electronic signals which can influence the sound quality, and some-

Figure 1. Spectral analysis of original signal (upper line) and signal recorded by Sure 58 microphone (lower line)

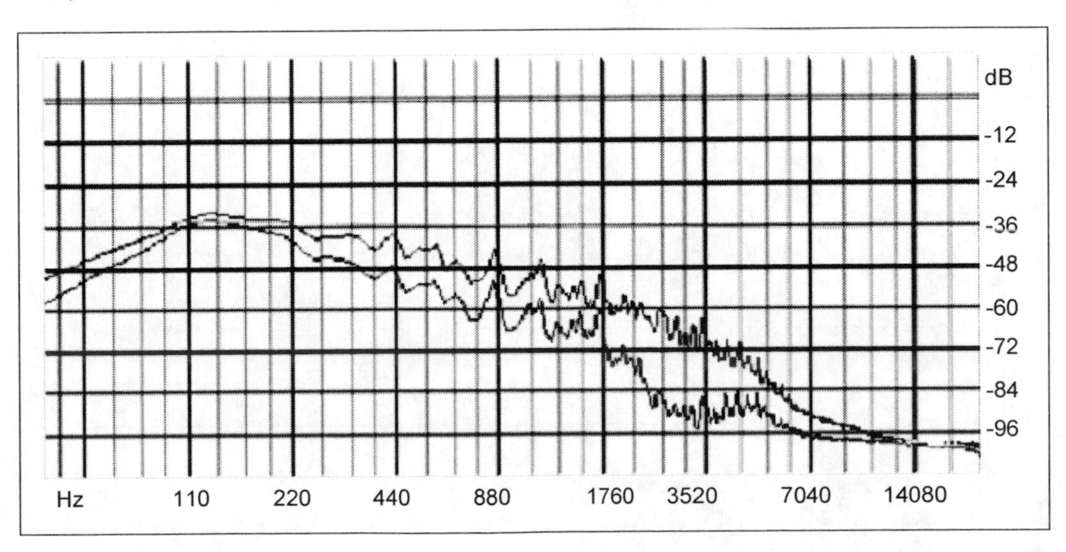

times the shared power source adds noise. Still, in a reasonably built state-of-the-art system, the distortion of the signal caused by the computer system should be minimal compared to the effects by microphone recordings or similar stages.

ACOUSTICS: REVERBERATION AND ECHO

To describe the effects of analogue environments on watermarked audio material in more detail, in this section we briefly introduce the research area of room acoustics as one example for the broad area of acoustics. Acoustic effects occur especially with microphone recordings. Reverberation and echo are acoustic effects caused by an audio signal reflected by the environment. The result is a trace of the original signal audible for some time after the original signal stopped. The time the reverb is audible is called reverberation time and is one of the most important characteristics of rooms with respect to acoustics. Reverb and echo share the same origin. If individual reflections can be distinguished by the listener, the effect is called echo, otherwise it is called reverberation. Reverberation is influenced by the geometrics of the room, its size, the reflecting material, and the presence of objects within the room. This modifies the spectral characteristics of the reflected signal, the delay time until the first reflection is audible, and the amount of time until the amplitude of the reflected sound fades out to become inaudible. The complete audio signal is usually divided into the direct signal, the early reflections caused by walls directly reflecting the sound to the listener, and the subsequent reverberations of follow-up reflections (Figure 2). For a more detailed discussion of reverberation effects and their simulation we suggest (Zölzer, 1997). Fundamentals of room acoustics are discussed for example, by Kuttfruf (2000).

Reverberation can be challenging for audio watermarking as it mixes the characteristics of audio frames with following frames by adding the sound of the reflection of the previous frame to the sound of the frame currently played.

PREVIOUS WORK

While numerous audio watermarking algorithms robust to DA/AD processing have been introduced in literature, only few articles on DA/AD benchmarking exist. In this section we summarise the results of two related articles we presented together with a number of colleagues. One by Steinebach, Lang, Dittmann, and Neubauer (2002) is on the topic of DA/AD robustness; the other by Steinebach, Zmudzinski, and Lang (2003) discusses the robustness against analogue radio transmission and also introduces a model for simulating this attack.

Figure 2. Room acoustics: Reverberation is added to the original signal when played back in a natural environment. The resulting signal is a mix of the direct signal, the early reflections and the subsequent reverberations.

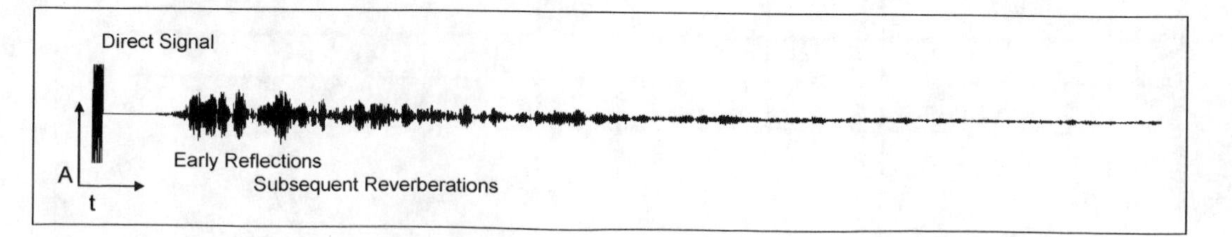

One important result of the first publication is the dependence of watermarking robustness on microphone types. A number of microphones have been compared in our tests, and a professional all-round recording microphone, the Sure SM-58, provided the best test results. But also a cheap sound card add-on enabled very good detection rates, while another cheap microphone for video recording caused worse results.

Another aspect is the influence of the audio material on the test results. Speech performs worse than music with respect to watermark detection. A reason for this could be the gaps between the spoken words which are filled with background noise in microphone recordings.

We also analyzed the influence of the distance between speaker and microphone on the detection results. While a greater distance leads to worse results, at a very close range of 30 cm we also found a decrease in detection rate. This can be caused by distortions occurring at close range with high volume playback.

In one of our articles (Steinebach et al., 2003) we analyzed the robustness of audio watermarking against FM radio transmission. While a strong watermark is robust against radio transmission and low quality line-in analogue recording, even a watermark of medium strength inaudible to most listeners survives radio transmission if good recording equipment is used for digitisation via line-in.

We also show that the attack scenario can be simulated by adding pink noise to the signal by measuring the influence of an FM transmission and recording to a set of test signals. This encourages the simulation of DA/AD attacks in benchmarking systems like Stirmark Benchmark for Audio (Steinebach, Petitcolas, Raynal, Dittmann, Fontaine, Seibel, et al., 2001).

DA/AD EVALUATION: MICROPHONE RECORDINGS

We now provide a number of recent research results for audio watermarking robustness against DA/AD attacks. For this, we use an algorithm developed by us and introduced by Steinebach (2004) based on common watermarking concepts (Bender, Gruhl, Morimoto, & Lu, 1996).

The watermarking technique works in the spectrum of uncompressed PCM audio data and follows a spread spectrum approach. For a better understanding of the test results and the behavior of the algorithm, its basic features are explained in the following:

1. As a first step, the algorithm watermark transforms the PCM source signal into the frequency spectrum using a fast Fourier transform (FFT). Segments of the source signal are created, also known as windows or frames.
2. In the next step, the watermark is embedded by deliberately modifying the energies of pseudo-randomly selected individual frequency bands. The selected frequencies are assigned to one of two groups, A and B. The process is controlled by a secret key.
3. The bit value embedded in the frame is defined by the relation of the energy in A and B. To embed a '0,' A must be greater than B. For a '1,' B must be greater than A. To enforce the correct relation of A and B, slight modifications on the energy of the individual frequency bands in A and B are done by the algorithm. The watermark is spread across the frequency spectrum in this way. The strength of the changes in the frequency bands can vary. The greater the change, the more reliably the watermark

can be retrieved; but at the same time, the perceived quality of the watermarked material deteriorates. An acoustic model is used, which simulates human perception and does not allow a modification to the material if it causes a perceptible distortion.

4. In the final step, the algorithm converts the data back into PCM format. In order to prevent noticeable gaps between the segments, windowing mechanisms are used to fade between marked and unmarked parts of the audio.

In addition, the technique uses a secret key with which the frequency bands which are to be altered are selected; we therefore speak of a quasi-random selection. Without knowledge of the key it cannot be determined which frequencies were changed without the original being present for comparison.

One important aspect of the evaluation is the relation of robustness and embedding strength. We therefore apply watermarks of different strength values called *a* to *h*. The strength of the watermarking algorithm is defined by the distance in dB of the watermark embedding artefacts to the point of inaudibility calculated by a psychoacoustic model. Negative values mean overriding the threshold of inaudibility and therefore a lower perceived quality.

For evaluating the robustness vs. microphone recording, we embedded a 16 bit watermark in

Table 1. Label: Signal-to-noise ratio

Level	SNR in dB	Level	SNR in dB
a	3	e	-4.5
b	0	f	-6
c	-1.5	g	-9
d	-3	h	-12

Figure 3 . Spectrum of white noise before (left) and after (right) microphone recording with a frequency range from 20 Hz to 22 kHz (y axis, logarithmic) and duration of 200 milliseconds (x axis)

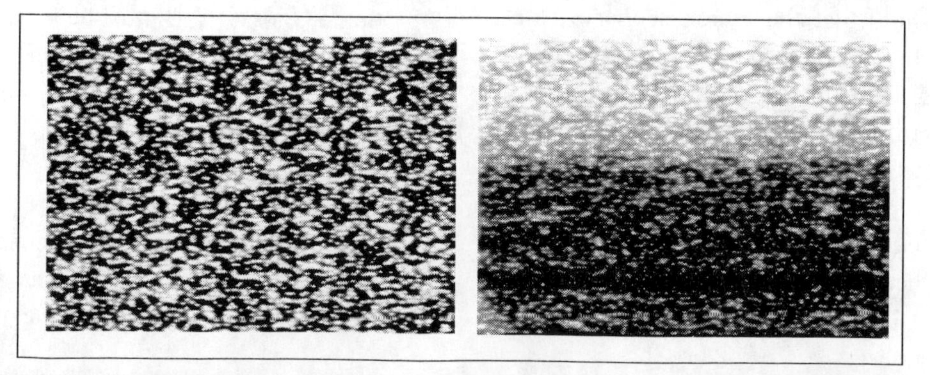

Note: The darker the area, the more energy is present at the corresponing frequency. One can see how the equal and random distribution of the energy before the recording is changed to a filtered and biased distribution after the microphone recording. Energy in high frequencies is almost completely removed and a clear focus on the mid range can be seen.

17 test files of one minute length with strength *a* to *h*. The watermark was repeated over the whole length of the file providing redundancy for increased robustness. On average, 20 instances of the 16 bit watermark could be embedded within one minute of the test material. The audio material consisted of music, sounds, atmospheres, and spoken words. All 136 files were played through a Yamaha MSP5 speaker featuring a rather neutral sound and rerecorded with a microphone supplied with a Creative Soundblaster sound card set up two meters away from the speaker. A cheap

Figure 4. Detection of watermarks in audio files after re-recording with a microphone given in percentage of successful detections

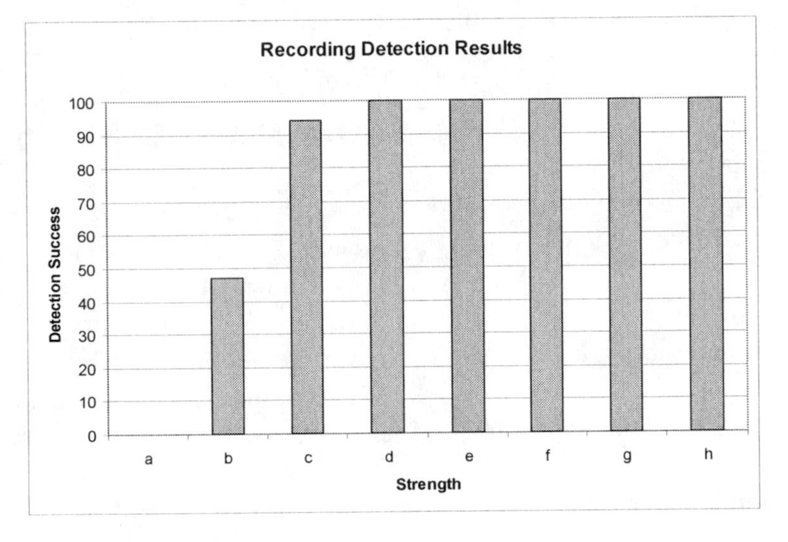

Figure 5. Number of watermarks in audio files after re-recording with a microphone given in percentage of originally embedded watermarks

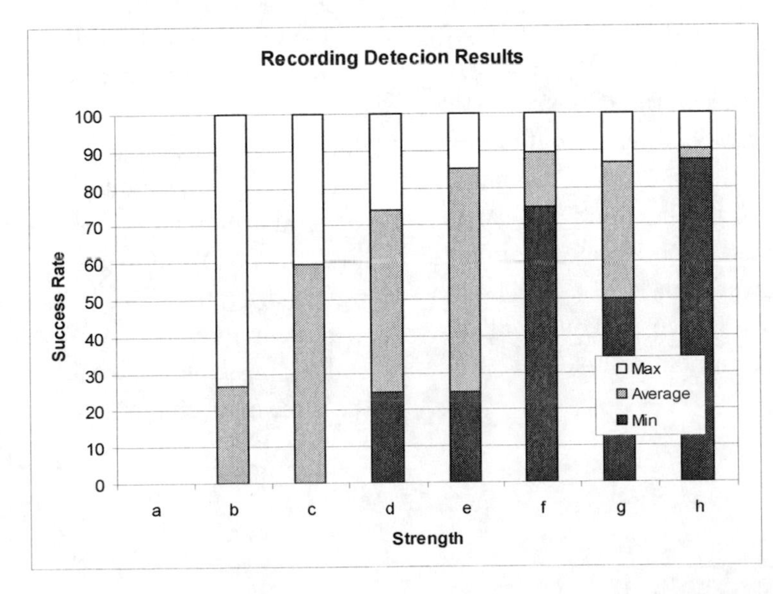

microphone was used intentionally in order to simulate a bad recording environment.

Both figures entitled 'Recording Detection Results' show clearly that the embedded watermarks are still present in microphone recordings.

Figure 4 shows that the retrieval process was successful. If at least one watermark could be successfully retrieved from a file, then embedding and retrieval were deemed successful. At level *a* no watermark could be detected, but at level *c* at least one watermark could be found in over 90% of the samples. At level *d*, which was classified in audio tests as imperceptible, in *every* file at least one watermark could be successfully retrieved.

Figure 5 shows how many of the embedded watermarks could be detected in 17 different samples. Here the minimum, maximum, and average values can be seen; for example, using level *b*:

- In some samples no watermark could be detected.
- On average approx. 25% of the watermarks could be detected.
- In at least one sample all watermarks could be detected.

As can be seen there are significant differences between the three values. Accordingly, the chances of detecting a watermark depend on the marked material. In some cases level *b* is sufficient, in others at least level *d* is needed to guarantee successful retrieval. At the high embedding strength levels *f* to *h*, the average success rate stays at a similar level while the minimal success rate significantly increases at level *h*. The success rate minimum of level *g* is much lower than level *f*, which can be caused by a single unsuccessful retrieval due to an increase in environmental noise.

DA/AD ROBUSTNESS OPTIMISATION STRATEGIES

In this section we discuss strategies to improve the robustness of audio watermarking algorithms against DA/AD conversion. This includes the embedding and the retrieval process. In both cases, understanding the potential effects of DA/AD conversion can help to improve the robustness of the embedded watermarks against such effects. We provide two examples for embedding and retrieval strategies

DOWNMIX ROBUSTNESS

When a multichannel signal is recorded by a monophonic microphone, all audible channels are mixed down into one single channel. To ensure the presence of a watermark in the resulting channel, some rules must be followed. It is important to ensure that, if the watermark is embedded in more than one channel, the parallel watermarks do not interfere with each other in the case of a downmix. This means that usually multiple channels cannot be used to increase the watermarking payload, but identically synchronised copies of the watermark must be embedded in all marked channels.

However, not all available channels need to be watermarked. Especially in surround sound recordings, some channels happen to be silent during some passages of a recording. In this case, the channel cannot and also does not need to be watermarked. The silent track will not have any effect on the downmix and therefore on the watermark retrieval as long as some other channels are marked. In Figure 6 we show a typical embedding strategy for a 5.1 surround signal; only the three main channels left, right, and centre are used for watermarking. The subwoofer and effect channels are ignored. The watermark is synchronised in the three main channels and redundantly embedded over the whole duration of the audio signal. Every time watermark embedding starts anew, it is decided for each individual channel if a watermark is embedded or if the audio content is not suitable for watermarking at this time. In this case, a gate function stops watermarking for the duration of the current watermark embedding. Alternatively this can also be done for each bit of the watermark.

Playback Speed Robustness

Playback speed changes can occur in conjunction with or without pitch changes. In analogue environments, usually both playback speed and pitch changes take place at the same time. This may lead to a strong decrease in audio watermarking detection rates. But various strategies exist to counter this. Often the watermarks are not removed by the playback speed change, but the synchronisation algorithm of the retrieval process is not able to find the starting point of the embedded watermark. Therefore playback speed changes are a synchronisation challenge. In recent experiments (Steinebach & Zmudzinski, 2006) we did show that watermark detection rates after pitch shifting and time stretching dramatically increase when the watermark synchronisation is re-established. As example detection rates of our watermarking algorithm after a 30% playback, time increase changed from undetectable without optimised resynchronisation to a success rate of about 85% percent. This improvement was achieved by reversing the effects of the previous attacks by pitch shifting and time stretching algorithms. Follow-up developments and tests show that even higher retrieval rates can be achieved by optimised watermark scanners trying to autosynchronise the watermarks after attacks.

For practical usage this means that various strategies are possible depending on the knowledge about the environments. In cases where the resulting speed changes can be estimated, the retrieval process can include a step to reverse the speed change by specialised algorithms. This only requires a small computational overhead and improves detection rates drastically. When the amount of speed changes can not be predicted, scanning for the embedded watermark may be necessary. This can become computationally intensive as the often complex and time consuming synchronisation phase of the watermark is improved by an even more complex search

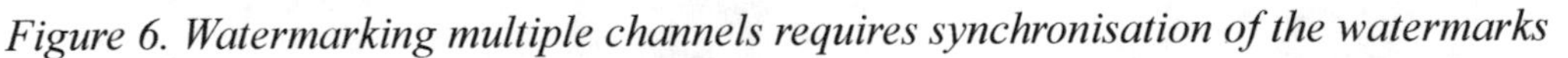
Figure 6. Watermarking multiple channels requires synchronisation of the watermarks

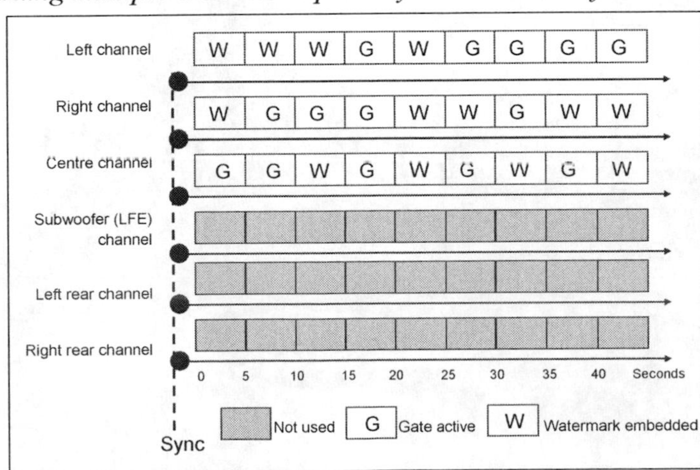

mechanism. In our test implementation, we scanned a defined range of neighbour frequencies for desynchronised watermarks to fight pitch shifting attacks. In average, this doubled the retrieval time of the watermark.

SUMMARY AND CONCLUSION

Robustness to DA/AD conversion is an important aspect of digital audio watermarking. On the one hand, it is required for a number of application scenarios; on the other hand it is challenging due to its multiple aspects. It combines the addition of noise, the removal of high and low frequencies, changes of the spectrum, playback speed changes, and pitch shifting. The watermark needs to be robust against the sum of these attacks to reliably resist DA/AD conversion. And only then it can offer a solution to the challenge of the 'analogue hole,' providing a copyright mechanism which stays within the content even after leaving digital and DRM controlled environments.

REFERENCES

Bender, W., Gruhl, D., Morimoto, N., & Lu, A. (1996). Techniques for data hiding. *IBM Systems Journal, 35*(3-4), 313-336.

Kuttruf, H. (2000, October). *Room acoustics* (1st ed.). Spon Press.

Shannon C. (1949). Communication in the presence of noise. *Proceedings Institute of Radio Engineers, 37*(1), 10-21.

Steinebach, M. (2004). *Digitale Wasserzeichen für Audiodaten*. Shaker Verlag Aachen.

Steinebach, M., Lang, A., Dittmann, J., & Neubauer, C. (2002, April 8-10). Audio watermarking quality evaluation: Robustness to DA/AD processes. Paper presented at the *International Conference on Information Technology: Coding and Computing, ITCC*, Las Vegas, Nevada (pp. 100-103). Piscataway, NJ: IEEE Computer Society.

Figure 7. The analogue hole is a challenge for many DRM approaches. Even a direct cable connection of soundcard input and output can be sufficient to disable security measures. Digital watermarking robust to DA/AD conversion can help to improve content security.

Steinebach, M., Petitcolas, F.A.P., Raynal, F., Dittmann, J., Fontaine, C., Seibel, S., et al. (2001, April 2-4). StirMark benchmark: Audio watermarking attacks. Paper presented at the *International Conference on Information Technology: Coding and Computing (ITCC 2001),* Las Vegas, Nevada (pp. 49-54).

Steinebach, M., & Zmudzinsiki, S. (2006). Robustheit digitaler Audiowasserzeichen gegen Pitch-Shifting und Time-Stretching, *Sicherheit 2006. Sicherheit - Schutz und Zuverlässigkeit. Beiträge der 3. Jahrestagung des Fachbereichs Sicherheit der Gesellschaft für Informatik e.V. (GI),* ISBN 3885791714, Kölln Verlag.

Steinebach, M., Zmudzinski, S., & Lang, A. (2003). Robustheitsevaluierung von digitalen Audiowasserzeichen im Rundfunkszenario, *CD-Rom zum 2. Thüringer Medienseminar der FKTG zum Thema Rechte Digitaler Medien Intellectual Properties & Content Management*, Erfurt

Steinebach, M., Zmudzinski, S., & Neichtadt, S. (2006). *Robust-audio-hash synchronized audio watermarking.* Paper presented at the 4th International Workshop on Security in Information Systems – (WOSIS 2006) (pp. 58-66). Paphos, Cyprus.

Zölzer, U. (1997, August). *Digital audio signal processing*, John Wiley & Sons.

Chapter XIII
Subjective and Objective Quality Evaluation of Watermarked Audio

Michael Arnold
Thomson, Germany

Peter Baum
Thomson, Germany

Walter Voeßing
Thomson, Germany

ABSTRACT

Methods for evaluating the quality of watermarked objects are detailed in this chapter. It will provide an overview of subjective and objective methods usable in order to judge the influence of watermark embedding on the quality of audio tracks. The problem associated with the quality evaluation of watermarked audio data will be presented. This is followed by a presentation of subjective evaluation standards used in testing the transparency of marked audio tracks as well as the evaluation of marked items with intermediate quality. Since subjective listening tests are expensive and dependent on many not easily controllable parameters, objective quality measurement methods are discussed in the section, Objective Evaluation Standards. The section Implementation of a Quality Evaluation presents the whole process of testing the quality taking into account the methods discussed in this chapter. Special emphasis is devoted to a detailed description of the test setup, item selection, and the practical limitations. The last section summarizes the chapter.

Evaluation of Watermarked Audio Data

The primary goal for evaluation techniques is to conduct valid and reliable tests which provide data for research, development, and quality control during deployment of the developed algorithms. Quality assessment of watermarked audio tracks is a significant challenge. No single objective metric to quantify the quality of an audio track carrying a watermark is currently available (Cox, Miller & Bloom, 2002).

Nevertheless, in the case of audio data, the problem of evaluating watermarked audio tracks is very similar to the problem of evaluating the quality of perceptual codecs. Both signal processing applications introduce distortions in the output signal which should be inaudible by exploiting psychoacoustic phenomena. Due to the underlying algorithm and the used psychoacoustic model, the types of distortions are:

- Artefacts according to block boundaries, since nearly all audio watermarking algorithms perform block wise processing of the audio signal (Arnold, Schmucker, & Wolthusen, 2003).
- Pre-echoes resulting from the smearing of noise in temporal domain due to block wise transformation from frequency to time.
- Increased roughness and modulation artefacts of the signals due to a lack of masking noise in silent regions of the audio signal
- Artificial signals by addition of nonexistent frequency components.
- Localization artefacts by changing delays in different channels.

Correspondingly, quality benchmarking principles and test methods used along the development of perceptual codecs can be applied in investigating the effects of audio watermarking encoding algorithms. Nevertheless, care has to be taken if one uses quality assessment methods for perceptual codecs (see the section, Implementation of a Quality Evaluation).

The different test procedures can be roughly classified into human subjective measurements and objective evaluation methods. A further distinction in the selection of the right test procedure is based on the definition of the quality evaluation problem:

- Testing the transparency of an audio watermarking codec.
- Comparing different audio watermarking algorithms regarding the quality of encoded tracks.
- Rating the quality of watermarked audio tracks.
- Testing watermarked tracks with intermediate quality.

Subjective Evaluation Standards and Methods

Performing subjective listening tests are still the ultimate evaluation procedures to judge the codec quality evaluation. Standardized test procedures have been developed to maximize the reliability of the results of subjective testing. The next two sections describe standards and general methods developed to perform subjective quality testing of coded audio data.

The subjective evaluation procedures will be generally distinguished regarding testing of the transparency of the watermarked items, rating of the quality of the processed items with respect to the reference signal, and testing of watermarked items with intermediate quality.

Testing Transparency

The ABX Double Blind Test

If the impairments introduced by the coding procedure are very small, one can assume transparency of the watermarked signal. In this case a subjective evaluation test for nontransparency can be conducted by the so-called ABX Double Blind Test or brief ABX Test. This is in contrast to an additional rating of the quality of watermarked tracks (see the section, The ITU-R BS.1116 Standard).

In the ABX test, the listener has access to three tracks labelled A, B, and X. The tracks A and B are the references. A is the original track whereas track B is the watermarked one. Track X is the unknown track which is selected randomly from A and B. Therefore both the person designing and doing the testing and the listener do not know if X is A or B[1]. During the test the subject is asked whether X is A or B by listening to the three items. In turn the listener will be either correct or wrong in judging whether X is A or B. Since the listener's answer in a trial is classified into one out of two categories it is a so-called Bernoulli trial. In a test with a number of n Bernoulli trials it is important that the response on each trial is independent on any other Bernoulli trial. If the requirement of independence is satisfied, the probabilities of wrong decisions in the listening test can be calculated as described below. This independence of the different trials imposes requirements on the setup of a listening test (see the section, Implementation of a Quality Evaluation).

Statistical hypotheses: In a listening study with n trials where the listener has to select one out of two alternatives, H_0 and H_1 are statistical hypotheses where the n trials constitute a random sample out of a conceptually infinite population of trials. Correspondingly the following null and alternative hypotheses are formulated:

H_0: Differences between original and watermarked tracks are not audible. The proportion of correct identifications p_0 in the population of trials is 0.5 and the listener will perform at chance.

H_1: Differences between original and watermarked item are audible. The proportion of correct identifications p_1 in the population of trials is greater than 0.5 and the listener will perform above chance.

A test for nontransparency H_1 is performed by trying to reject the transparency hypotheses H_0 is a so-called directional hypothesis. In this case listeners who are able to correctly identify X out of A and B with $p_1 > 0.5$. This is in contrast to a nondirectional hypothesis $p_1 \neq 0.5$ where correct identifications will be more or less by chance ($p_0 = 0.5$) which would not reflect the ability of the listeners to hear differences.

In practice the true proportion of correct identifications p_1 is never known. From the test $p_1 = c/n$ with c the number of correct identifications and n the sample size is measured. However it is not known which of the c identifications are due to the ability of the listener to hear differences and which are by chance. If p_1 were known the whole listening test, which was conducted to test hypothesis about p_1, it would be superfluous. The designer of the test has to define a value for p_1 (for example $p_1 = 0.6$) which represents such a small improvement over chance performance that it will not be used as a confirmation for audible differences.

Correct decisions and errors: In the ABX test the ability to detect differences between the original and the watermarked object is tested. A correct decision is called a hit, so a subject produces a result of the form "k hits out of a number of n Bernoulli trials," with k being the test variable. The probability of obtaining exactly *k* hits out of n with detection probability *p* can be calculated by the binomial distribution function:

Equation 1. Binomial distribution

$$P(k,n,p) = \binom{n}{k} p^k (1-p)^{n-k}$$

Under the hypothesis H_0 the probability to get *k* hits out of *n* with $p_0 = 0.5$ is:

$$P_0(k,n,p_0) = \binom{n}{k} p_0^k (1-p_0)^{n-k} = \binom{n}{k} 0.5^n$$

The alternative hypothesis H_1 asserts that the proportion of correct identifications $p_1 > 0.5$ and the probability to get *k* hits out of *n* is:

$$P_1(k,n,p_1) = \binom{n}{k} p_1^k (1-p_1)^{n-k}$$

For large n the direct calculation of the binomial distribution with equation (1) becomes troublesome. The individual probabilities for $k = 1, \ldots, n$ can be easily calculated via the following recursion formula starting from $P(0,n,p) = (1\text{-}p)^n$:

$$P(k,n,p) = P(k-1,n,p)\frac{n-k+1}{k}\frac{p}{1-p}$$

As an example, the two probability density functions for $n = 20$, $p_0 = 0.5$ and $p_1 = 0.7$ are plotted in Figure 1 (see Table 1).

A decision to reject or not to reject H_0 has to be based on a certain threshold for k resulting in different probabilities for correct decisions and errors. Due to the overlap of the two probability density functions, four different decisions are possible according to the defined threshold (see Table 2 and Figures 2 and 3).

Figure 1. Distribution functions for n = 20 trials and $p_0 = 0.5$, $p_1 = 0.7$

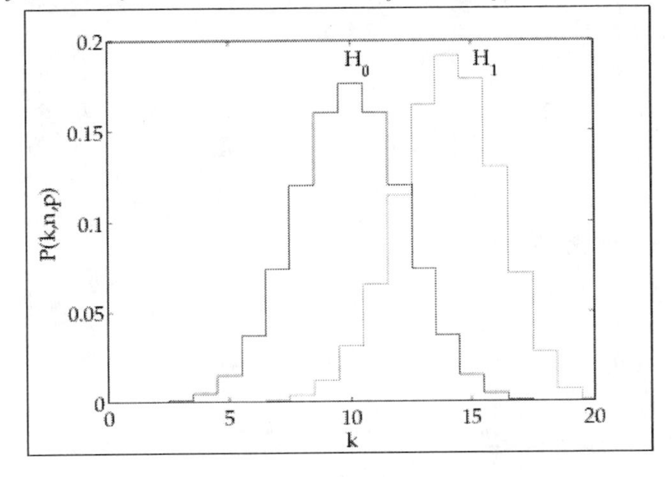

Table 1. Probabilities for H_0 and H_1

k	$p(k, 20, 0.5)$	$p(k, 20, 0.7)$
7	0.073	0.001
8	0.120	0.003
9	0.160	0.012
10	0.176	0.030
11	0.160	0.065
12	0.120	0.114
13	0.073	0.164
14	0.036	0.191
15	0.014	0.178
16	0.004	0.130
17	0.001	0.071

Table 2. True states, decisions and corresponding probabilities

		True State	
		H_0 is true (inaudible)	H_0 is false (audible)
Decision	H_0 is accepted (inaudible)	Correct ($1-\alpha$)	Wrong acceptance β
	H_0 is rejected (audible)	Wrong rejection α	Correct ($1-\beta$)

Figure 2. Correct decision $1-\alpha$ and error region α for threshold $T = 15$

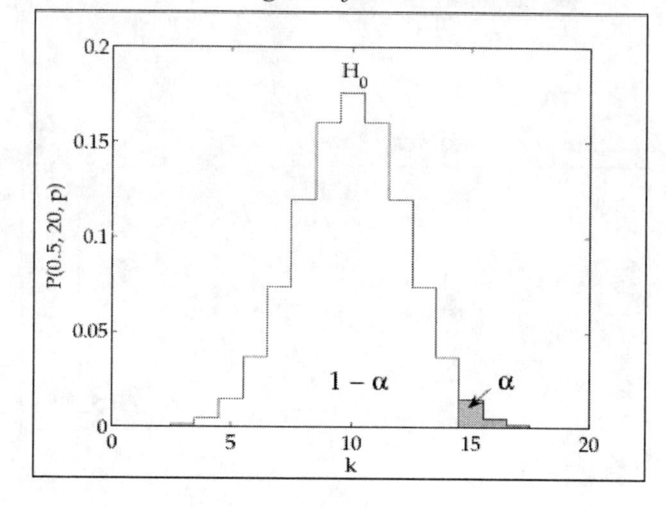

Figure 3. Correct decision $1-\beta$ and error region β for threshold $T = 15$

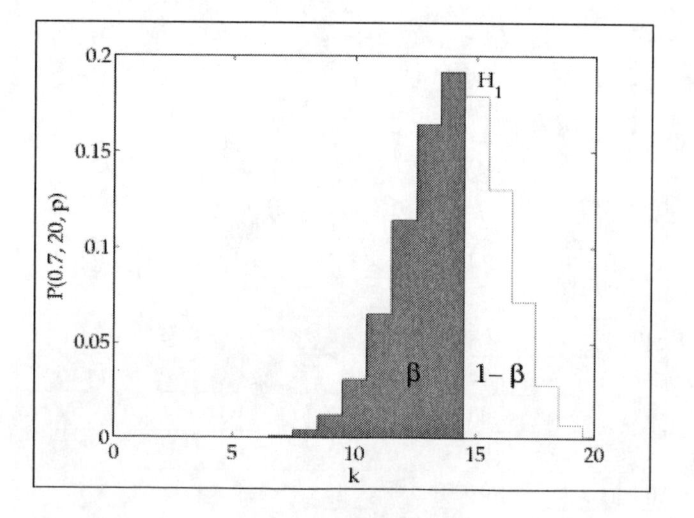

The two errors are:

- Wrongly rejecting (with error probability α) the null hypothesis and concluding that inaudible differences are audible; an error a watermarking system developer would like to avoid (type I error)
- Wrongly accepting (with error probability β) the null hypothesis concluding that audible differences are inaudible; an error an expert listener would like to avoid (type II error)

Significance Level and Type I and II Errors: The type I (α) and type II (β) error for threshold $0 \leq T \leq n$ can be determined by calculating the areas in Figure 2 and Figure 3 according to the equations:

Equation 2. Type I and II error calculation

$$\alpha(T,n,p_0) = \sum_{k=T}^{n} P(k,n,p_0)$$

$$\beta(T,n,p_1) = \sum_{k=0}^{T-1} P_1(k,n,p_1)$$

The threshold T is the lower bound of the so-called critical region B. B is the range of values which occur with a probability P by choosing a level of significance α.

Equation 3. Definition of critical region B

$$P(k \in B \mid H_0) \leq \alpha$$

For example, using a significance level $\alpha = 0.05$ and applying equation (3) for $n = 20$ trials using equation (2) leads to a threshold of $T = 15$. The critical region $B = 15, \ldots, 20$, that is, if a subject has more then 14 hits, the error probability of wrongly rejecting the null hypothesis, that is, concluding that the watermark is audible despite it being imperceptible, is lower or equal 5 %.

The corresponding type II error, that is, concluding that the watermark is inaudible despite it is audible, according to equation (2) is $\beta(15,20,0.7) = 0.5836$. In this example the traditional significance level of 0.05 used in statistical analysis results in a high risk of type II error β. Thus a wrong acceptance of the inaudibility hypothesis would be critical from the listener position due to the high error probability β included in the decision process.

The statistical test is called powerful if there is a high probability (which is equal to $1-\beta$) for correctly rejecting H_0 if H_0 is false. Therefore increasing the power of the test is equivalent to decreasing type II error (corresponding to the shaded region in Figure 3).

The power of the test can be increased in several ways. Both hypotheses are binomial distributions with corresponding mean $E[k] = np_{(0|1)}$ and variance $\text{var}[k] = np_{(0|1)}(p_{(0|1)} - 1)$. If the number of trials n gets bigger, the mean of k grows n and the standard deviation of k grows proportional to the root of the number of trials $\sqrt{n}$. Thus the probability distributions become more concentrated and the overlapping regions and corresponding errors are decreased. Another approach is to shift the probability distribution for H_1 to bigger values of the mean np_1 by increasing the detection ability p_1. The proportion of correct identifications p_1 can be increased by using expert listeners, a careful selection of the test material, and equipment which will be detailed in the section, Implementation of a Quality Evaluation. Due to the interdependence of type I and II errors on the threshold T (see equation (2)), type II error β can be decreased by increasing the type I error α as a last resort. Whether type I or type II error is more important depends on the application. If the relative importance is not further specified, the listening test should be designed with approximately equal error rates (Leventhal, 1986). The required sample size n can be calculated by choosing values for the error rates and hypothized detection proportion p by using the normal approximation to the binomial distribution.

Normal approximation: If the sample sizes n are large enough[2], the distributions of k and the sample proportion p are approximately normal according to the central limit theorem. The corresponding mean and variance for the approximately normal distribution of k are identical to the mean np and variance $np(p-1)$of the binomial distribution.

The significance level α and type II error β with continuous correction for small sample sizes are calculated as follows:

$$z_\alpha = \frac{(T-0.5)-np_0}{\sqrt{np_0(1-p_0)}}$$

$$z_\beta = \frac{np_1-(T-0.5)}{\sqrt{np_1(1-p_1)}}$$

The following figure depicts the continuous correction for $n = 20$, threshold $T = 12$, and the proportion of correct detection $p_0 = 0.5$ and $p_1 = 0.7$.

Figure 4. Continuous correction for H_0 at threshold $T = 12$

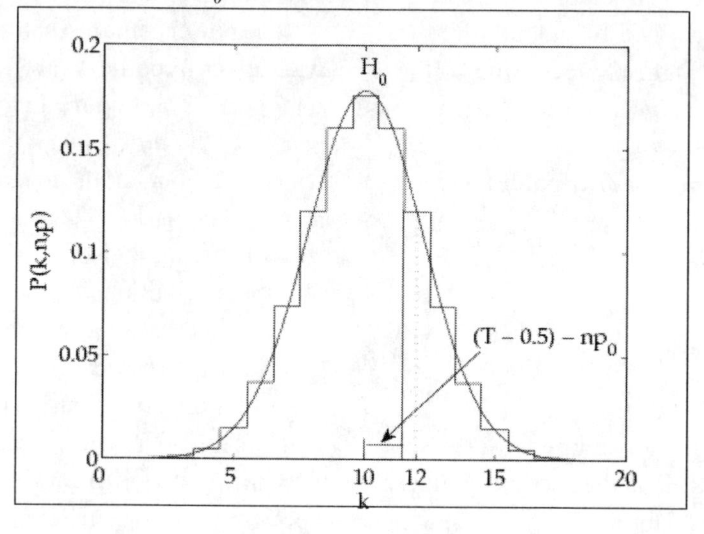

Figure 5. Continuous correction for H_1 at threshold $T = 12$

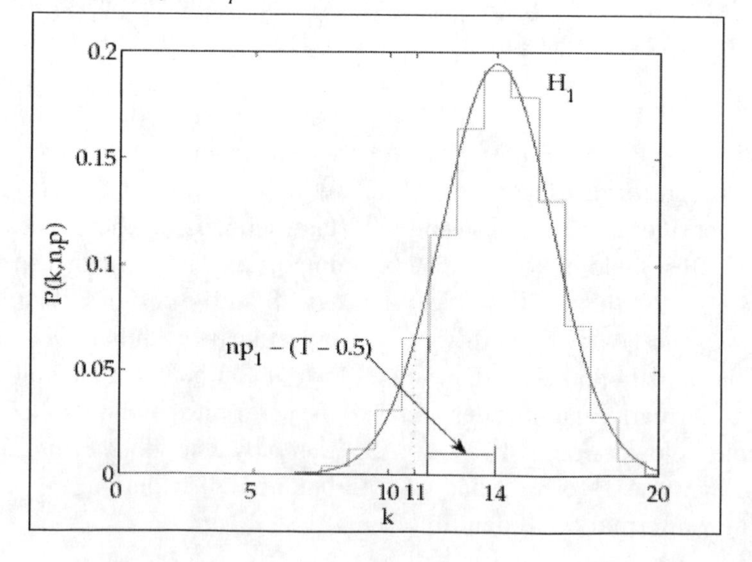

Calculation of Sample Size: Using the same threshold T in calculation of both approximations, the sample size n can be calculated as a function of the two proportions of identifications $(p_{01} p_{A})$ and the z-values for the defined error probabilities (z_{α}, z_{β}) (Burstein,1988).

Equation 4. Calculation of sample size n as a function of error probabilities and detection probabilities.

$$n(\alpha, \beta, p_0, p_1) = \left(\frac{z_{\alpha}\sqrt{p_0(1-p_0)} + z_{\beta}\sqrt{p_1(1-p_1)}}{p_1 - p_0} \right)$$

The sample size $n(\alpha, \beta, 0.5, p_1)$ as a function of the detection probability p_1 for equal error ($\alpha = \beta$) rates are plotted in Figure 6.

The person designing the test has to decide which value of p_1 is a negligible improvement in comparison to just guessing with $p_0 = 0.5$ that it will not be used as a confirmation for audible differences. By defining the significance level α with the condition of equal error rates ($\alpha = \beta$), one can determine the sample size n from Figure 6. As can be deduced the sample size increases considerably if the error rates are halved.

The evidence of the statistical significant result depends on the defined error risks. A result at:

- A risk of α and β below 0.1% suggests extremely strong evidence about the audibility respectively inaudibility of a difference.
- A risk of α and β between 0.1-1% suggests very strong evidence about the audibility respectively inaudibility of a difference.
- A risk of α and β between 1-5% suggests strong evidence about the audibility respectively inaudibility of a difference.
- A risk of α and β between 5-10% suggests moderate evidence about the audibility respectively inaudibility of a difference.

Design of the ABX test: The design of the ABX test requires the specification of the number of correct identifications p_1 and the error risks (α, β). The required number of trials n can in turn be calculated from these figures using equation (4). With the significance level α and the number of trials n, the threshold T can be derived from equation (2).

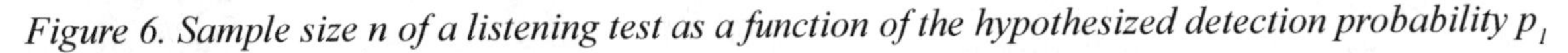

Figure 6. Sample size n of a listening test as a function of the hypothesized detection probability p_1

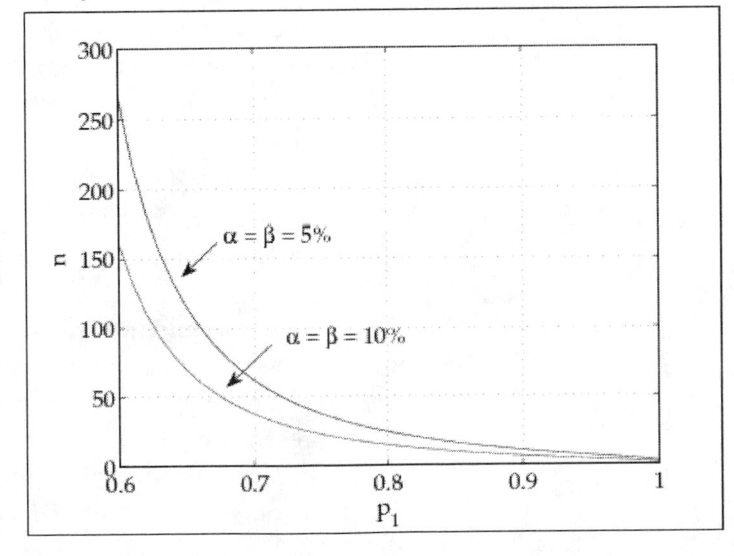

Analysis and interpretation of results: For each audio track, count the number of correct identifications k out of n trials. Comparison of the number of hits k with threshold T leads to the interpretation:

$k < T$ The watermark is inaudible.

$k \geq T$ The watermark is audible.

In a table, the (yes/no) decision for accepting the inaudible or audible hypothesis should be presented for each audio track (Arnold, 2002).

Rating the Quality

The ITU-R BS.1116 Standard

This section describes the standard for subjective evaluations of small impairments of high quality perceptual audio coders as specified in the International Telecommunication Union (ITU-R) Recommendation BS.1116 (ITU-R, 1997a)[3]. It has been designed to assess the degree of annoyance any degradation of the audio quality causes to the audio track. A continuous grading scale with the fixed points derived from the ITU-R subjective difference grade (SDG) scale (Recommendation ITU-R BS.1284, ITU-R, 1997b) listed in Table 3 is used.

The test procedure is a so-called double-blind A-B-C triple-stimulus hidden reference comparison test. Stimuli A always contains the reference signal, whereas B and C are pseudo randomly selected from the coded and the reference signal. After listening to all three items, the subject has to grade either B or C according to the above mentioned grading scale. As soon as the subject has completed the grading of a trial it should be possible to proceed to the next trial. Since one of them is the hidden reference, there are two possible outcomes of the grading process. First, the coded signal and the hidden reference are correctly identified. In this case the hidden reference will be assigned the default grade of 5.0 and its impairment is graded accordingly. Second, the reference signal is identified as being coded, which is an indication that no difference was audible. The results of the listening tests are reported with so-called subjective difference grade metric (see Table 3). The SDG value is derived from the rating results by subtracting the scores of the actual hidden reference signal from the score of the actual coded signal:

$$SDG = Score_{\text{Signal Under Test}} - Score_{\text{Reference Signal}}$$

A SDG value of 0 corresponds to an inaudible watermark whereas a value of -4.0 indicates an audible very annoying watermark.

Table 3. ITU-R five-grade impairment scale

Impairment	Grade	SDG
Imperceptible	5.0	0.0
Perceptible, but not annoying	4.0	-1.0
Slightly annoying	3.0	-2.0
Annoying	2.0	-3.0
Very annoying	1.0	-4.0

Design of the test: The standard (ITU-R, 1997a) specifies 20 subjects as an adequate size for the listening panel if the technical conditions can be strictly controlled. Per grading session, 10-15 trials should be scheduled. The test signals should be approximately 10-20 seconds long. To avoid listener fatigue the rest periods between grading sessions should be greater or equal to the session time.

Analysis and interpretation of results: The SDG values represent the data which should be used for the statistical analysis. In a graphical representation the mean SDG value and the 95% confidence interval should be plotted as a function of the different audio tracks (Lemma,

Aprea, Oomen, & Kerkhof, 2003) for the different coding algorithms to clearly reveal the distance to transparency (SDG = 0). In order to be able to interpret the average performance of each of the watermark codecs, the reliability of the differences between all SDG means have to be analyzed statistically.

In a first stage, an analysis of variance (ANOVA) should be applied to test the hypothesis that there are no differences between the SDG means.

H_0: The SDG means of the different watermarking codecs are equal

H_1: H_0 is incorrect

If any significant effects are found by applying ANOVA it tells only that two or more SDG means are unequal but does not tell which means are different. Therefore subsequent statistical tests have to be applied to find out which of the means are different from others. These so-called post hoc comparisons are Tukey's HSD[4], Neuman-Keul's, and Scheffé's tests. These tests can make all pair wise comparisons between the SDG means.

For an example experiment applied to subjective testing of multichannel audio systems see Kirby (1995).

Testing Intermediate Quality

The ITU-R BS.1534 Standard

In certain watermarking applications it might be reasonable to use an intermediate audio quality. If for example a MP3 encoded audio track with a bit rate of 48 kBit/s, which corresponds to a quality level well below transparency, shall be marked for forensic tracking, it is not necessary that the watermark is not audible. In this case it is sufficient if audio quality is not further decreased by the watermark encoding process.

As noted above, the ITU BS.1116 recommendation and its objective counterpart BS.1387 are not intended to be used in this scenario. The comparatively new subjective listening tests termed multi stimulus with hidden reference anchors (MUSHRA) ITU-R BS.1534 (ITU-R, 2001)[5], should be used to evaluate items with intermediate quality. In contrast to the former standards, the outcome of this test is not to see whether the marked audio signal and the original signal could be distinguished by the listener, but to rate the quality of the marked signal relative to other test signals of a reference and a hidden anchor. The five-interval continuous quality scale (CQS), which is divided into five intervals as shown in Table 4, is used for this grading.

Design of the test: As it is the case for the ITU-R BS.1116 test, the standard (ITU-R, 2001) specifies 20 subjects as an adequate size for the listening panel if the technical conditions can be strictly controlled. Each trial contains three signals (1 reference + 1 hidden reference + 1 hidden anchor) plus the impaired signals from the watermarking codecs. No more than 15 signals corresponding to 12 impaired signals should be presented in any trial. The test signals should be approximately 10 - 20 seconds long.

Analysis and interpretation of results: The normalized scores in the range 0 - 100 represent the data which should be used for the statistical

Table 4. Five-interval continuous quality scale (CQS)

Quality	Grade	Internal numerical representation
Excellent	5.0	100
Good	4.0	80
Fair	3.0	60
Poor	2.0	40
Bad	1.0	20

analysis. In a graphical representation the mean score value and the 95 % confidence interval should be plotted for fixed test condition and audio track for the different coding algorithms. The mean and standard deviation is obtained by averaging over the listeners and is used to compare the different codecs.

For an example experiment applied to subjective testing of MPEG Layer II see Soulodre and Lavoie (2004).

Objective Evaluation Standards

The ITU-R BS.1387 Standard

The ultimate goal of the objective measurement algorithms is to substitute the subjective listening tests by modelling the listening behaviour of human beings. The output of the algorithms for objective measurements is the so-called objective difference grade (ODG) consisting of a single number to describe the audibility of the introduced distortions like SDG used in subjective listening tests.

This objective measurement method of the perceived audio quality (PEAQ) is an international standard, ITU-R BS.1387 (ITU-R, 1998). It should replace the described ITU-R BS.1116 and is therefore only useful in detecting small distortions in high quality audio material.

The PEAQ method for objective measurement of audio quality fit into the principle architecture according to Figure 7.

A difference measurement technique is used to compare the reference (original) signal and the test (processed, i.e., compressed or watermarked) signal. Both the reference signal and the signal under test are processed by an ear model, which calculates an estimate for the audible signal components. The PEAQ method includes a basic and an advanced version. The basic version of ITU-R BS.1387 uses an ear model based on the FFT. The advanced version uses the ear model of the basic version and one based on a filter bank.

The signal outputs from both ear models represent the two different internal representations of the reference and the test signal. These components can be regarded as the representation of the signals in the human auditory system. The internal representation is often related to the masked threshold, which in turn is based on psychoacoustic experiments performed by Zwicker and Fastl (1999). These signals being functions of time and frequency are further processed into functions of time. A subsequent averaging in time results in a number of single values of the model output variables (MOVs).

In the basic version the eleven model output variables are derived from the FFT model. The advanced version calculates five model output variables which are partially derived from the FFT

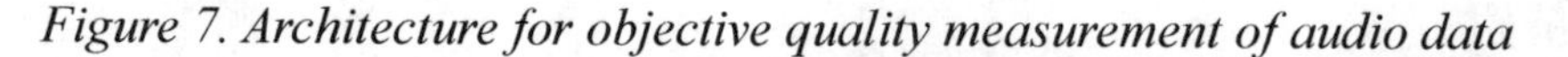

Figure 7. Architecture for objective quality measurement of audio data

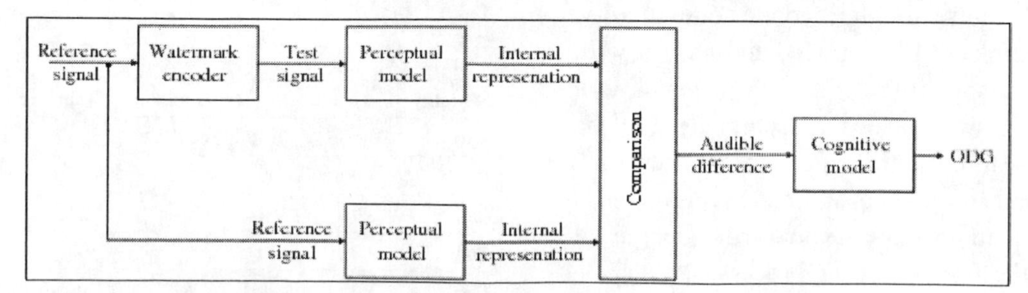

Note: Reproduced from Arnold (2003) by permission of Artec House.

and the filter bank model. Because the results of the listening tests are judged with a single SDG value, the corresponding ODG value has to be derived from the audible difference. The calculated MOVs will be the input of a neural network in order to calculate the ODG which measures the quality degradation of the test signal with respect to the reference. To provide good matches to the corresponding subjective impairment scale (see Table 3), the neural network has been trained exhaustively. This neural network models the processing of the signals by the human brain during the listening tests.

Design of the test: Since the intention of ITU-R BS.1387 standard is to replace the ITU-R BS.1116 standard, the design of this test method reflects the corresponding test method of ITU-R BS.1116.

Analysis and interpretation of results: The ODG values can be computed with the freely available implementation by Peter Kabal (2003). In a graphical representation, the ODG value should be plotted as a function of the different audio tracks for the different coding algorithms to clearly reveal the distance to transparency (ODG = 0).

For an example experiment applied to investigate the influence of audio watermarking on the audio quality with PEAQ and a comparison with the results of subjective listening tests, see Neubauer and Herre (1998).

Practical Limitations

As it is the case for the peak-signal-to-noise-ratio (PSNR) in evaluating the influence of compression on images, the objective measurement metric namely ODG does not always correlate very well with the result from the subjective listening tests (see Keiler, 2006).

The distortions introduced during lossy compression of audio data by adding quantization noise are similar to the watermarking encoding process if the masking threshold is used for shaping the distortion signal. In this case the ODG value correlates quite well with the SDG value. Therefore the ITU-R BS.1387 standard has to be used with care in assessing the quality of watermarked audio tracks. It can be used as a preselection mechanism for critical items regarding the watermarking embedding process. However a final judgement regarding quality has to be based on subjective listening tests.

Implementation of a Quality Evaluation

This section will provide a practical guideline for a quality evaluation for watermarked audio tracks. To perform a reliable testing, the whole test includes the following tasks which will be detailed below:

- Project objective (project leader)
- Test objective (test designer and project leader)
- Selection of test items
- Design of the test
- Conducting the test
- Analyzing the test results
- Interpretation and reporting of test results

Project Objective

Defining the research project objective is the most important requirement in order to select the right test. Does the robustness improvement of an existing audio watermarking algorithm influence the quality of watermarked audio tracks? Does the new developed audio watermarking algorithm produce watermarked audio tracks with transparent quality? Can an audio watermarking algorithm be a substitution of another one to permit cost reduction? Which of a set of audio watermarking algorithms produce the best quality of watermarked tracks? What quality setting is sufficient for the anticipated application? Which of the samples are critical to the specific audio watermarking technology?

Test Objective

If the project objective can be clearly stated, the test objective in terms of quality can be determined; overall difference (transparency), relative preference (rating), acceptability (intermediate quality), and so forth. In general one has to distinguish between testing the transparency, rating the quality, or testing watermarked tracks with intermediate quality. A good approach is to clearly write down the project objective, the test objective, and a brief statement of how the test results will be interpreted and used.

Selection of Items

Special emphasis has to be devoted to a careful selection of the test material used. Since it is not possible to test thousand of sound files in listening tests, the test material should be a representative sample from the infinite population of all possible tracks for the intended application. If specific types of sound signals like speech signals are used within the application, selecting a representative sample is a manageable task. In general it will not be possible to explicitly narrow the test material due to the great variety of sound material. In this case a standardized set of sound material to test audio watermarking algorithms should be used.

This ensures that:

- The evaluation results of different watermarking systems are comparable.
- Subsequent tests between different versions of a watermarking algorithm are comparable.
- The used sound files represent critical tests cases.

The standard test material used in evaluating perceptual audio codecs is from the sound quality assessment material (SQAM) material (EBU, 1988). These signals are chosen to reveal to the listener small impairments which can happen in analogue and digital audio systems. The collection of signals represents test signals for different tasks. A further subselection of audio test samples from this material is driven by the algorithm of the watermarking codec to be tested. Since the goal of a low bit-rate codec and a watermarking encoder of ensuring the quality of processed signal is the same, the selection of the test signals can be based on these items (see ITU-R, 1998).

Attributes of the audio signal which can cause problems to the watermark encoder are:

1. Attacks in the time domain which can result in smearing of noise in the temporal domain due to block wise transformation from frequency to time domain (pre-echoes).
2. Tonal signals are sensitive against noise insertion and increase in roughness due to a lack of masking signals hiding the additional watermark signal.
3. Natural speech signals which consist of a lot of interval times where no watermarks can be embedded.
4. Small bandwidth signals where watermark encoder may add additional frequencies components during embedding which are not masked by the original signal.

Table 5 provides an example list of test signals with comments about the special attributes of each signal.

Design of the Test

Based on the definition of the research project, the corresponding test objective, and the screening of

Table 5. Subset of the test signals suitable for assessing watermark encoder

Item	Attribute
Castanets	1.
Flute	2.
Glockenspiel	2.,4.
Marimba	2.,4.
Speech Female English	3.
Speech Male English	3.
Speech Male German	3.
Triangle	2.,4.
Xylophone	2.

the appropriate sound material, the test has to be designed. The design of the test involves:

- Selection of the test technique (see the sections, Subjective Evaluation Standards and Methods - Objective Evaluation Standards).
- Selecting and training subjects.
- Recording of test results.
- Description of test equipment, setup and implementation.

Selection of Test Technique

The selection of the test technique is based on the test objective:

- Use ABX to determine whether an overall difference between original and watermarked tracks exists.
- Use standard ITU-R BS.1116 to compare different watermarking codecs by rating the quality.
- Use standard ITU-R BS.1116 to compare same watermarking codec with quality parameter varied.
- Use standard ITU-R BS.1534 to evaluate watermark items with intermediate quality.

Selecting and Training Subjects

Just as well as a use of a standardized test set, a careful selection of the listeners ensures that the most sensitive evaluation of the test objective is obtained. If designated expert listeners are participating in the listening test the correct identification proportion can be increased and correspondingly the errors involved in the test decreased (see the section, The ABX Double Blind Test). The ITU-R BS.1116 standard (ITU-R, 1997a) (see Section 3 and Appendix 1 of Annex 1) specifies a procedure for the selection of the listeners.

Before conducting the test, the listeners have to be familiarized with the test equipment. Prepared instructions including a description of the presentation of sound material should let them understand the test methodology. In the training phase (see Section 4.1 and Appendix 3 of Annex 1 of the ITU-R BS.1116 standard (ITU-R, 1997a)) they should be exposed to all selected sound material that they will be listening to during the actual test.

Number of Subjects

Furthermore, the type of listeners influences the number of subjects that should do the listening test. For the ABX test, the calculation of the necessary number of subjects to ensure specified decision errors was presented in section The ABX Double Blind Test. The ITU-R standards BS.1116 (see the section, The ITU-R BS.1116 standard) and BS.1534 (see the section, The ITU-R BS.1534 Standard) include guidelines on the number of subjects to be used in the listening test. Table 6 contains a summary on the number of the subjects for the different presented test methods.

Table 6. Number of listeners for different test methods

Test name	Number of subjects
ABX Double Blind Test	According to decision errors and detection probability
ITU-R BS.1116	20
ITU-R BS.1534	20
ITU-R BS.1387	No listeners required

Nevertheless the number of subjects can be reduced significantly if experts are performing the listening test in contrast to average listeners.

Recording Results

A score sheet has to be designed for each test to record the results of each trial. An automated presentation of the test samples and recording software of the results is beneficial. It minimizes errors occurring during presentation of sound files and the additional transferring of the results from the score sheet in the database for later evaluation.

Listening Environment

The purpose of creating a specific listening environment is to define reference environments for certain types of listening tests. The listening rooms specified in the standards ensure comparability of the listening results obtained from research work over different laboratories.

The listening room has to control the following acoustic characteristics:

- Distribution of low-frequency standing waves
- Room reverberation
- Background noise
- Interferences from outside (acoustic and vibration)

The ITU-R BS.1116 standard (ITU-R, 1997a) designed to assess small impairments of high quality audio includes the specification of a reference listening room (see Section 8 of Annex 1 in the standard).

Testing multichannel sound systems requires the usage of a reference room for performing the listening test. The influence of embedding watermarks into mono and stereo signals can also be tested using headphones. In this case the influence of the room characteristics can be neglected. Nevertheless in order to minimize disturbance from outside, the listening test should be performed in a separate cabin to reduce noise and vibration.

Test Equipment, Setup, and Implementation

To present the test material optimally, high quality equipment should be used (see standard ITU-R BS.1116, ITU-R, 1997a). The listener should be alone in the listening room. Social interaction with other listeners during listening time will violate the independence assumption for different trials. In general no listening time limit should be imposed during the training and test phase. However due to getting tired with a long listening test the time should be limited to 30 minutes. In order to guarantee the independence of the individual trials the presented sound material should be randomized for each trial and each listener.

Conducting the Listening Test

A test stage should precede the actual listening test in order to be sure that everything works properly and all requirements of the test design are met. Variations have to be recorded in order to take them into account in the analyzing step.

Analyzing Results

The specific statistical analysis of the results is determined by the test design which specified the appropriate test method. The result data should be analyzed for the test objective.

Interpretation and Reporting

The obtained data from the statistical analysis have to be reviewed in relation to the project and test objectives and expressed in terms of the stated objectives. This should be written in a test report that summarizes the project and test objectives, the test items used, the listeners expertise, the test performed, the data obtained, and the conclusion drawn from the statistical analysis.

SUMMARY

This chapter described quality evaluation methods of watermarked audio tracks. The application of the procedures presented was motivated by the similar problem in the evaluation of the performance of compression codecs with regard to the quality of lossy compressed audio data. Like compression codecs, which add quantization noise during the encoding process, audio watermarking systems alter the audio in a way, which has to be kept inaudible. Therefore both subjective listening tests and objective measurement methods can be transferred to the problem of evaluating the quality of watermarked audio tracks with small impairments. Furthermore the problem of judging watermarked audio tracks with intermediate quality by applying the so-called MUSHRA test was addressed. Subjective listening test standards and classical methods namely the ABX test were presented in detail. To overcome the problem of time-consuming and cost-intensive subjective testing, the appropriate objective test standard namely BS.1387 was presented. A guideline for the practical implementation of a quality evaluation procedure was presented in a separate section.

FUTURE RESEARCH DIRECTIONS

Researchers, developers, broadcasters, and other customers want to assess different implementations of watermarking algorithms. Objective measurement methods are superior to subjective listening tests especially during the development phase of new audio watermarking algorithms, because the former requires, due to its ability to be automated, a lot less effort, time, and money than latter.

The presented objective measurement method (PEAQ) the ODG correlates quite well with the SDG when averaged on a large collection of audio tracks (Treurniet & Soulodre, 2000). Nevertheless the individual audio tracks show a great deviation between the ODG and its subjective counterpart (Treurniet & Soulodre, 2000).

Hence one important research questions is how the objective measurement methods can be improved so that it can be applied to measure the quality of the watermarked audio tracks.

The objective measurement model consists of a psychoacoustic and a cognitive model (Thiede, Treurniet, Bitto, Schmidmer, Sporer, Beerends, et al., 2000). An improvement of objective measurement methods should be achieved if both parts of the model advance. There is a need to integrate more sophisticated human auditory models, which give a better approximation of the perceptual thresholds in order to determine the detectability of stimuli near those thresholds (Baumgarte, 1998; Dau, 1996a; Dau, Püschel,

& Kohlrausch, 1996b). The cognitive model in the PEAQ algorithm is tuned by using databases which consists of audio files processed by different codecs and corresponding SDG values. The calibration of the model was done by minimizing the difference between ODG and the distribution of the mean SDG. Therefore the model implicitly mirrors the artefacts due to lossy compression (Thiede et al., 2000).

The available watermarking algorithms exploit different psychoacoustic phenomena and therefore will introduce varying artefacts. This is quite the same problem for newer codecs using, for example, spectral band replication (SBR) (Dietz, Lijeryd, Kjorling, & Kunz, 2002). Whether the cognitive part of the PEAQ algorithm trained with the results from the subjective listening tests performed on compressed audio tracks is useful for evaluating the quality of watermarked tracks, it has yet to be proven. Otherwise, calibration of the model to the artefacts introduced by watermarking methods may lead to a better correlation. Therefore, subjective listening tests and objective measurement have to be made in order to provide insight in the applicability of the presently available evaluation tools.

Moreover, objective methods for evaluating quality of multichannel sounds have not been addressed so far. An interesting research questions explores how the embedding of watermark affects the presentation of multichannel sounds.

REFERENCES

Arnold, M. (2002, December). Subjective and objective quality evaluation of watermarked audio tracks. In C. Busch, M. Arnold, P. Nesi, & M. Schmucker (Eds.), *Proceedings of the Second International Conference on WEB Delivering of Music (WEDELMUSIC 2002)* (Vol. 4675, pp. 161-167). Darmstadt, Germany: IEEE Computer Society Press.

Arnold, M., Schmucker, M., & Wolthusen, S. (2003). *Techniques and applications of digital watermarking and content protection*. Boston: Artech House.

Baumgarte, F. (1998). Evaluation of a physiological ear model considering masking effects relevant to audio coding. In *Proceedings of the 105th Audio Engineering Society Convention*. San Francisco: AES.

Burstein, H. (1988, November). Approximation formulas for error risk and sample size in *ABX* testing. *AES Journal, 36*(11), 879-883.

Cox, I.J., Miller, M.L., & Bloom, J.A. (2002). *Digital watermarking*. San Francisco: Morgan Kaufmann Publishers.

Dau, T. (1996). *Modeling auditory processing of amplitude modulation*. Unpublished doctoral thesis, University Oldenburg, Oldenburg, Germany.

Dau, T., Püschel, D., & Kohlrausch, A. (1996). A quantitative model of the effective signal processing in the auditory system. I. Model structure. *AES Journal, 99*(6), 3615-3622.

Dietz, M., Lijeryd, L., Kjorling, K., & Kunz, O. (2002). *Spectral band replication, a novel approach in audio coding*. In *Proceedings of the 112th Audio Engineering Society Convention*. Munich, Germany: AES.

EBU. (1988, April). *Sound quality assessment material recordings for subjective tests*. Bruxelles, Belgium.

ITU-R. (1997a). *Recommendation BS.1116-1, methods for subjective assessement of small impairments in audio systems including multichannel sound systems*. International Telecommunications Union Radiocommunication Assembly.

ITU-R. (1997b). *Recommendation BS.1284-1, general methods for the subjective assessement of audio quality.* International Telecommunications Union Radiocommunication Assembly.

ITU-R. (1998). *Recommendation BS.1387, method for objective measurements of perceived audio quality (PEAQ).* International Telecommunications Union Radio Communication Assembly.

ITU-R. (2001). *Recommendation BS.1534, method for the subjective assessment of intermediate quality level of coding systems.* Paper presented by the International Telecommunications Union Radiocommunication Assembly.

Kabal, P. (2003, December). *An examination and interpretation of ITU-R BS.1387: Perceptual evaluation of audio quality* (Tech. Rep.). Montreal, Canada: McGill University, Department of Electrical & Computer Engineering.

Keiler, F. (2006, May). Real-time subband-ADPCM low-delay audio coding approach. In *Proceedings of the 120th Audio Engineering Society Convention.* Paris: AES.

Kirby, D.G. (1995, October). ISO/MPEG subjective tests on multichannel audio systems. In *Proceedings of the 99th Audio Engineering Society Convention.* New York: AES.

Lemma, A.N., Aprea, J., Oomen, W., & Kerkhof, L. van de. (2003, April). A temporal domain audio watermarking technique. *IEEETASSP: IEEE Transactions on Signal Processing, 51*(4), 1088–1097.

Leventhal, L. (1986, June). Type 1 and type 2 errors in the statistical analysis of listening tests. *AES Journal, 34*(6), 437-453.

Neubauer, C., & Herre, J. (1998, September). Digital watermarking and its influence on audio quality. In *Proceedings of the 105th Audio Engineering Society Convention.* San Francisco: AES.

Soulodre, G.A., & Lavoie, M.C. (2004, October). Subjective evaluation of MPEG layer II with spectral band replication. In *Proceedings of the 117th Audio Engineering Society Convention.* San Francisco: AES.

Thiede, T., Treurniet, W.C., Bitto, R., Schmidmer, C., Sporer, T., Beerends, J.G., et al. (2000). Evaluation of the ITU-R objective audio quality measurement method. *AES Journal, 48*(1/2), 3-29.

Treurniet, W.C., & Soulodre, G.A. (2000). Evaluation of the ITU-R objective audio quality measurement method. *AES Journal, 48*(3), 164-173.

Zwicker, E., & Fastl, H. (1999). *Psychoacoustics: Facts and models* (2nd ed.). Heidelberg, Germany: Springer-Verlag.

ENDNOTES

1. Hence the name double blind test
2. A good rule of thumb is to use the normal approximation only if $np > 15$ and $n(p-1) > 15$
3. Published in 1994 and updated in 1997
4. HSD stands for honestly significant difference
5. Published in 2001 and updated in 2003

Chapter XIV
Watermarking Security

Teddy Furon
INRIA, France

François Cayre
LIS /INPG, France

Caroline Fontaine
CNRS/IRISA, France

ABSTRACT

Digital watermarking studies have always been driven by the improvement of robustness. Most of articles of this field deal with this criterion, presenting more and more impressive experimental assessments. Some key events in this quest are the use of spread spectrum, the invention of resynchronisation schemes, the discovery of side information channel, and the formulation of the embedding and attacking strategies as a game. On the contrary, security received little attention in the watermarking community. This chapter presents a comprehensive overview of this recent concept. We list the typical applications which require a secure watermarking technique. For each context, a threat analysis is proposed. This presentation allows us to illustrate all the certainties the community has on the subject, browsing all key papers. The end of the chapter is devoted to what remains not clear, intuitions, and future trends.

WATERMARKING SECURITY

So far, this book has presented watermarking as the art of hiding metadata in content in a robust manner. 'Hiding' has unfortunately many meanings. Some understand that the embedding of metadata does not cause any perceptual distortion. Watermarking is then the art of creating a communication channel inside a piece of content without spoiling its entertainment. Others cast a security requirement in the word 'hiding.' This surprisingly happened at the very beginning of the digital watermarking story as detailed below.

The objectives of this chapter are to motivate the need of a new concept of 'watermarking security' through an historical perspective. This concept then needs to be clearly defined especially with respect to robustness. A second objective is to build a methodology to assess the security level of watermarking schemes. The analysis is twofold; a first theoretical study is based on the amount of information about the secret key which leaks from observations made by the attacker. The second study is more practical as a kind of proof of concept. Its goal is to show that the exploitation of these information leakages stems in a practical estimation of the secret key.

BACKGROUND

Birth of a New Concept

As a first element of the background section, we would like to motivate the birth of security, a new concept in watermarking, by an historical point of view.

In the analog age, content was protected by copyright laws included in intellectual property treaties dating back from the 50s (Maillard & Furon, 2004). There was a balance between conflicting issues like the copyright holders interests and the user-friendly usage of content. The digital age and the merging of formats from the entertainment and computer industries broke this balance in the 90s, spoiling copyright holders. Technical barriers have been created to enforce the copyright laws[1]. As cryptography leaves insecure protected content once decrypted by users, a recent technology named digital watermarking was perceived as the last line of defence. It allows to firmly bound content with metadata such as the copyright holder identity (copyright protection (Craver, Memon, Yeo, & Yeung, 1998) or the copy status (copy protection) (Bloom, Cox, Kalker, Linnartz, Miller, & Traw, 1999). At that time, the naïve rationale was, 'If you cannot see it, and if it is not removed by common processing, then watermarking must be secure.'

Unfortunately, digital watermarking was too young a science to support such an adventurous assertion. The technique was even lacking sufficient robustness to fulfil the requirements of these first applications. Defeats happened very soon (Stern & Craver, 2001), so that the watermarking community envisaged applications where security is not an issue (e.g., content enhancement). On the front of copyright and copy protection, new laws have been promoted in the 2000s forbidding the circumvention of a digital right management (DRM) system (Maillard & Furon, 2004). In a way, this new legal framework patches the security flaws of technical barriers, including digital watermarking. There are now three walls of defence; new laws protect the technical barriers which protect the enforcement of old copyright law which protect content's use and exploitation. On the other hand, absolute security does not exist (not even in cryptography) and a high security level has a cost which nobody wants to pay for (copyright holders, device manufacturers, and users). The goal of the entertainment industry is not to erase piracy but to maximise their incomes. To this end, weak security is better than no security (Cox & Miller, 2001), and a slightly secure but cheap protection system is enough to 'keep honest people honest.'

This historical point of view shows that security of digital watermarking has clearly lost interest in real life applications. However, it becomes a hot issue in the watermarking community (Barni & Perez-Gonzalez, 2005; Bartolini, Barni & Furon, 2002). We believe that researchers have stretched the limit of robustness to almost its maximum so that new attacks pertain more to security than classical robustness. Because a secure but nonrobust watermarking technique would be useless, robustness is the weakest link and it was the priority to be fixed. Huge improvements have been done in this field, and security now appears as the next issue on the list. Even if it is less important for real applications, it is also theoretically challenging because very few certainties are known about watermarking security.

Elements of Definition

Does a short and concise definition of watermarking security exist? This question stems from two facts: watermarking security has different implications according to the targeted application, and security is too close to robustness to be clearly distinguished (Doërr & Dugelay, 2005). Note that, so far, we have discussed security, understanding it as security of robust watermarking. It is time now to broaden our scope. In copy protection, copyright protection, and fingerprinting, we need to assess that dishonest users cannot remove the watermark signal. However, note that in copy protection, a pirate should not be able to change content status to a less restrictive one (e.g., from 'Copy Never' to 'Copy Once') (Bloom et al., 1999). In fingerprint, a collusion (group of pirates) should not frame an innocent user, that is, they should not change their hidden messages to the identifier of an honest user (Trappe, Wu, Wang & Liu, 2003). In copyright protection, authors should not copy and paste their watermark (possibly issued by a trusted third party) in content they did not create (by embedding a watermark signal or by creating a fake original) (Craver et al., 1997; Kutter, Voloshynovskiy & Herrigel, 2000). This ruins the identification of the owner because two watermarking channels interfere in the same piece of content. In authentication, the goal of the pirate is not to remove the authenticating watermark signal but to sign content in place of the secret key holder (Wong & Memon, 2001). In steganography, the pirate does not remove watermark signals but detects the presence of hidden data, and the watermarking technique used for it (Anderson & Petitcolas, 1998). The oracle attack is a threat whenever the opponent has access to a watermarking detector as in copy protection for consumer electronics devices (Cox & Linnartz, 1998). The attacker first estimates the secret key, testing the detection process on different pieces of content (Linnartz & van Dijk, 1998); this disclosure then helps the attacker forge pirated content. Note that in this last case, the number of detection tries is of utmost importance.

In a will to decouple the concept of security from the application layout, the following list suggests criteria to make a clear cut between robustness and security.

- **Intention:** In security, there obviously exists a pirate. In robustness, a classical content processing made without any malicious intention might delude the watermark decoder.
- **Specific hack:** Robustness usually considers classical content processing which can be dangerous for whatever watermarking technique. In security, pirates apply malicious attacks dedicated to one watermarking technique.
- **Removal:** In robustness, the effect of the attack is to delude the watermarking decoder. The attack succeeds in removing enough watermarking energy or it has desynchronised the embedder and the decoder. In security, we have seen that pirates' goals are different according to the targeted application.
- **Number of Steps:** In robustness, the pirate applies a processing to the watermarked piece of content. This is a single step process. In security, the pirate observes several watermarked pieces of content and gains from these observations some information about the watermarking technique and the secret key in use. Then, with this 'stolen' knowledge, the pirate attacks protected content. This is a two-step process. Some say the pirate is not fair, in the sense that the pirate is not contented with the official instruction (e.g., the watermarking technique according to the Kerckhoffs' principle), but the pirate tries to access all the information which may

be of any help for the goal (e.g., the secret key) (Barni, Bartolini & Furon, 2003).

- **Probability of success:** In robustness, an attack is usually not always successful, but it leads to a given bit error rate (decoding) or probability of a miss (detection). In security, a successful hack is almost granted when the pirate has an accurate estimation of the secret key (if this is the goal).

However, Kalker (2001) formulated very elegant definitions of robustness and security. These may not encompass all cases, but they are the only concise attempts we are aware of. 'Robust watermarking is a mechanism to create a communication channel that is multiplexed into original content [...]. It is required that, firstly, the perceptual degradation of the marked content [...] is minimal and, secondly, that the capacity of the watermark channel degrades as a smooth function of the degradation of the marked content. [...]. Watermarking security refers to the inability by unauthorised users to have access to the raw watermarking channel. [...] to remove, detect and estimate, write or modify the raw watermarking bits.'

Articles proposing a complete analysis of robust watermarking security are rare but becoming more and more popular. The authors are only aware of the pioneer work by Mittelholzer (1999), where two digital modulation schemes achieve perfect secrecy, and more recent works sketching a general framework for security analysis (Barni et al., 2003; Furon & Duhamel, 2003). The main idea is here to adapt Shannon's definition of cryptography security to watermarking. The remaining part of this chapter explains and illustrates this view of watermarking security. It has been developed in the seminal paper of Cayre, Fontaine, and Furon (2005c), and strengthened and generalised to quantisation-based watermarking schemes by Comesaña, Pérez-Freire, and Pérez-González (2005), Pérez-Freire, Comesaña, and Pérez-González (2005), and Pérez-Freire, Pérez-González, Furon, and Na (in press).

Notation

Let us introduce some notational conventions used in this chapter. Vectors are set in bold font, matrices in calligraphic font, and sets in black board font. Data are written in small letters, and random variables in capital ones. The length of the vectors considered in this chapter is N_v; $\mathbf{x}(i)$ is the i-th component of vector $\mathbf{x}$. The probability density function of random variable $\mathbf{X}$ (or its probability mass function if $\mathbf{X}$ is discrete) is denoted by $p_{\mathbf{X}}(.)$. Hidden messages have N_c bits and secret keys are usually composed of N_c elements (e.g., N_c secret carriers in the spread spectrum case). Finally, N_o vectors are considered; $\mathbf{x}^{N_o} = \{\mathbf{x}_j\}_{j=1}^{N_o}$ represents a collection of vectors and $\mathbf{x}_j$ is the vector $\mathbf{x}$ associated to the j-th observation.

Figure 1. Global point of view of the embedding process

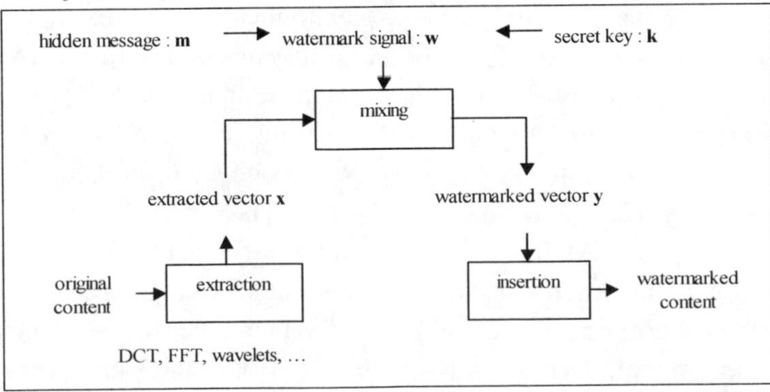

A Theory for Watermarking Security

This section deals with the above-mentioned attempts to find a theory of watermark security. What is the role of this theory? Before its existence, the security assessment of a watermarking technique was fuzzy in the sense that the analysts had to think about an attack and to see how dangerous it was. In other words, the security assessment was clearly dependent on the cleverness of the analysts. Maybe, later on, one will discover a more powerful attack which will lower the security level of the watermarking technique. In a way, the role of a watermarking security theory is to assess the security level once for all.

A Methodology Inspired by Cryptanalysis

The methodology presented in this chapter is clearly inspired by the cryptanalysis. It is based on three key articles in cryptology: Kerckhoffs (1883), Shannon (1949) and Diffie and Hellman (1976). We first briefly present these ideas, before formalising them in the following subsections.

Kerckhoff's Principle. Kerckhoffs (1883) stated that keeping an encryption algorithm secret for years is not realistic, and this principle is now used in any cryptographic study. In watermarking, the situation is similar, and it is assumed that the opponent knows the watermarking algorithm. Hence, for a given design and implementation of an algorithm, the security stems from the secrecy of the key. The designer's challenge is, 'Am I sure that an opponent will not exploit some weaknesses of the algorithm to disclose the secret key?' Watermarking processes are often split into three functions. The first one extracts some features from content (issued by a classical transform, such as DCT, wavelet, FFT, Fourier Mellin, etc.), which are stored in a so-called extracted vector x. The second one mixes the extracted vector with the secret watermark signal, giving a watermarked vector y. Then, an insertion function reverses the extraction process to come back in the original world, putting out the watermarked document. Figure 1 illustrates the embedding process. The detection follows an analogous process as sketched in Figure 2. According to the Kerckhoff's principle, the opponent knows all the involved functions. The opponent thus observes the watermarked vectors from contents there is access to, because the extraction function has no secret parameter.

Shannon's Approach. The methodology that Shannon (1949) exposed for studying the security of encryption schemes is here transposed to watermarking. The embedder has randomly picked up a secret key, and used it to watermark several pieces of content. The opponent observes these pieces of watermarked content, all related to the same secret key but hiding different messages. The watermarking technique is perfectly secure if and only if no information about the secret key leaks from the observations. If it is not the case, the security level is defined as the number of observations which are needed to disclose the secret key. The bigger the information leakage is, the smaller the security level of the watermarking scheme will be. In a way, a perfectly secure watermarking scheme has a security level equaling infinity. What does it mean in real life? The core idea is that pirates observing watermarked content can derive some knowledge about the secret key. In other words, information about secret parameters leaks from one watermarked content. Suppose an application with explicit requirements where the list of potential robustness attacks is finite. Suppose there exists a watermarking technique fulfilling all these requirements. Is this secure? When one watermarked content is released, the answer is yes thanks to the robustness of the watermarking technique, and also because the amount of leakage is certainly too small.

When several pieces of content are watermarked with the same secret key and released, then the question remains. Even if the original documents are independent, the embedding has rendered the watermarked documents dependent on the same variable, the secret key. This mutual dependency might leave prints revealing the value of the secret. When pirates observe many pieces of content watermarked with the same key, each of them leaking some information, then their knowledge increases. It reaches a point where an accurate estimation of the secret key allows powerful attacks.

This magnitude of order defines in a way the security level of the watermarking technique as a number of contents; if one watermarks with the same secret key more pieces of content than allowed by the security level, then a pirate can disclose it later on. In a way, this approach is pessimistic because the pirate, in practice, will extract less information about the secret than assessed by the theory, but this guarantees a lower bound.

Diffie-Hellman's Terminology. Diffie and Hellman (1976) wrote one of the most famous articles in cryptography as it strikes the creation of new directions such as public key cryptography and digital signature. It is also, as far as the authors know, the first time where several contexts of attack are envisaged according to the kind of data observed by the opponent. In order to avoid a security analysis for each application, these contexts allow to group analyses having common features, assessing a security level for a given context. In watermarking, the adversary has at least access to watermarked content, but, in some cases, the adversary might also observe the hidden messages (for instance, the name of the author in copyright protection or the status of a movie in copy protection) or the original data (for instance, DVD movies are watermarked for copy protection; but original version of old movies were not protected). Here are some contexts:

- The watermarked only attack (WOA), in which the opponent only has access to N_o watermarked vectors $\mathbf{y}^{N_o}$
- The known message attack (KMA), in which the opponent has access to N_o watermarked vectors and the associated messages $(\mathbf{y},\mathbf{m})^{N_o}$
- The known original attack (KOA), in which the opponent has access to N_o watermarked vectors and the corresponding original ones $(\mathbf{y},\mathbf{x})^{N_o}$

The reader might be surprised that the KOA context deserves any attention. Seemingly, there is no need to attack watermarked content when one has the original version. The pirate does not hack these pieces of content, but the goal is to gain information about the secret key, in order to later on hack different pieces of content watermarked with the same key.

Other contexts (not studied here) will certainly deserve a proper study in the future.

- The estimated original attack, in which the opponent has access to original content but

Figure 2. Global point of view of the decoding process

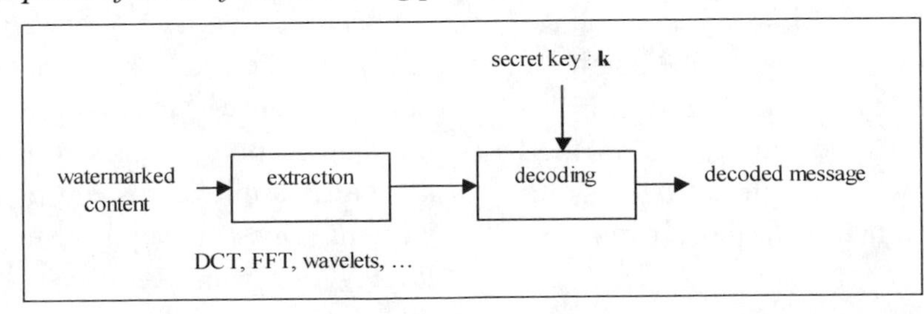

at a lower quality than the watermarked versions. Are small pictures in thumbnail gallery or movies trailers watermarked? Another possibility is that the opponent denoises watermarked content to estimate its original version.

- The constant message attack, in which the opponent observes pieces of content watermarked with the same key and the same unknown message.
- The multiple embedding single original attack, in which the opponent has access to several watermarked versions of the same content with different hidden messages. Collusion in fingerprinting and tracing traitors applications is targeted here. However, the collusion attack (i.e., the process made by the group of colluders) shall not be reduced here to a simple average of the multiple watermarked versions. A proper study must reveal whether a more powerful attack exists in order to assess the security level of fingerprinting schemes.
- The multiple embedding multiple original attack, in which the opponent has access to several watermarked contents (each) of some originals. Fingerprinting of movies (i.e., several video blocks) is targeted here.

Measure of Information Leakages and Physical Interpretation

The recent attempts of setting a methodology for watermarking security assessment are based on the adaptation of the fundamental work of Shannon. There is nothing new here except the adaptation. The key idea of this theory is to measure the amount of information leakage. Shannon mutual information (Cayre, Fontaine & Furon, 2005b; Comesana et al., 2005) or Fisher information matrix (Cayre et al., 2005b) are tools used for this purpose. We will not address the differences between these two tools (Comesana et al., 2005). There are pros and cons, and even other ways to measure information (Kullback Leibler distance or Renyi information) (Cachin, 1997), and also relations between them (Amari, 1985); the Fisher information matrix (FIM) defined below is a Riemann metric of the Kullback-Leibler divergence. What is of utmost importance is that the measurement tool provides a physical interpretation.

Shannon's Measure. In the case where the secret key $\mathbf{K}$ is a discrete random variable, the entropy $H(\mathbf{K})$ measures the uncertainty of the opponent on the true value of $\mathbf{k}$. For instance, suppose $\mathbf{K}$ is a vector composed of N_v antipodal (i.e., ± 1) samples randomly picked up by the embedder among the 2^{N_v} potential keys (i.e., a random vector with uniform probability mass distribution: $P(\mathbf{k}_i) = 2^{-N_v}, \forall \mathbf{k}_i \in \mathbb{K}$), the entropy then equals:

$$H(\mathbf{K}) = -\sum_{i=1}^{2^{N_v}} P(\mathbf{k}_i) \log_2 P(\mathbf{k}_i) = N_v \qquad (0.1)$$

and the uncertainty of the opponent is of N_v bits, as we take the logarithm with base 2 in the calculus of the entropy.

When the opponent makes N_o observations[2] $\mathbf{O}^{N_o}$, uncertainty is now evaluated through a conditional entropy, which Shannon named *equivocation*: $H(\mathbf{K} \mid \mathbf{O}^{N_o}) = H(\mathbf{K}) - I(\mathbf{K}; \mathbf{O}^{N_o})$. The information leakage is measured by the mutual information between the observations and the secret key. The bigger the information leakage is, the smaller the uncertainty of the opponent is. Equivocation is a nonincreasing function with N_o. It goes from $H(\mathbf{K})$ ideally down to 0. When the equivocation becomes null, this means that the opponent has enough observations to uniquely determine the secret key. Shannon defined the unicity distance as the first value of N_o for which the equivocation becomes null, meaning that the set of all possible keys is now reduced to only one element. This is a way to measure the security level N_o^* of a primitive.

Unfortunately, this interpretation is not suitable for any watermarking scheme, as it only works for discrete secret key **K**. It is well known that differential entropy $h(\mathbf{K})$ (or conditional differential entropy) of a continuous random variable **K** does not measure a quantity of information; one can easily check that, contrary to its discrete counterpart, differential entropy can become negative. Mutual information $I(\mathbf{K};\mathbf{O}^{N_o})$ is always pertinent as a measure of information leakages; but the physical interpretation of the equivocation as the remaining uncertainty does not hold when the secret key is regarded as a continuous random variable. For instance, the equivocation can take positive or nonpositive values, going to $-\infty$ when the attacker has a perfect knowledge, ruining the concept of unicity distance. But, equivocation can still be used with another interpretation:

$$\sigma_{\hat{\mathbf{K}}}^2 \geq \frac{1}{2\pi e} e^{2h(\mathbf{K}|\mathbf{O}^{N_o})} \tag{0.2}$$

where $\sigma_{\hat{\mathbf{K}}}^2$ is the minimum error variance in the estimation of the key (remember that the Gaussian distribution yields the maximum entropy at any given variance). This variance goes to zero as the equivocation goes to $-\infty$. The security level N_o^* is redefined as the number of observations to meet equality in equation (0.2) for a given estimator error variance. In a way, the rationale would be that to have a given probability of success in the attack, the opponent needs to know the secret key with a given accuracy. This accuracy, translated in terms of minimum estimation error variance, corresponds to a minimum number of observations N_o^* (Comesana et al., 2005).

Fisher's Measure. For continuous random variable secret key, the calculus of mutual information is always very complex, usually numerical simulations are mandatory, and sometimes they are not even tractable (Comesana et al., 2005). This is the reason why another information measure is proposed. In statistics, Fisher was one of the first to introduce the measure of the amount of information supplied by the observations about an unknown parameter to be estimated. Suppose observation **O** is a random variable with a probability distribution function depending on a parameter vector $\boldsymbol{\theta}$. The Fisher information matrix concerning $\boldsymbol{\theta}$ is defined as:

$$FIM(\boldsymbol{\theta}) = E\psi\,\psi' \quad with \quad \psi = \nabla_{\boldsymbol{\theta}} \log p_{\mathbf{O}}(\mathbf{o};\boldsymbol{\theta}) \tag{0.3}$$

where E is the mathematical expectation operator and $\nabla_{\boldsymbol{\theta}}$ is the gradient vector operator defined by $\nabla_{\boldsymbol{\theta}} = (\partial/\partial\theta(1),\ldots,\partial/\partial\theta(N_{\boldsymbol{\theta}}))'$. The Cramèr-Rao theorem gives a lower bound of the covariance matrix of an unbiased estimator of parameter vector $\boldsymbol{\theta}$ whenever the FIM is invertible:

$$\mathbf{R}_{\hat{\boldsymbol{\theta}}} \geq FIM(\boldsymbol{\theta})^{-1} \tag{0.4}$$

in the sense of non-negative definiteness of the difference matrix. In our framework, the parameter vector can be the watermark signal or the secret key. Theorem (0.4) provides us a physical interpretation; the bigger the information leakage is, the more accurate the estimation of the secret parameter can be.

The FIM is also an additive measure of the information, provided the observations are statistically independent. Suppose that the watermark signal has been added in N_o pieces of content whose extracted vectors are independent and identically distributed as $\mathbf{X} \sim N(\mathbf{0}, R_{\mathbf{X}})$. The observations are N_o watermarked signals. Then:

$$\log p_{\mathbf{O}}(\mathbf{o};\mathbf{w}) = -1/2\sum_{j=1}^{N_o}(\mathbf{y}_j - \mathbf{w})R_{\mathbf{X}}^{-1}(\mathbf{y}_j - \mathbf{w})' + const$$

Calculation readily gives $FIM(\mathbf{w}) = N_o R_{\mathbf{X}}^{-1}$. This models applications where we detect the presence of (and not decode) watermarks, or also template signals which resynchronise content transformed by a geometric attack.

The mean square error $E\{\|\hat{\boldsymbol{\theta}} - \boldsymbol{\theta}\|^2\}$ is the trace of $R_{\hat{\boldsymbol{\theta}}}$, and thus its lower bound decreases in N_o^{-1}.

However, the rate $N_o^* = N_o tr(FIM(\boldsymbol{\theta})^{-1})$ depends on the statistical model and consequently the nature of observations of the contexts. It means that the estimation is significantly more accurate when the number of independent observations increases at an order of N_o^*. The bigger N_o^* is, the more difficult is the disclosure of the secret key. This notion is close to the unicity distance of the above subsection. This is the reason why we use the same notation N_o^* (although, once again, absolutely not defined in the same way).

Examples of the Theoretical Security Analysis

This section provides an example of a theoretical security analysis applied to the substitutive watermarking scheme. A summary of the main results for the spread spectrum watermarking schemes is given at the end of this section. Moreover, Pérez-Freire et al. (2005, 2006) recently analysed the security level of lattice quantisation based watermarking scheme.

Mathematical Model of Substitutive Watermarking Security

In a substitutive watermarking scheme, a binary vector $\mathbf{x} = (x(1),..., x(N_v))'$ is extracted from the content. For instance, in the famous technique invented by Burgett, Koch, and Zhao (1998), N_v pairs of absolute values of DCT coefficients of an image are compared. The message to be hidden is a binary vector $\mathbf{m} = (m(1),..., m(N_c))'$. The secret key is a list of N_c integers $\mathbf{k} = [k(1),..., k(N_c)]$ with $1 \leq k(l) \leq N_v$ and $k(l) \neq k(l')$ if $l \neq l'$. The embedding process copies $\mathbf{x}$ in $\mathbf{y}$ and then substitutes the $k(l)$-th bit of $\mathbf{y}$ by the l-th bit of the message to be hidden: $\mathbf{y}(\mathbf{k}(l)) = \mathbf{m}(l)$. The inverse extraction function maps back the watermarked vector $\mathbf{y}$ into the content. The decoding simply reads the bits whose indices are given by the secret key.

Example 1. $N_v = 8$ and $N_c = 4$:

$\mathbf{m} = (1101)$, $\mathbf{k} = [2,8,5,3]$

$\mathbf{x} = (01001011)$, $\mathbf{y} = (0\mathbf{1100}01\mathbf{1})$

The uncertainty of the opponent is given by the entropy of the secret key that the embedder has randomly selected among $N_v!/(N_v - N_c)!$ possible keys. Thus:

$$H(\mathbf{K}) = \log_2 \frac{N_v!}{(N_v - N_c)!} \quad (0.5)$$

Perfect Covering

We show here that a substitutive watermarking scheme provides perfect covering in the sense that observing a watermarked vector $\mathbf{y}$ does not help the attacker in refining knowledge about the watermark signal $\mathbf{w}$. Formally, perfect covering is granted when $p_{\mathbf{W}}(\mathbf{w}) = p_{\mathbf{W}}(\mathbf{w} \mid \mathbf{y})$. It means that the original vector and the watermark vector are strongly mixed during embedding such that it is not possible to split the watermarked vector back to them.

We can model the substitutive watermarking as follows: let $\mathbf{x}$ be a binary N_v-length random vector, whose probability mass function is uniform and equal to 2^{-N_v} and $\mathbf{w}$ be a binary N_v-length vector whose bits equal to 1 indicate the bits to be flipped. Hence, we have $\mathbf{y} = \mathbf{x} \oplus \mathbf{w}$, giving:

$$p_{\mathbf{Y}}(\mathbf{y}) = \sum_{\mathbf{w} \in \mathbf{W}} p_{\mathbf{Y}}(\mathbf{y} \mid \mathbf{w}) p_{\mathbf{W}}(\mathbf{w}) = \sum_{\mathbf{w} \in \mathbf{W}} p_{\mathbf{X}}(\mathbf{y} \oplus \mathbf{w}) p_{\mathbf{W}}(\mathbf{w})$$

$$= 2^{-N_v} \sum_{\mathbf{w} \in \mathbf{W}} p_{\mathbf{W}}(\mathbf{w}) = 2^{-N_v}$$

$$p_{\mathbf{Y}}(\mathbf{y} \mid \mathbf{w}) = p_{\mathbf{X}}(\mathbf{y} \oplus \mathbf{w}) = 2^{-N_v}$$

The Bayes rule, $p_{\mathbf{Y}}(\mathbf{y} \mid \mathbf{w}) p_{\mathbf{W}}(\mathbf{w}) = p_{\mathbf{W}}(\mathbf{w} \mid \mathbf{y}) p_{\mathbf{Y}}(\mathbf{y})$, then gives $p_{\mathbf{W}}(\mathbf{w}) = p_{\mathbf{W}}(\mathbf{w} \mid \mathbf{y})$.

Watermarked Only Attack

With the substitutive method providing perfect covering, it is then very easy to show that $I(\mathbf{Y};\mathbf{W}) = 0$. It is obvious that $I(\mathbf{Y};\mathbf{W}) \geq I(\mathbf{Y};\mathbf{K}) \geq 0$ which implies that $I(\mathbf{Y};\mathbf{K}) = 0$. There is no information leakage, and the equivocation is equal to $H(\mathbf{K})$ whatever the number of observations. In a way, one can say that security level $N_o^* = +\infty$.

Known Message Attack

If the opponent observes only one watermarked content $\mathbf{y}_1$ and its hidden message $\mathbf{m}_1$, the indices i such that $\mathbf{y}_1(i)=\mathbf{m}_1(l)$ are possible values of $k(l)$. Denote $\mathbb{S}_1(l)$ this set. As $P(\mathbf{y}_1(i) = \mathbf{m}_1(l) \mid i \neq \mathbf{k}(l)) = 1/2$, there are in expectation $1+(N_v-1)/2$ elements in this set.

Now assume that the opponent observes several contents $\mathbf{y}^{N_o}$ and their hidden messages $\mathbf{m}^{N_o}$. Set $\mathbb{S}_{N_o}(l)$ is now defined by $S_{N_o}(l) = \{i : \mathbf{y}_j(i) = \mathbf{m}_j(l)\ \forall j, 1 \leq j \leq N_o\}$. The probability that $\mathbf{y}_j(i) = \mathbf{m}_j(l)\ \forall j$ knowing that $i \neq \mathbf{k}(l)$ is $1/2^{N_o}$. Thus, in expectation, $|\mathbb{S}_{N_o}| = 1+(N_v-1)/2^{N_o}$ and the equivocation about $\mathbf{k}(l)$ is equal to $\log_2(1+2^{-N_o}(N_v-1))$. However, there might be some overlaps between the N_c sets $\mathbb{S}_{N_o}(l)$, and the total equivocation is smaller than the sum of the equivocations about $\mathbf{k}(l)$. As the calculus is quite complex, we stay with this approximation:

$$H(\mathbf{K} \mid (\mathbf{Y},\mathbf{M})^{N_o}) \leq N_c \log_2(1+2^{-N_o}(N_v-1)) \qquad (0.6)$$

As Shannon (1949) did for cryptanalysis, we can approximate this equivocation by $N_c(\log_2(n-1)-N_o)$ when $N_o \ll \log_2(N_v-1)$, and by $2^{-N_o} N_c (N_v-1)/\log(2)$ when $N_o \gg \log_2(N_v-1)$. These approximations are show in Figure 3. The unicity distance is approximately given by $N_o^* = \log_2 N_v$.

Known Original Attack

If the opponent observes only one watermarked content $\mathbf{y}_1$ and its original version $\mathbf{x}_1$, the indices i such that $\mathbf{x}_1(i) \neq \mathbf{y}_1(i)$ are possible values for the key samples. There are in expectation $N_c/2$ of such indices, as $p(\mathbf{x}_1(\mathbf{k}(l)) = \mathbf{m}_1(l)) = 1/2$. When the opponent observes j pairs, the set $\mathbb{S}_j = \{l : \exists j', 1 \leq j' \leq j, \mathbf{x}_{j'}(l) \neq \mathbf{y}_{j'}(l)\}$ grows up. However, the event that an index revealed by a new pair was already known happens with a probability $|\mathbb{S}_{j-1}|/N_c$. This leads to the following series:

$$|\mathbb{S}_j| = |\mathbb{S}_{j-1}| + N_c(1-|\mathbb{S}_{j-1}|/N_c)/2 = N_c(1-2^{-j}) \qquad (0.7)$$

Yet, it is not possible to assign a key sample to one of these indices. The equivocation is then the sum of two terms; one is due to the $N_c - |\mathbb{S}_{N_o}|$ undisclosed indices to be picked up randomly among the remaining candidates, the second one is due to the $N_c!$ possible permutations of the chosen indices, (see equation (0.8))

The security level (in the unicity distance sense) is not defined as the equivocation is always greater than zero. This is due to the term $\log_2(N_c!)$ reflecting the ambiguity in the order of the estimated key samples. We preferably consider that within a number of observations greater than

Equation 0.8.

$$H(\mathbf{K} \mid (\mathbf{Y},\mathbf{X})^{N_o}) = \log_2\left(\frac{(N_v - \lceil |\mathbb{S}_{N_o}| \rceil)!}{(N_v - N_c)!(N_c - \lceil |\mathbb{S}_{N_o}| \rceil)!}\right) + \log_2(N_c!)$$

$N_o^* = \log_2 N_c$, the opponent learns all the indices stored in the secret key. This information is helpful for watermark jamming. The opponent can also notice if two hidden messages are the same. Yet, the ambiguity prevents the opponent reading the hidden messages (cannot put the hidden bits in the right order), and writing hidden messages.

Figure 3 gives a good synthesis of the results. In the WOA case, the equivocation is constant; the opponent cannot get any information on the key. In the KMA case, the opponent is able to completely disclose the key, and then will be able to read, erase, write, or modify hidden messages. In the KOA case, the equivocation decreases down to a positive value; the opponent is able to recover the components of the key but up to a permutation, and then will be able to erase the hidden message, but not to read or write a proper one.

Spread Spectrum Watermarking Schemes

We will not detail the theoretical analysis of the security of the spread spectrum watermarking schemes because the mathematical developments are more cumbersome. The main known results are summarised below.

The mathematical model is very simple. Denote by $\mathbf{x}$ a vector of N_v samples extracted from original content. The embedding is the addition of the watermark signal which is the modulation of N_c private carriers $\{\mathbf{u}_l\}, 1 \le l \le N_c$:

$$\mathbf{w} = \frac{\gamma}{\sqrt{N_c}} \sum_{l=1}^{N_c} \mathbf{a}(l)\mathbf{u}_l \qquad (0.9)$$

where $\gamma > 0$ is a small gain fixing the embedding strength, and $\|\mathbf{u}_l\| = 1, 1 \le l \le N_c$. The watermark

Figure 3. Substitutive watermarking. Equivocations for WOA, KMA and KOA, against the number of observations. The triangle and the square respectively mark the security level for KMA and KOA.

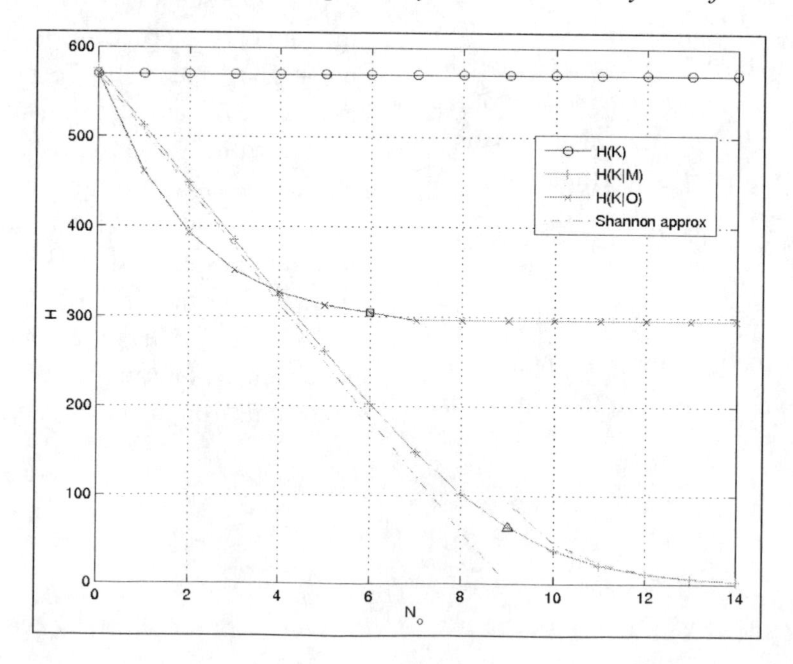

to content power ratio (WCR) equals $\gamma^2\sigma_{\mathbf{a}}^2/\sigma_{\mathbf{x}}^2$ (or $10\log_{10}(\gamma^2\sigma_{\mathbf{a}}^2/\sigma_{\mathbf{x}}^2)$ if expressed in dB). An inverse extraction function puts back vector $\mathbf{y}=\mathbf{x}+\mathbf{w}$ into the media to produce the watermarked content. Symbol vector $\mathbf{a}$ represents the message to be hidden/transmitted through content. In the case of a direct sequence spread spectrum (DSSS), the modulation is a simple BPSK: $\mathbf{a}(l)=(-1)^{\mathrm{m}(l)}, 1\leq l\leq N_c$ and $\sigma_{\mathbf{a}}^2=1$.

For security reason, the carriers are private and issued by a pseudo-random generator fed by a seed. Many people think the secret key is the seed. This is not false as the disclosure of the seed obviously gives the carriers and allows the access to the watermarking channel. However, the knowledge of the carriers is sufficient and the pirate usually has no interest in getting back to the seed. Hence, in this analysis, the secret key, defined as the object the opponent is keen on revealing, is constituted by the carriers.

According to the methodology exposed above, the theoretical security analysis considers several watermarked vectors $\mathbf{y}_j$, $1\leq j\leq N_o$, with different embedded symbols $\mathbf{a}_j=(\mathbf{a}_j(1),...,\mathbf{a}_j(N_c))'$ being linearly mixed by the $N_v\times N_c$ matrix $\mathfrak{U}=(\mathbf{u}_1...\mathbf{u}_{N_c})$. This matrix is the secret to be disclosed. In the sequel, we try to express the Fisher information matrix concerning the N_vN_c long vector $\mathfrak{U}^v=(\mathbf{u}_1'...\mathbf{u}_{N_c}')'$. To cancel intersymbol interferences at the decoding side, carriers are two-by-two orthogonal vectors; $\mathfrak{U}'\mathfrak{U}=\mathfrak{I}_{N_c}$, where $\mathfrak{I}_N$ is the $N\times N$ identity matrix. Index i denotes the i-th sample of a signal, whereas j indices the different signals. Thus, there are N_o watermarked vectors given by:

$$\mathbf{y}_j=\mathbf{x}_j+\frac{\gamma}{\sqrt{N_c}}\mathfrak{U}\mathbf{a}_j \tag{0.10}$$

or, equivalently, concatenating N_o vectors $\mathbf{x}_j$ (respectively $\mathbf{y}_j$ or $\mathbf{a}_j$) column-wise in the $N_v\times N_o$ matrix $\mathfrak{X}$ (respectfully $\mathfrak{Y}$) or the $N_c\times N_o$ matrix $\mathfrak{A}$):

$$\mathfrak{Y}=\mathfrak{X}+\frac{\gamma}{\sqrt{N_c}}\mathfrak{U}\mathfrak{A} \tag{0.11}$$

Known Message Attack. In this subsection, the opponent has access to (watermarked signals/hidden messages) pairs: $(\mathbf{y},\mathbf{a})^{N_o}$. Assume, for simplicity reason, that each occurrence of random vector $(\mathbf{x})$ is independently drawn from $N(\mathbf{0},\sigma_{\mathbf{x}}^2\mathfrak{I}_{N_v})$. The FIM has then the following inverse:

$$FIM(U^v)^{-1}=\frac{N_c\sigma_{\mathbf{x}}^2}{N_o\gamma^2\sigma_{\mathbf{a}}^2}\mathfrak{I}_{N_vN_c} \tag{0.12}$$

with a BPSK modulation, $\sigma_{\mathbf{a}}^2=1$. The information leakage is linear with the number of observations thanks to the assumption of independence, and the rate is given by the watermark to content power ratio per carrier $\gamma^2/N_c\sigma_{\mathbf{x}}^2$. The security level of spread spectrum based watermarking techniques against KMA is $N_o^*=N_vN_c^2\sigma_{\mathbf{x}}^2/\gamma^2$ of (watermarked signals/hidden messages) pairs.

Known Original Attack. The opponent observes $(\mathbf{y},\mathbf{x})^{N_o}$. The vector difference of each observation j gives the source signals $\mathbf{a}_j$ being linearly mixed by the $N_v\times N_c$ matrix $\mathfrak{U}$:

$$\mathbf{d}_j=\mathbf{y}_j-\mathbf{x}_j=\frac{\gamma}{\sqrt{N_c}}\mathfrak{U}\mathbf{a}_j \tag{0.13}$$

Assume that $N_o\geq N_c$ and that there are at least N_c linearly independent messages. The difference matrix $\mathfrak{D}=\mathfrak{Y}-\mathfrak{X}\propto\mathfrak{U}\mathfrak{A}$ (where $\propto$ means 'proportional to') is then full rank, and $Span(\mathfrak{D})=Span(\mathfrak{U})$. The observation of difference vectors discloses the secret subspace $Span(\mathfrak{U})$, provided symbol matrix $\mathfrak{A}$ is full rank. However, this does not reveal the private carriers. This is the blind source separation (BSS) problem, and Comon (1994) proved that it is possible to identify the correct basis, that is $\mathfrak{U}$, but up to a permutation and scale ambiguity. The scale ambiguity is indeed a sign ambiguity in our problem, as we set $\mathfrak{U}'\mathfrak{U}=\mathfrak{I}$. In conclusion, at best the mixing matrix is identified by $\hat{\mathfrak{U}}=\Pi\Sigma\mathfrak{U}$ with Π a permutation matrix and Σ a diagonal matrix whose elements are ±1. At best for the opponent, the secret carriers are identified up to a signed permutation (i.e., matrix $\Pi\Sigma$) ambiguity.

The asymptotic accuracy of the estimations is known to be only dependent on the symbols distribution, and especially on its non-Gaussianity. Indeed, as in our case, when symbols are discrete with bounded support symbols, the trace of Cramèr-Rao Bound decreases at a faster rate than $1/N_o$ (Cardoso, 1998; Gamboa & Gassiat, 1997).

Watermarked Only Attack. In this section, the sources are unknown and can then be regarded as nuisance parameters, which render the estimation of $\mathfrak{U}$ less accurate (Amari & Cardoso, 1997). But, the situation is even worse here as the FIM becomes singular preventing us from applying the Cramèr-Rao bound. This problem stems from two facts. First, we did not integrate some constraints during our derivation. Especially, we know that $\mathbf{u}_l^t \mathbf{u}_k = \delta_{l,k}$. Stoica and Ng (1998) give an alternative expression for the bound in the case where the unconstrained problem is unidentifiable and the FIM noninvertible. However, the integration of the above-mentioned constraints in the derivation of the FIM is not sufficient for $N_c > 1$. The second fact is that an ambiguity remains about the order and sign of the carriers.

We prefer to approximate the information leakage about carriers by a FIM whose inverse is:

$$FIM(\mathfrak{U}^v)^{-1} = \frac{N_c \sigma_{\mathbf{x}}^2}{N_o \sigma_{\mathbf{a}}^2 \gamma^2} \mathfrak{U}^{\perp} \mathfrak{U}^{\perp t} \quad (0.14)$$

where $\mathfrak{U}^{\perp}$ is a basis of the complementary subspace of $Span(\mathbf{u}_1)$ in $\mathbb{R}^{N_v}$. The security level is then $N_o^* \geq N_v N_c^2 \sigma_{\mathbf{x}}^2 / \gamma^2$. This result is quite surprising because the security level is almost the same against KMA and WOA. Yet, the estimation of the secret carriers remains up to a signed permutation in the WOA.

Possible Hacks. The conclusion of this security analysis stands in the different possibilities to forge pirated content:

- The pirate discloses secret subspace $Span(\mathfrak{U})$ The pirate can now focus attack's noise in this subspace to jam the communication far more efficiently, and can also nullify the watermarked signals projection in this subspace to remove the watermark.
- The pirate discloses the secret carriers up to a signed permutation. The abovementioned hacks are still possible. Besides, the pirate can detect whether two watermarked pieces of content share the same hidden message. The pirate can also flip some randomly chosen bits. Moreover, the accidental knowledge of hidden messages in few watermarked pieces of content might remove this ambiguity.
- The pirate discloses the secret carriers. The pirate has full access to the watermarking channel to read, write, or erase hidden messages.

Of course, the quality of the pirated pieces of content depends on the accuracy of the pirate's estimation. Cayre, Fontaine, and Furon (2004) deal with this aspect.

FUTURE TRENDS

This section questions the methodology presented in this chapter, showing that the assessment of watermarking security is still at its infancy. New tracks have to be explored to bring practical solutions more closely related to what really matters in real life applications.

Careful Comparison with Cryptography

In cryptography, unless a symmetric encryption scheme is broken, a N_v bit secret key ensures a security level of N_v bits. While introducing the methodology based on the adaptation of Shannon's approach to watermarking security, a small example gave the value of the equivocation when the secret key is an antipodal career of length N_v, as could be used in spread spectrum: $H(\mathbf{K}) = N_v$ bits. Hence, it seems that in watermarking also, the secret's strength is given by N_v bits. The following sections mitigate this comparison.

Weak Secret Keys. Not all antipodal sequences are convenient for spread spectrum watermarking. For instance, sequences with zero average are much more preferable in order to get rid of the direct component of the host signal and not to add more distortion than needed. Other sequences might be considered as 'weak' because they yield a bias in the detection statistic. Therefore, the number of secret keys is indeed $\binom{N_v}{N_v/2}$. A Stirling's approximation shows that the equivocation is now:

$$H(\mathbf{K}) = \log_2 \binom{N_v}{N_v/2} \simeq N_v - \frac{1}{2}\log_2\left(\frac{N_v \pi}{2}\right) \quad bits \qquad (0.15)$$

which is smaller that the first estimation.

Importance of the Error Variance. A second issue is the accuracy of the estimated key. Contrary to cryptographic encryption schemes, an exact disclosure is not mandatory, and powerful attacks can be launched based on a rough estimation of the secret key. For instance, in a zero-bit watermarking scheme based on spread spectrum, the important factor is the normalised correlation $\rho = \hat{\mathbf{u}}'\mathbf{u} / N_v$. Watermarked signals can then be written as $\mathbf{y} = \mathbf{x} + \gamma_e \mathbf{u}$. The attacker has an estimated career sequence $\hat{\mathbf{w}}$. To remove the watermark, the attacker forms $\mathbf{z} = \mathbf{y} - \gamma_a \hat{\mathbf{u}}$. The attack will therefore be successful if the correlation is below the threshold $\tau : \mathbf{x}'\mathbf{u} + N_v(\gamma_e - \gamma_a \rho)$. If ρ is close to one, then the attack has a power $N_v^{-1} \|\mathbf{y} - \mathbf{z}\|^2 = \gamma_a^2$ similar to the embedding power γ_e^2. Moreover, the distortion between the original work and the attacked signal decreases with $\rho : \|z - x\|^2 = N_v(\gamma_e^2 + \gamma_a^2 - 2\gamma_a\gamma_e\rho)$. Simulations on real images by Cayre et al. (2005c) show that a minimum of $\rho_{\min} = 0.4$ is enough to remove watermarks with a good perceptual quality.

We would like to translate this specific feature of spread spectrum watermarking in terms of equivocation in Shannon's framework. If the attacker succeeds to correctly estimate k samples of the true key, then the normalised correlation between the two antipodal sequences is $\rho = (2k - N_v)/N_v$. Thus, the attacker needs at least $k_{\min} = N_v(\rho_{\min} + 1)/2$ correct samples to have a 'good' estimated key $\hat{\mathbf{u}}$ (i.e., sufficiently correlated). For a given secret key chosen by the embedder, the attacker has indeed $\binom{N_v}{k_{\min}}$ possible good estimated keys with exactly $k_{\min}$ correct samples. However, we do not need exactly $k_{\min}$ but at least $k_{\min}$ correct samples. Hence, the number of good estimated keys is indeed:

$$|\hat{\mathbb{U}}| = \sum_{k=k_{\min}}^{N_v} \binom{N_v}{k}$$

Assuming that the size of this set is dominated by the first term of the sum, and using Stirling's approximation, we have the following bound, (see Box 1).

Box 1.

$$\log_2(|\hat{\mathbb{U}}|) \geq N_v \left(\rho_{\min} \log\left(\frac{1-\rho_{\min}}{1+\rho_{\min}}\right) - \log(1-\rho_{\min}^2) \right) / 2\log(2) + N_v$$

Figure 4: A collection of watermarked signals with ( ) with the Spread Spectrum technique (left) and the Spread Transform Scalar Costa Scheme (right). Red circles (resp. green crosses) represent signals hiding symbol '0' (resp. '1'). The grey area (resp. white) is the decoding region associated to symbol '0' (resp. '1'). The main axis of the ellipses are given by the eigenvalues of the covariance matrix of the watermarked signals.

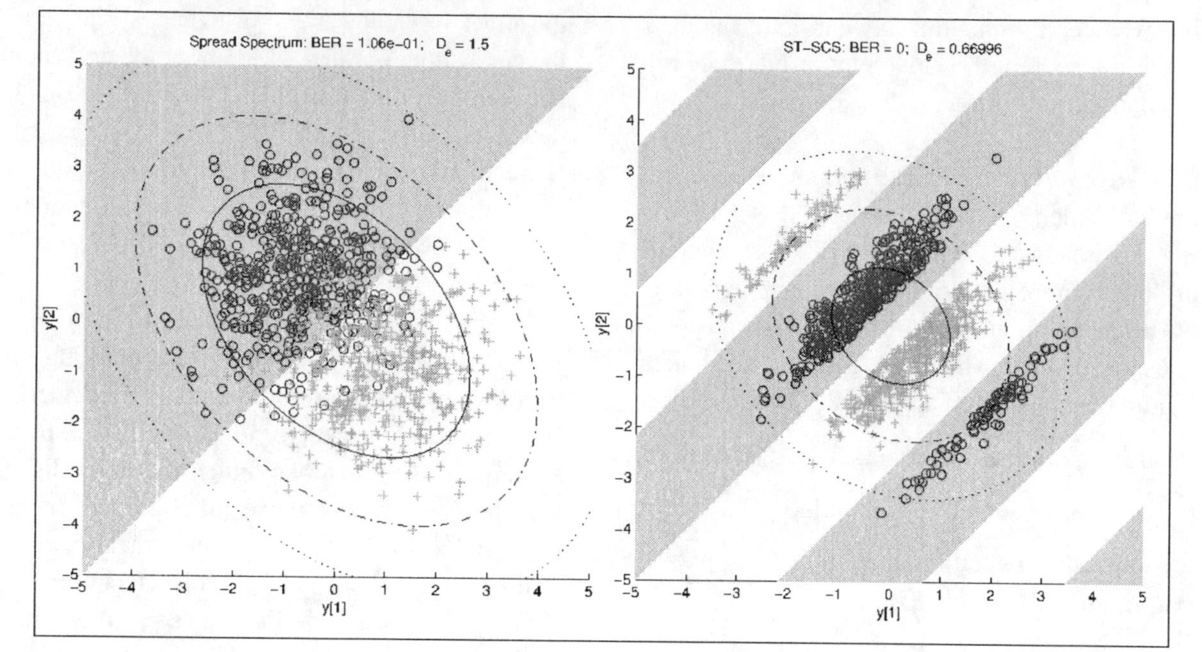

Box 1). For $\rho_{min} = 0.4$, the bound equals $0.79N$ bits whereas numerical calculations give $\approx 0.84N_v$ bits. In Shannon's framework, $\log_2(|\hat{\mathbf{U}}|)$ corresponds to the a posteriori equivocation $H(\mathbf{U}\,|\,\mathbf{O}^{N_o})$. In cryptography, the attacker has to disclose the exact value of the secret key bit by bit. The unicity distance is then defined by $H(\mathbf{K}\,|\,\mathbf{O}^{N_o^*}) = 0$. In watermarking, an estimation of the secret key is sufficient. The unicity distance in this small example should be defined by $H(\mathbf{U}\,|\,\mathbf{O}^{N_o^*}) = 0.84N_v$ bits. To conclude, we draw the following comparison. In cryptography, unless an encryption scheme is broken, a N_v bit secret key ensures a security level of N_v bits; the equivocation starts at N_v bits and the attacker needs to reduce it down to 0. In spread spectrum watermarking, the equivocation starts at $N_v - \log_2(\sqrt{N_v})$, and the attacker needs to reduce it down to $0.84\,N_v$. The security level is around 15% of the length of the carrier!

Practical Approaches

The methodology presented so far does not say anything on the way how to extract and exploit the information leakage. Hence, the practical part of the security analysis is of utmost importance. It gives a proof of concept that there exists at least one estimation algorithm. It may not be the most efficient in terms of computing power, and it may not be statistically efficient in the sense that its variance is greater than the Cramèr-Rao bound. It just proves that the attack is a real threat. This is the reason why most articles based on the methodology proposed in this chapter, such as Cayre et al. (2005c), also encompass a practical part.

Whereas the theory is just the adaptation of Shannon's security model, practice of watermarking security consists in inventing original and efficient algorithms. Their cores are often known signal processing tools from fields which have a priori nothing in common with watermarking. This extremely interesting part of the job proves once again that watermarking is a multifield science.

Here are some useful signal processing tools to hack watermarking schemes.

Maximum Likelihood Estimator. Let us consider the watermarking embedder as a system to be identified. Hence, in the known message attack, we observe pairs of input (i.e., message $\mathbf{m}$) and output (watermarked content $\mathbf{y}$). In other words, we have the framework of an input-output identification. If it is possible to write the likelihood $p(\mathbf{y}^{N_o}, \mathbf{m}^{N_o} \mid \mathbf{K})$, the opponent can use the maximum likelihood estimator (MLE) which finds $\hat{\mathbf{K}}$ by maximizing the likelihood or nullifying its derivative. The MLE is known to be unbiased and consistent, that is, it asymptotically achieves the Cramèr-Rao bound mentioned above. Cayre, Fontaine, and Furon (2005a) have applied it to spread spectrum schemes.

Expectation Maximisation Algorithm. Unknown messages can be considered as hidden data. The MLE based on $p(\mathbf{y}^{N_o}, \mathbf{m}^{N_o} \mid \mathbf{K})$ is not usually practical in this case. But, the EM algorithm approaches it by the iteration of a two-step process:

- **Expectation**. Having an estimation of the key $\hat{\mathbf{K}}(i)$, we estimate the messages $\hat{\mathbf{m}}^{N_o}(i)$. This basically corresponds to the decoding algorithm, known by the pirate according to Kerckhoffs' principle.
- **Maximisation**. Having an estimate of the message $\hat{\mathbf{m}}^{N_o}(i)$, we upgrade our estimation of the secret key $\hat{\mathbf{K}}(i+1)$. This step uses for instance the MLE seen above.

Cayre et al. (2005a) have used EM algorithm for spread spectrum schemes.

Principal Component Analysis. Many watermarking schemes use projection onto N_c orthonormal private vectors or carriers $\{\mathbf{u}_i\}$ in order to increase the SNR at the decoding side. In general, one can write $\mathbf{y} = \mathbf{x} + \mathbf{w}$, with $\mathbf{x}$ the host signal, and $\mathbf{w} = \sum_{i=1}^{N_c} \gamma \mathbf{a}(i) \mathbf{u}_i$. The coefficients $\gamma \mathbf{a}(i)$ carry the message to be hidden and tackle the perceptual constraint. We assume they are independent from $\mathbf{x}$, i.i.d and centered. It means that the watermark signal $\mathbf{w}$ lives in a small subspace whose dimension is N_c, whereas $\mathbf{x}$ belongs to $\mathbb{R}^{N_v}$. This leaves clues for the pirate as the energy of the watermark is focused on a small subspace. For instance, if $\mathbf{x}$ is a white noise, the covariance matrix of $\mathbf{y}$ is $R_\mathbf{y} = \sigma_\mathbf{x}^2 \mathfrak{I} + \sum_{i=1}^{N_c} \gamma^2 E\{\mathbf{a}_i^2\} \mathbf{u}_i \mathbf{u}_i^t$ whereas $R_\mathbf{x} = \sigma_\mathbf{x}^2 \mathfrak{I}$. This means that $R_\mathbf{x}$ has one eigenvalue $\sigma_\mathbf{x}^2$ with order N_v, whereas $R_\mathbf{y}$ has one eigenvalue $\sigma_\mathbf{x}^2$ with order $N_v - N_c$, and N_c other eigenvalues equal to $\sigma_\mathbf{x}^2 + \gamma^2 E\{\mathbf{a}_i^2\}$. Moreover, these N_c biggest eigenvalues are related to eigenvectors $\mathbf{u}_i$. Consequently, it is very easy for the pirate to estimate these private carriers: (a) estimate the covariance matrix $R_\mathbf{y}$ with $\hat{R}_\mathbf{y} = \sum_{i=1}^{N_o} \mathbf{y}(i)\mathbf{y}(i)^t / N_o$, (b) make an eigen-decomposition of this matrix, and (c) isolate the eigen-vectors corresponding to the N_c biggest eigenvalues. Figure 4 illustrates this for toy examples. Ellipses show that the watermarked signals are no more white signals. Cayre et al. (2005a) and also Doërr and Dugelay (2004) apply principal component analysis to break spread spectrum based schemes.

Independent Component Analysis. In the case where $E\{\mathbf{a}_i^2\} = cst$, then $R_\mathbf{y}$ has one eigenvalue $\sigma_\mathbf{x}^2 + \gamma^2 E\{\mathbf{a}_i^2\}$ with associated subspace of dimension N_c. When successful, the PCA reveals this subspace and gives a basis, which is not the one used by the embedder: $\{\mathbf{u}_1, \ldots, \mathbf{u}_{N_c}\}$. The pirate can focus the attack noise on this subspace, or remove the watermark signal by nullifying the projection

Figure 5. PCA vs. ICA. PCA finds the secret carriers up to a rotation, whereas ICA succeeds to align the estimated carriers with the original ones. An ambiguity remains in their order (permutation) and orientation (sign).

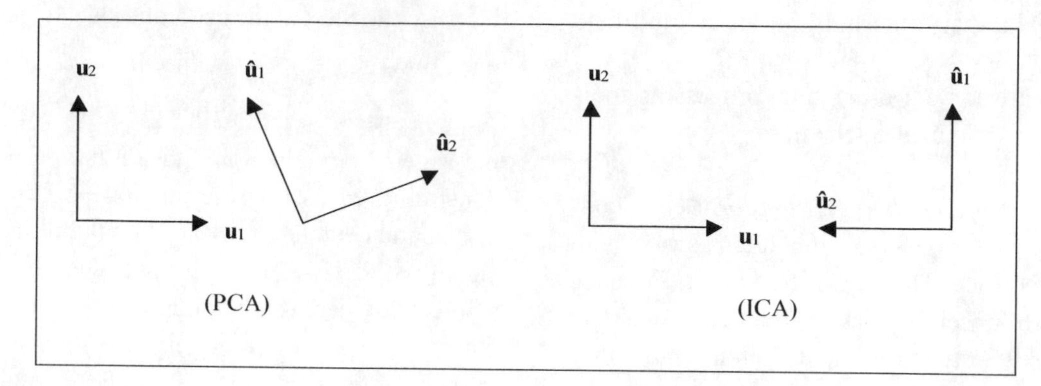

of **y** onto this subspace. Yet, the attacker cannot have a read and write access on the watermarking channel.

If symbols $\mathbf{a}_i$ are statistically independent, an independent component analysis (ICA) rotates the PCA basis until the estimated symbols 'look like' independent. When successful, the ICA yields estimated carriers which correspond to the real basis up to permutation $\pi(.)$ and change of sign: $\hat{\mathbf{u}}_i = \pm\mathbf{u}_{\pi(i)}$. Figure (5) illustrates this ambiguity. This ambiguity prevents the pirate from embedding/decoding messages, but the pirate can check if two watermarked contents have the same hidden message or the pirate can flip bits of hidden messages. Cayre et al. (2005a) apply ICA to analyse practically the security level of spread spectrum based schemes.

Clustering. Doërr and Dugelay (2004) have tested clustering tools to break a video watermarking technique. This technique randomly embeds one of N_c watermark signals in one video frame. An average attack does not work as it only estimates a mixture of these N_c signals. However, if a spatial filter succeeds to isolate enough watermark energy, the pirate obtains noisy estimations of the N_c watermark signals. The pirate's goal is now to split this set of estimations into N_c clusters of estimations corresponding each to one watermark signal, and whose centroids would be good estimates of the N_c watermark signals. This is a typical task for the *k*-means algorithm.

Vector Quantisation. A closely related tool is the vector quantisation which is used for replacement attack. The pirate has a database of signal blocks and wishes to replace a block in a watermarked content by a similar block of the database. The word similar is here important. The vector quantisation is used to find in the database the most similar 'codeword' (i.e., block) in the sense of the Euclidean distance. This tool is used for attacking video watermarking techniques (Doërr & Dugelay, 2004) or block-based authentication schemes (Holliman & Memon, 2000).

Computational Security Assessment

Another crucial point is the complexity of the algorithm disclosing the secret key. Shannon (1949) already denoted this notion of complexity by the term 'work'. It might be theoretically possible to extract enough information in order to estimate the secret key, while, in practice, demanding too much computing power. Cryptographers distinguish unconditional security (it is proven that no information leaks from the observations) and computational security (the best known algorithm requires an unreasonable amount of computing power) (Menezes, Van Oorschot, & Vanstone, 1996). Unconditional security is exceptional is cryptology, so that most of cryptanalyses are computational oriented.

Katzenbeisser (2005) adapts such computational cryptanalysis framework to watermarking. First, an opponent is defined as a Turing machine with polynomial-time complexity. It is also probabilistic in the sense that the opponent can make random decisions in the algorithm. Then, a challenge is proposed to the attacker in order to measure the advantage, that is, the attacker's ability of disclosing 'some property' about the secret key. As the attacker is probabilistic, the probability of success to the challenge needs to be evaluated and compared to a probability of success for a random guess.

For instance, Katzenbeisser (2005) proposes the following challenge: a judge picks up two secret keys $\mathbf{K}_1$ and $\mathbf{K}_2$, and will use one of them to watermark a piece of content $\mathbf{x}_{test}$. The challenge is to disclose which key is used. A random guess gives 1/2 as a probability of success. The attacker is given two oracles, $O_{\mathbf{K}_1}$ and $O_{\mathbf{K}_2}$, each one watermarking content with one of the two secret keys. The attacker can test the oracles during a fixed amount of time; the attacker watermarks originals $\mathbf{x}_1, \mathbf{x}_2, ..., \mathbf{x}_{N_o}$ using the oracles, observing $\mathbf{y}_1, \mathbf{y}_2, ..., \mathbf{y}_{N_o}$. The attacker is free to process these data provided the complexity is polynomial-time bounded. When this phase is finished, the attacker has no longer access to the oracles. The judge then produces the watermarked content $\mathbf{y}_{test}$ using key $\mathbf{K}_{test} \in \{\mathbf{K}_1, \mathbf{K}_2\}$, gives the opponent $\mathbf{y}_{test}$, and asks whether $\mathbf{K}_{test} = \mathbf{K}_1$. The advantage of the attacker is measured by the probability of success minus 1/2. This advantage should vanish exponentially as the length of the key increases.

This methodology for watermarking security assessment is certainly closer to what matters in reality, especially since it does not provide the attacker an unbounded computer power as before. However, for the moment, no watermarking schemes have been analysed with this framework.

FUTURE RESEARCH DIRECTIONS

The future of watermarking security is quite simple as it consists in discovering whether secure and robust watermarking does exist. However, the answer will be certainly not that simple. First, robustness and security are matters of quantity. The question is not whether a technique is robust or secure but how much it is. For the time being, researchers have been analysing the security of existing watermarking techniques. But, as security will become more and more popular, watermarkers in the future will have to present a security level assessment beside the perceptual distortion, the embedding rate, and the robustness of their proposed schemes.

It is likely that researchers in this field will succeed in slowly increasing the security levels, but will they be able to keep a constant robustness? Our point of view is that the fundamental issue is indeed the following one: is there a fundamental trade-off between robustness and security? So far, there is one theoretical answer, but this trade-off has been observed for some watermarking schemes.

For instance, counter-measures to geometric attacks either use invariant coefficients to embed the watermark signal. However, there is a low number of these coefficients, resulting in small secret key space. Either they use periodic templates to resynchronise the watermark decoder. But, a direct application of this chapter shows that pirates will easily estimate these templates.

Another question concerns the tools presented in this chapter. In one hand, Shannon's information theory has been the support of digital communications for the past 60 years. As watermarking is above all a communication with side-information problem, it is not surprising to resort to Shannon's tools to assess its security. On the other hand, these information theoretical concepts are not common in cryptanalysis. Shannon has applied his concepts to secrecy systems in 1948, but they are not used anymore. Cryptographic security levels are nowadays assessed via a complexity basement; what computing power is required to disclose the secret key of this cryptosystem?

CONCLUSION

As in cryptanalysis, measurement of information leakages is the fundamental principle underlying the theoretical framework for robust watermarking security assessment presented in this chapter. A watermarking technique, even robust, is not secure if the opponent has refine knowledge on the presumably secret key while pieces of content are watermarked with the same key. The security level is then defined by the number of observations the opponent needs in order to accurately estimate the secret key.

The conclusion of this chapter is not that spread spectrum based watermarking techniques or substitutive schemes are broken. The goal is to warn the watermarking community that security is a crucial issue. Designers should not only control the imperceptibility and the robustness of their schemes but also assess their security levels. Depending on the application designers are targeting (and especially on the observations available to the pirate), watermarking several pieces of content with the same key might bring threats. This potentially causes difficulties on the key management. For instance, it is not clear how a blind watermarking decoder will be informed of the secret key if this later one is to be changed according to the security levels assessed in this chapter.

REFERENCES

Amari, S. I. (1985). *Differential-geometrical methods in statistics.* Lectures notes in statistics. Berlin: Springer-Verlag.

Amari, S.I., & Cardoso, J.F. (1997). Blind source separation: Semiparametric statistical approach. *IEEE Transactions on Signal Processing, 45*(11).

Anderson, R., & Petitcolas, F. (1998, May). On the limits of steganography. *IEEE Journal of Selected Areas in Communications, 16*(4), 474-481.

Barni, M., Bartolini, F., & Furon, T. (2003, October). A general framework for robust watermarking security. *Signal Processing, 83*(10), 2069-2084.

Barni, M., & Perez-Gonzalez, F. (2005, January). Special session on watermarking security. In E. J. Delp & P. W. Wong (Eds.), *Security, steganography, and watermarking of multimedia contents vii* (Vol. 5681, pp. 685-768). San Jose, CA: SPIE.

Bartolini, F., Barni, M., & Furon, T. (2002, September). Special session on watermarking security. In *Proceedings of the 11th European Signal Processing Conference (EUSIPCO)* (Vol. 1, pp. 283-302, 441-461). Toulouse, France.

Bloom, J., Cox, I., Kalker, T., Linnartz, J.P., Miller, M., & Traw, C. (1999, July). Copy protection for DVD video. In *Proceedings of the IEEE, 87*(7), 1267-1276.

Burgett, S., Koch, E., Zhao, & J. (1998, March). Copyright labeling of digitized image data. *IEEE Communications Magazine, 36*(3), 94-100.

Cachin, C. (1997). *Entropy measures and unconditional security in cryptography* (Vol. 1). Hartung-Gorre Verlag.

Cardoso, J.F. (1998, October). Blind signal separation: Statistical principles. In *Proceedings of the IEEE, 90*(8), 2009-2026.

Cayre, F., Fontaine, C., & Furon, T. (2004, October). Watermarking attack: Security of WSS techniques. In I. Cox, T. Kalker & H.K. Lee (Eds.), *Proceedings of the International Workshop on Digital Watermarking* (Vol. 3304, pp. 171-183). Seoul, Korea: Springer-Verlag.

Cayre, F., Fontaine, C., & Furon, T. (2005a, January). Watermarking security part II: Practice. In E. J. Delp & P. W. Wong (Eds.), *Proceedings of spie-is&t electronic imaging, spie* (Vol. 5681, pp. 758-768). San Jose, California.

Cayre, F., Fontaine, C., & Furon, T. (2005b, January). Watermarking security part I: Theory. In E.J. Delp & P.W. Wong (Eds.), *Proceeding of spie-is&t electronic imaging, spie* (Vol. 5681, pp. 746-757), San Jose, California.

Cayre, F., Fontaine, C., & Furon, T. (2005c, October). Watermarking security: Theory and practice. *IEEE Transactions on Signal Processing, 53*(10).

Comesana, P., P′erez-Freire, L., & P′erez-Gonzalez, F. (2005, June). Fundamentals of data hiding security and their application to spread-spectrum analysis. In *Proceedings of the 7th Information Hiding Workshop (IH05),* Barcelona, Spain. Springer-Verlag.

Comon, P. (1994). Independent component analysis, a new concept? *Signal Processing, 36*(3), 287-314.

Cox, I., & Linnartz, J.P. (1998, May). Some general methods for tampering with watermarks. *IEEE Journal on Selected Areas in Communications, 16*(4), 587-93.

Cox, I., & Miller, M. (2001, October). Electronic watermarking: The first 50 years. In J.L. Dugelay & K. Rose (Eds.), *Proceedings of the Fourth Workshop on Multimedia Signal Processing (MMSP)* (pp. 225-230). Cannes, France.

Craver, S., Memon, N., Yeo, B.L., & Yeung, M.M. (1997, October). On the invertibility of invisible watermarking technique. In *Proceedings of International Conference on Image Processing* (pp. 540-543). Washington, DC.

Craver, S., Memon, N., Yeo, B.L., & Yeung, M. (1998, May). Resolving rightful ownership with invisible watermarking techniques: Limitations, attacks, and implications. *IEEE Journal of Selected Areas in Communications, 16*(4), 573-587.

Diffie, W., & Hellman, M. (1976, November). New directions in cryptography. *IEEE Transactions on Information Theory, 22*(6), 644-654.

Döerr, G., & Dugelay, J.L. (2004, October). Security pitfalls of frame-by-frame approaches to video watermarking. *IEEE Trans. Sig. Proc., Supplement on Secure Media, 52*(10), 2955-2964.

Döerr, G., & Dugelay, J.L. (2005, January). Collusion issue in video watermarking. In E. J. Delp & P. W. Wong (Eds.), *Security, steganography, and watermarking of multimedia contents* (Vol. 5681, pp. 685-696). San Jose, CA: SPIE.

Furon, T., & Duhamel, P. (2003, April). An asymmetric watermarking method. *IEEE Transactions on Signal Processing, 51*(4), 981-995.

Gamboa, F., & Gassiat, E. (1997, December). Source separation when the input sources are discrete or have constant modulus. *IEEE Transactions on Signal Processing, 45*(12), 3062-3072.

Holliman, M., & Memon, N. (2000, March). Counterfeiting attacks on oblivious block-wise independent invisible watermarking schemes. *IEEE Transactions on Image Processing, 9*(3), 432-441.

Kalker, T. (2001, October). Considerations on watermarking security. In J. L. Dugelay & K. Rose (Eds.), *Proceedings of the Fourth Workshop on Multimedia Signal Processing (MMSP)* (pp. 201-206). Cannes, France.

Katzenbeisser, S. (2005). Computational security models for digital watermarks. In *Proceedings of the Workshop on Image Analysis for Multimedia Interactive Services (WIAMIS).*

Kerckhoffs, A. (1883, January). La cryptographie militaire. *Journal des Sciences Militaires, 9*, 5-38.

Kutter, M., Voloshynovskiy, S., & Herrigel, A. (2000, January). Watermark copy attack. In P. W. Wong & E. Delp (Eds.), *Security and watermarking of multimedia contents ii* (Vol. 3971). San Jose, CA: SPIE Proceedings.

Linnartz, J., & van Dijk, M. (1998, April). Analysis of the sensitivity attack against electronic watermarks in images. In D. Aucsmith (Ed.), *Proceedings of the Second International Workshop on Information Hiding* (Vol. 1525). Portland, OR: Springer-Verlag.

Maillard, T., & Furon, T. (2004, July). Towards digital rights and exemptions management systems. *Computer Law and Security Report, 20*(4), 281-287.

Menezes, A., Van Oorschot, P., & Vanstone, S. (1996). *Handbook of applied cryptography*. CRC Press.

Mittelholzer, T. (1999, September). An information-theoritic approach to steganography and watermarking. In A. Pfitzmann (Ed.), *Proceedings of the Third International Workshop on Information Hiding* (pp. 1-17). Dresden, Germany: Springer-Verlag.

Pérez-Freire, L., Comesana, P., & Pérez-González, F. (2005, June). Information-theoretic analysis of security in side-informed data hiding. In *Proceedings of the 7th Information Hiding Workshop (IH05).* Barcelona, Spain: Springer-Verlag.

Pérez-Freire, L., Pérez-González, F., Furon, T., & Na, P.C. (in press). Security of lattice-based data hiding against the known message attack. *IEEE Transactions on Information Forensics and Security.*

Shannon, C. (1949, October). Communication theory of secrecy systems. *Bell System Technical Journal, 28*, 656-715.

Stern, J., & Craver, S. (2001, October). Lessons learned from the SDMI. In J. L. Dugelay & K. Rose (Eds.), *Proceedings of the Fourth Workshop on Multimedia Signal Processing (MMSP)* (pp. 213-218), Cannes, France.

Stoica, P., & Ng, B.C. (1998). On the Cramér-Rao bound under parametric constraints. *IEEE Signal Processing Letters, 5*(7), 177-179.

Trappe,W., Wu, M., Wang, Z., & Liu, K. (2003, April). Anti-collusion fingerprinting for multimedia. *IEEE Transactions on Signal Processing, 51*(4), 1069-1087.

Wong, P. W., & Memon, N. (2001, October). Secret and public key image watermarking schemes for images authentication and ownership verification. *IEEE Transactions on Image Processing, 10*(10), 1593-1601.

ADDITIONAL READING

Cox, I., Doërr, G., & Furon, T. (2006, November). Watermarking is not cryptography. In *Proceedings of the International Workshop on Digital Watermarking, Invited Talk*, Jeju island, Korea.

Perez-Freire, L., Comesaña, P., Troncoso-Pastoriza, J. R., & Pérez-Gonzalez, F. (2006, October). Watermarking security: A survey. *Transactions on Data Hiding and Multimedia Security I, 4300*, 41-72.

Security Analysis of Practical Watermarking Technique:

Bas, P., & Hurri, J. (2005, September). Security of DM quantization watermarking scheme: A practical study for digital images. In *Proceedings of the*

International Workshop of Digital Watermarking (IWDW), Sienna, Italy.

Bas, P., & Loboguerrero, A. (2005, May). *First Wavila challenge: Several consideration on the security of a feature-based synchronisation scheme for digital image watermarking.* Paper presented at the First Wavilla Challenge, Barcelonna, Spain.

Doërr, G., & Dugelay, J. L. (2004). Security pitfalls of frame-by-frame approaches to video watermarking. *IEEE Transactions on Signal Processing, Supplement on Secure Media, 52*(10), 2955-2964.

Pérez-Freire, L., Pérez-Gonzàlez, F., Furon, T., & Comesaña, P. (2006, December). Security of lattice-based data hiding against the known message attack. *IEEE Transactions on Information Forensics and Security, 1*(4), 421-439.

Sensitivity Attack:

Comesaña, P., Pérez-Freire, L., & Pérez-Gonzàlez, F. (2006, September). Blind Newton sensitivity attack. *IEE Proceedings on Information Security, 153*(3), 115-125.

El Choubassi, M., & Moulin, P. (2005, Jan). A new sensitivity analysis attack. In *Proceedings of the SPIE Conference*, San Jose, California.

Lessons Learned from the BOWS Challenge:

Comesaña-Alfaro, P., & Pérez-Gonzàlez, F. (2007, February). Two different approaches for attacking BOWS. In *SPIE Proceedings of the Security, Steganography, and Watermarking of Multimedia Contents IX, San Jose, California.*

Craver, S.A., Atakli, I., & Yu, J. (2007, February). How we broke the BOWS watermark. In *SPIE Proceedings of the Security, Steganography, and Watermarking of Multimedia Contents IX*, San Jose, California.

Piva, A., & Barni, M. (2007, February). The first BOWS contest: Break our watermarking system. In *SPIE Proceedings of the Security, Steganography, and Watermarking of Multimedia Contents IX*, San Jose, California.

Potential Countermeasures:

Bas, P., & Cayre, F. (2006, September). Achieving subspace or key security for WOA using Natural or Circular Watermarking. In *Proceedings of the ACM Multimedia and Security Workshop*, Geneva, Switzerland.

Lin, E.T., & Delp, E.J. (2004, October). Temporal synchronization in video watermarking. *IEEE Transactions on Signal Processing: Supplement on Secure Media, 52*(10), 3007-3022.

Moulin, P., & Briassouli, A. (2004, October). A stochastic QIM algorithm for robust, undetectable image watermarking. In *Proceedings of the IEEE International Conference on Image Processing (ICIP),* Singapore.

Compilation of References

Abbate, A., Decusatis, C. M., & Das, P. K. (2002). *Wavelets and subbands, fundamentals and applications.* Boston: Birkhauser.

Abe, M., & Smith, J.O. (2004). *Design criteria for simple sinusoidal parameter estimation based on quadratic interpolation of FFT magnitude peaks.* Paper presented at the AES 117th Convention, San Francisco.

Achlioptas, D., Gomez, C., Kautz, H., & Selman, B. (2000). Generating satisfiable problem instances. In *AAAI.*

Ajtai, M. (1996). *Generating hard instances of lattice problems* (ECCC Report TR96-007). Electronic Colloquium on Computational Complexity.

Amari, S. I. (1985). *Differential-geometrical methods in statistics.* Lectures notes in statistics, Berlin: Springer-Verlag.

Amari, S.I., & Cardoso, J.F. (1997). Blind source separation: Semiparametric statistical approach. *IEEE Transactions on Signal Processing, 45*(11).

Anand, D., & Niranjan, U. C. (1998, October). *Watermarking medical images with patient information.* Paper presented at the IEEE EMBS Conference, Hong Kong, China.

Anderson, R., & Petitcolas, F. (1998, May). On the limits of steganography. *IEEE Journal of Selected Areas in Communications, 16*(4), 474–481.

Ansari, R., Malik, H., & Khokhar, A. (2004, May). *Data-hiding in audio using frequency-selective phase alteration.* Paper presented at the IEEE International Conference on Acoustics Speech and Signal Processing, Montreal, Canada.

Antonopoulou, H. (2002). A user authentication protocol based on the intractability of the 3-COLORING problem. *Journal of Discrete mathematical Sciences & Cryptography, 5*(1), 17-21.

Armeni, S., Christodoulakis, D., Kostopoulos, I., Kountrias, P. D., Stamatiou, Y. C., & Xenos, M. (2003). An information hiding method based on computational intractable problems. In Y. Manolopoulos & S. Evripidou (Eds.), *8th Panhellenic Conference on Informatics Advances in Informatics* (Vol. 2563, pp. 262-278). Springer-Verlag.

Armeni, S., Christodoulakis, D., Kostopoulos, I., Kountrias, P. D., Stamatiou, Y. C., & Xenos, M. (2003). Secure information hiding based on computationally intractable problems. *Journal of Discrete Mathematical Sciences & Cryptography, 6*(1), 21-33.

Arnold, M. (2000). Audio watermarking: Features, applications and algorithms. In *Proceedings of the IEEE International Conference on Multimedia and Expo* (Vol. 2, pp. 1013-1016).

Arnold, M. (2002, December). Subjective and objective quality evaluation of watermarked audio tracks. In C. Busch, M. Arnold, P. Nesi, & M. Schmucker (Eds.), *Proceedings of the Second International Conference on WEB Delivering of Music (WEDELMUSIC 2002)* (Vol. 4675, pp. 161-167). Darmstadt, Germany: IEEE Computer Society Press.

Arnold, M., & Kanka, S. (1999). *MP3 robust audio watermarking.* Paper presented at the DFG VIIDII Watermarking Workshop, Erlangen, Germany.

Arnold, M., Schmucker, M., & Wolthusen, S. (2003). *Techniques and applications of digital watermarking and content protection.* Boston: Artech House.

Arttameeyanant, P., Kumhom, P., & Chamnongthai, K. (2002). *Audio watermarking for Internet.* Paper presented at the IEEE ICIT'02 (Vol. 2, pp. 976-979).

Babai, L. (1985). *Trading group theory for randomness.* Paper presented at the 17th Annual ACM Symposium on Theory of Computing (pp. 421-420).

Barni, M., & Perez-Gonzalez, F. (2005, January). Special session on watermarking security. In E. J. Delp & P. W. Wong (Eds.), *Security, steganography, and watermarking of multimedia contents vii* (Vol. 5681, pp. 685-768). San Jose, CA: SPIE.

Barni, M., Bartolini, F., & Furon, T. (2003, October). A general framework for robust watermarking security. *Signal Processing, 83*(10), 2069-2084.

Barni, M., Bartolini, F., Rosa, A.D., & Piva, A. (2003). Optimal decoding and detection of multiplicative watermarks. *IEEE Transactions on Signal Processing, 51*(4).

Bartolini, F., Barni, M., & Furon, T. (2002, September). Special session on watermarking security. In *Proceedings of the 11th European Signal Processing Conference (EUSIPCO)* (Vol. 1, pp. 283-302, 441-461). Toulouse, France.

Bas, P., Chassery, J. M., & Macq, B. (2002). Geometrically invariant watermarking using feature points. *IEEE Transactions on Image Processing, 11*(9), 1014-1028.

Basia, V., Pitas, I., & Nikolaidis, N. (2001). Robust audio watermarking in the time-domain. *IEEE Transactions on Multimedia, 3*(2), 232-241.

Bassia, P., & Pitas, I.P. (1998, September 8-11). Robust audio watermarking in time domain. *EUSIPCO,* 25-28.

Bassia, P., Pitas, I., & Nikolaidis, N. (2001). Robust audio watermarking in the time domain. *IEEE Transactions on Multimedia, 3*(2), 232-241.

Baumgarte, F. (1998). Evaluation of a physiological ear model considering masking effects relevant to audio coding. In *Proceedings of the 105th Audio Engineering Society Convention.* San Francisco: AES.

Bender, W., Gruhl, D., & Morimoto, N. (1996). Techniques for data hiding. *IBM Systems Journal, 35*(3), 313-336.

Billingsley, P. (1995). *Probability and measure.* New York: Wiley.

BitTorrent. (2006). Retrieved March 26, 2007, from http://www.bittorrent.com

Black, M., & Zeytinoglu, M. (1995). Computationally efficient wavelet packet coding of wide-band stereo signals. In *Proceedings of the IEEE International Conference on Acoustics, Speech, and Signal Processing* (pp. 3075-3078).

Bloom, J., Cox, I., Kalker, T., Linnartz, J., Miller, M., & Traw, C. (1999). Copy protection for DVD video. *Proceedings of the IEEE, 87*(7), 1267-1276.

Boff, R.K., Kaufman, L., & Thomas, D.J. (1986). *Handbooks of perception and human performance* (Vol. I). John Wiley & Sons.

Boney, L., Tewfic, A., & Hamdy, K. (1996). *Digital watermarks for audio signals.* Paper presented at the IEEE International Conference on Multimedia Computing and Systems (pp. 473-480).

Bosi, M. (1997). Perceptual audio coding. *IEEE Signal Processing Magazine, 14*(5), 43-49.

Bosi, M., & Goldberg, R. E. (2003). *Introduction to digital audio coding and standards.* Kluwer Academic Publishers.

Boyarsky, A., & Scarowsky, M. (1979). On a class of transformations which have unique absolutely continues invariant measures. *Transactions of American Mathematical Society, 255,* 243-262.

Brandenburg, K. (1999). *MP3 and AAC explained.* Paper presented at the AES 17th International Conference on High Quality Audio Coding.

Burgett, S., Koch, E., Zhao, & J. (1998, March). Copyright labeling of digitized image data. *IEEE Communications Magazine, 36*(3), 94-100.

Burstein, H. (1988, November). Approximation formulas for error risk and sample size in *ABX* testing. *AES Journal, 36*(11), 879-883.

Cachin, C. (1997). *Entropy measures and unconditional security in cryptography* (Vol. 1). Hartung-Gorre Verlag.

Cano, P., Batlle, E., Kalker, T., & Haitsma, J. (2002, December). *A review of algorithms for audio fingerprinting.* Paper presented at the International Workshop on Multimedia Signal Processing, U.S. Virgin Islands.

Cardoso, J.F. (1998, October). Blind signal separation: Statistical principles. In *Proceedings of the IEEE, 90*(8), 2009-2026.

Carnero, B., & Drygajlo, A. (1999). Perceptual speech coding and enhancement using frame-synchronized fast wavelet packet transform algorithms. *IEEE Transactions on Signal Processing, 47*(6), 1622-1635.

Cayre, F., Fontaine, C., & Furon, T. (2004, October). Watermarking attack: Security of wss techniques. In I. Cox, T. Kalker & H.K. Lee (Eds.), *Proceedings of the International Workshop on Digital Watermarking* (Vol. 3304, pp. 171-183). Seoul, Korea: Springer-Verlag.

Cayre, F., Fontaine, C., & Furon, T. (2005a, January). Watermarking security part II: Practice. In E. J. Delp & P. W. Wong (Eds.), *Proceedings of spie-is&t electronic imaging, spie* (Vol. 5681, pp. 758-768), San Jose, California.

Cayre, F., Fontaine, C., & Furon, T. (2005c, October). Watermarking security: Theory and practice. *IEEE Transactions on Signal Processing, 53*(10).

Celik, M., Sharma, G., & Tekalp, A.M. (2005, March). *Pitch and duration modification for speech watermarking.* Paper presented at the IEEE International Conference on Acoustics Speech and Signal Processing, Philadelphia.

Chaitin, G.H. (1966). On the length of programs for computing finite binary sequences. *Journal of the ACM, 13,* 547-570.

Cheeseman, P., Kanefsky, B., & Taylor, W.M. (1991). *Where the really hard problems are.* Paper presented at the International Joint Conference on Artificial Intelligence (pp. 331-337).

Chen, B., & Wornell, G. (2001). Quantization index modulation: A class of provably good methods for digital watermarking and information embedding. *IEEE Trans. Information Theory, 47,* 1423-1443.

Chen, B., & Wornell, G.W. (2001). Quantization index modulation: A class of provably good methods for digital watermarking and information embedding. *IEEE Transactions on Information Theory, 47*(4).

Cheng, Q., & Sorensen, J. (2001, May). *Spread spectrum signaling for speech watermarking.* Paper presented at the IEEE International Conference on Acoustics Speech and Signal Processing, Salt Lake City, Utah.

Cheng, S., & Yu, H., & Xiong, Z. (2002). Enhanced spread spectrum watermarking of MPEG-2 AAC audio. In *Proceeding of the IEEE International Conference on Acoustics, Speech, and Signal Processing* (pp. 3728-3731).

Chenyu, W., Jie, Z., Zhao, B., & Gang, R. (2003). Robust crease detection in fingerprint images. In *Proceedings of the IEEE International Conference on Computer Vision and Pattern* Recognition (pp. 505-510). Madison, Wisconsin.

Chipcenter. (2006). *Tutorial on spread spectrum.* Retrieved March 18, 2007, from http://archive.chipcenter.com/knowledge_centers/digital/features/showArticle.jhtml?articleID=9901240

Chou, J., Ramchandran, K., & Ortega, A. (2001). Next generation techniques for robust and imperceptible audio data hiding. In *Proceedings of the IEEE International Conference on Acoustics, Speech, and Signal Processing* (pp. 1349-1352), Salt Lake City, Nevada.

Chvátal, V., & Reed, B. (1992). *Mick gets some (the odds are on his side).* Paper presented at the 33rd IEEE Annual Symposium on Foundations of Computer Science (pp. 620-627).

Coello, C.A., Veldhuizen, D.A., & Lamont, G.B. (2002). *Evolutionary algorithms for solving multi-objective problems.* Kluwer Academic Publishers.

Comesana, P., P´erez-Freire, L., & P´erez-Gonzalez, F. (2005, June). Fundamentals of data hiding security and their application to spread-spectrum analysis. In *Proceedings of the 7th Information Hiding Workshop (IH05),* Barcelona, Spain. Springer-Verlag.

Comon, P. (1994). Independent component analysis, a new concept? *Signal Processing, 36*(3), 287–314.

Cook, P., & Scavone, G. P. (1999). The synthesis toolKit (STK). In *Proceedings of the International Computer Music Conference*, Beijing, China. Retrieved March 19, 2007, from http://ccrma.stanford.edu/software/stk/papers/stkicmc99.pdf

Cook, S. (1971). *The complexity of theorem-proving procedures.* Paper presented at the 3rd Annual ACM Symposium on Theory of Computing (pp. 151-158).

Cooper, G.R., & McGillem, C.D. (1986). *Modern communications and spread spectrum.* New York: McGraw-Hill Book Company.

Costa, M. (1983). Writing on dirty paper. *IEEE Trans. Information Theory, IT, 29,* 439-441.

Cover, T., & Thomas, J. A. (1991). *Elements of information theory.* New York: Wiley-Interscience.

Cox, I. J., Kilian, J., Leighton, F., & Shamoon, T. (1999). Secure spread spectrum watermarking for multimedia. *IEEE Trans. Image Processing, 6,* 1673-1687.

Cox, I. J., Miller, M. L., & Bloom, J. A. (2002). *Digital watermarking.* San Francisco: Morgan Kaufmann.

Cox, I., & Linnartz, J.P. (1998, May). Some general methods for tampering with watermarks. *IEEE Journal on Selected Areas in Communications, 16*(4), 587-93.

Cox, I., & Miller, M. (2001). Electronic watermarking: The first 50 years. In *Proceedings of the IEEE Workshop on Multimedia Signal Processing* (pp. 225-230). Cannes, France.

Cox, I., Kilian, J., Leighton, F., & Shamoon, T. (1997). Secure spread spectrum watermarking for multimedia. *IEEE Transactions on Image Processing, 6*(12), 1673-1687.

Cox, I., Miller, M., & Bloom, J. (2002). *Digital watermarking.* San Francisco: Morgan Kaufmann Publishers.

Cox, I.J. (1998). Spread spectrum watermark for embedded signalling. *United States Patent 5,848,155.*

Cox, I.J., Kilan, J., Leighton, T., & Shamoon, T. (1997). Secure spread spectrum watermarking for multimedia. *IEEE Transactions on Image Processing, 6*(12), 1673-1687.

Craver, S., & Stern, J. (2001). Lessons learned from SDMI. In *Proceedings of the IEEE International Workshop on Multimedia Signal Processing* (pp. 213-218). Cannes, France.

Craver, S., Memon, N., Yeo, B.L., & Yeung, M. (1998, May). Resolving rightful ownership with invisible watermarking techniques: Limitations, attacks, and implications. *IEEE Journal of Selected Areas in Communications, 16*(4), 573-587.

Craver, S., Memon, N., Yeo, B.L., & Yeung, M.M. (1997, October). On the invertibility of invisible watermarking technique. In *Proceedings of International Conference on Image Processing* (p. 540-543), Washington, DC.

Craver, S.A., Wu, M., Liu, B., Stubblefield, A., Swartzlander, B., & Wallach, D. S. (2001). Reading between the lines: Lessons from the SDMI challenge. In *Proceeding of the 10th USENIX Security Symposium.*

Cvejic, N. (2004). *Algorithms for audio watermarking and steganography.* Unpublished doctoral thesis, University of Oulu, Department of Electrical and Information Engineering.

Cvejic, N., & Seppänen, T. (2003). Robust audio watermarking in wavelet domain using frequency hopping and patchwork method. In *Proceedings of the 3rd International Symposium on Image and Signal Processing and Analysis* (pp. 251-255).

Cvejic, N., & Seppänen, T. (2003, September). *Robust audio watermarking in wavelet domain using frequency hopping and patchwork method.* Paper presented at the 3rd International Symposium on Image and Signal Processing and Analysis, Rome, Italy.

Cvejic, N., & Seppänen, T. (2004). Spread spectrum audio watermarking using frequency hopping and attack characterization. *Signal Processing, 84,* 207-213.

Cvejic, N., & Seppänen, T. (2005). Increasing robustness of LSB audio steganography by reduced distortion LSB coding. *Journal of Universal Computer Science, 11*(1), 56-65.

Cvejic, N., Keskinarkaus, A., & Seppanen, T. (2001). *Audio watermarking using m-sequences and temporal masking.*

Paper presented at the IEEE Workshop on Applications of Signal Processing to Audio and Acoustics (pp. 227-230).

Czerwinski, S., Fromm, R., & Hodes, T. (1999). *Digital music distribution and audio watermarking* (Project Report UCB IS 219).

Das, T. K., & Maitra, S. (2004). Cryptanalysis of correlation based watermarking schemes using single watermarked copy. *IEEE Signal Processing Letters, 11*, 446-449.

Das, T. K., Kim, H. J., & Maitra, S. (2005). Security evaluation of generalized patchwork algorithm from cryptanalytic viewpoint. *Lecture Notes in Artificial Intelligence, 3681*, 1240-1247.

Dau, T. (1996). *Modeling auditory processing of amplitude modulation.* Unpublished doctoral thesis, University Oldenburg, Oldenburg, Germany.

Dau, T., Püschel, D., & Kohlrausch, A. (1996). A quantitative model of the effective signal processing in the auditory system. I. Model structure. *AES Journal, 99*(6), 3615-3622.

Delannay, D., & Macq, B. (2000). Generalized 2D cyclic patterns for secret watermark generation. In *Proceedings of the IEEE International Conference of Image Processing* (pp. 72-79).

Deller, J.R., Jr., Hansen, J.H.L., & Proakis, J.G. (2000). *Discrete time processing of speech signals* (2nd ed.). New York: IEEE Press.

Depovere, G., Kalker, T., & Linnartz, J. P. (1998). Improved watermark detection reliability using filtering before correlation. In *Proceedings of the IEEE International Conference on Image Processing* (Vol. 1, pp. 430-434).

Depovere, G., Kalker, T., Haitsma, J., Maes, M., de Strycker, L., Termont, P., et al. (1999). The viva project: Digital watermarking for broadcast monitoring. In *Proceedings of the IEEE International Conference on Image Processing* (pp. 202-205). Kobe, Japan.

Dietz, M., Lijeryd, L., Kjorling, K., & Kunz, O. (2002). *Spectral band replication, a novel approach in audio coding.* In *Proceedings of the 112th Audio Engineering Society Convention.* Munich, Germany: AES.

Diffie, W., & Hellman, M. (1976, November). New directions in cryptography. *IEEE Transactions on Information Theory, 22*(6), 644-654.

Dittmann, J., Steinebach, M., & Steinmetz, R. (2001). *Digital watermarking for MPEG audio layer 2.* Paper presented at the Multimedia and Security Workshop at ACM Multimedia.

Do¨err, G., & Dugelay, J.L. (2004, October). Security pitfalls of frame-by-frame approaches to video watermarking. *IEEE Trans. Sig. Proc., Supplement on Secure Media, 52*(10), 2955-2964.

Do¨err, G., & Dugelay, J.L. (2005, January). Collusion issue in video watermarking. In E. J. Delp & P. W. Wong (Eds.), *Security, steganography, and watermarking of multimedia contents* (Vol. 5681, pp. 685-696). San Jose, CA: SPIE.

Dong, X., Bocko, M. F., & Ignjatovic, Z. (2004). Data hiding via phase manipulation of audio signals. In *Proceedings of the IEEE International Conference on Acoustics, Speech, and Signal Processing* (pp. 377-380). Montréal, Canada.

Dong, X., Bocko, M., & Ignjatovic, Z. (2004, May). *Data hiding via phase modification of audio signals.* Paper presented at the IEEE International Conference on Acoustics Speech and Signal Processing, Montreal, Canada.

EBU. (1988, April). *Sound quality assessment material recordings for subjective tests.* Bruxelles, Belgium.

edonkey. (2006). Retrieved March 26, 2007, from http://www.edonkey2000.com

Eggers, J., Bäuml, R., Tzschoppe, R., & Girod, B. (2003). Scalar Costa scheme for information embedding. *IEEE Trans. Signal Processing, 51*, 1003-1019.

Erkucuk, S., Krishnan, S., & Zeytinoglu, M. (2006). A robust audio watermark representation based on linear chirps. *IEEE Transactions on Multimedia, 8*(5), 925-936.

Esmaili, S., Krishnan, S., & Raahemifar, K. (2003). Audio watermarking time-frequency characteristics. *Canadian Journal of Electrical and Computer Engineering, 2*(28), 57-61.

Faundez-Zanuy, M., Hagmuller, M., Kubin, G., & Kleijn, W. B. (2005, April). *The Cost-277 speech database.* Paper presented at the 13th European Signal Processing Conference, Barcelona, Spain.

Foo S.W., Xue, F. & Li M. (2005). *A blind audio watermarking scheme using peak point extraction.* Paper presented at the IEEE International Symposium on Circuits and Systems (pp. 4409-4412).

Foo, S.W., Ho, S.M., & Ng, L.M. (2004). *Audio watermarking using time-frequency compression expansion.* Paper presented at the IEEE International Symposium on Circuits and Systems (pp. 201-204.

Foo, S.W., Yeo, T.H., & Huang, D.Y. (2001). An adaptive audio watermarking system. *IEEE Tencon.*, 509-513.

Furon, T., & Duhamel, P. (2003, April). An asymmetric watermarking method. *IEEE Transactions on Signal Processing, 51*(4), 981-995.

Furui, S. (1992). *Digital speech processing, synthesis, and recognition.* Marcel Dekker Inc.

Gamboa, F., & Gassiat, E. (1997, December). Source separation when the input sources are discrete or have constant modulus. *IEEE Transactions on Signal Processing, 45*(12), 3062-3072.

Gang, L., Akansu, A.N., & Ramkumar, M. (2001). *MP3 resistant oblivious steganography.* Paper presented at the IEEE International Conference on Acoustics, Speech and Signal Processing (Vol. 3, pp. 1365-1368).

Garas, J., & Sommen, P. (1998). Time/pitch scaling using the constant-q phase vocoder. In *Proceedings of the STW's 1998 Workshops CSSP98 and SAFE98* (pp. 173-176).

Garcia R.A. (1999) *Digital watermarking of audio signals using a psychoacoustic auditory model and spread spectrum theory.* Paper presented at the 107th Convention on Audio Engineering Society (preprint 5073).

Gardner, G.W. (1992). *The virtual acoustic room.* Unpublished master's thesis, MIT Computer Science and Engineering.

Garey, M.R., & Johnson, D.S. (1979). *Computers and Intractability, a guide to the theory of NP-completeness.* W.H. Freeman and Company.

Ghoraani, B., & Krishnan, S. (in press). Chirp-based image watermarking schemes. *IEEE Transactions on Information Forensics and Security.*

Giannoula, A., Tefas, A., Nikolaidis, N., & Pitas, I. (2003). Improving the detection reliability of correlation-based watermarking techniques. In *Proceedings of the IEEE International Conference on Multimedia and Expo* (Vol. 1, pp. 209-212).

Girin, L., & Marchand, S. (2004). Watermarking of speech signals using the sinusoidal model and frequency modulation of the partials. In *Proceedings of the IEEE International Conference on Acoustics, Speech, and Signal Processing* (pp. I633-636). Montréal, Canada.

Goerdt, A. (1996). A threshold for unsatisfiability. *Journal of Computer and System Sciences, 53*, 469-486.

Goldreich, O. (1995). Randomness, interactive proofs and zero-knowledge: A survey. In R. Herken (Ed.), *The universal turing machine: A half-century survey.* Springer-Verlag.

Goldreich, O., Micali, S., & Widgerson, A. (1991). Proofs that yield nothing but their validity or all languages in NP have Zero-Knowledge proof systems. *Journal of the ACM, 38*(1), 691-729.

Goldwasser, S., Micali, S., & Rackoff, C. (1985). *The knowledge complexity of interactive proof-systems.* Paper presented at the 17th Annual ACM Symposium on Theory of Computing (pp. 291-304).

Gomes, L. de C. T. (2001). Resynchronization methods for audio watermarking. In *Proceedings of the 111th AES Convention* (Preprint 5441). New York.

Green, M.D., & Swets, A.J. (1998). *Signal detection theory and psychophysics.* Peninsula Publishing.

Gruhl, D., & Bender, W. (1996). Echo hiding. In *Proceedings of the Information Hiding Workshop* (pp. 295-315).

Gruhl, D., & Lu, A., & Bender, W. (1996). Echo hiding. In *Proceedings of the Information Hiding Workshop* (pp. 295-315). University of Cambridge.

Gurijala, A., & Deller, J.R., Jr. (2001, May). *Robust algorithm for watermark recovery from cropped speech.* Paper presented at the IEEE International Conference on Acoustics Speech and Signal Processing, Salt Lake City, Utah.

Gurijala, A., & Deller, J.R., Jr. (2003, September). *Speech watermarking by parametric embedding with an fidelity criterion.* Paper presented at the Interspeech Eurospeech, Geneva, Switzerland.

Gurijala, A., & Deller, J.R., Jr. (2005, July). *Detector design for parametric speech watermarking.* Paper presented at the IEEE International Conference on Multimedia and Expo, Amsterdam, The Netherlands.

Gurijala, A., Deller, D. D. R., Seadle, M. S., & Hansen, J. H. L. (2002). Speech watermarking through parametric modeling. In *Proceedings of the International Conference on Spoken Language Processing* (CD-ROM), Denver, Colorado.

Hagmüller, M., Horst, H., Kröpfl, A., & Kubin, G. (2004, September). *Speech watermarking for air traffic control.* Paper presented at the 12th European Signal Processing Conference, Vienna, Austria.

Harris, F. J. (1978). On the use of windows for harmonic analysis with the discrete Fourier transform. *Proceedings IEEE, 66*, 51-83.

Hartung, F., & Kutter, M. (1999). Multimedia watermarking techniques. In *Proceedings of the IEEE, 87*(7), 1709-1107.

Hartung, F., Su, J. K., & Girod, B. (1999). Spread spectrum watermarking: Malicious attacks and counterattacks. In *Proceedings of the SPIE Security and Watermarking of Multimedia Contents* (pp. 147-158).

Hatada, M., Sakai, T., Komatsu, N., & Yamazaki, Y. (2002, July). Digital watermarking based on process of speech production. In *Proceedings of the SPIE: Multimedia Systems and Applications V*, Boston.

Haykin, S. (1996). *Adaptive filter theory* (3rd ed.). NJ: Prentice Hall.

He, X., & Scordilis, M. (2005). Improved spread spectrum digital audio watermarking based on a modified perceptual entropy psychoacoustic model. In *Proceedings of the IEEE Southeast Conference* (pp. 283-286).

He, X., & Scordilis, M. (2006). A psychoacoustic model based on the discrete wavelet packet transform. *Journal of Franklin Institute, 343*(7), 738-755

He, X., Iliev, A. I., & Scordilis, M. S. (2004). A high capacity watermarking technique for stereo audio. In *Proceedings of the IEEE International Conference on Acoustics, Speech, and Signal Processing* (pp. 393-396). Montréal, Canada.

Heeger, D. (1997). Signal detection theory. *Teaching handout.* Department of Psychology, Stanford University.

Herken, R. (1995). *The universal turing machine: A half-century survey.* Springer-Verlag.

Hernandez, J.J., Amado, M., & Perez-Gonzalez, F. (2000). DCT-domain watermarking techniques for still images: Detector performance analysis and a new structure. *IEEE Transactions on Image Processing, 9*(1).

Holliman, M., & Memon, N. (2000, March). Counterfeiting attacks on oblivious block-wise independent invisible watermarking schemes. *IEEE Transactions on Image Processing, 9*(3), 432-441.

Hong, Z., Wu, M., Wang, Z., & Liu, K. (2003). Nonlinear collusion attacks on independent fingerprints for multimedia. In *Proceedings of the IEEE Computer Society Conference on Multimedia and Expo* (613-616). Baltimore.

Hsieh, C.T., & Tsou, P.Y. (2002). *Blind cepstrum domain audio watermarking based on time energy features.* Paper presented at the 14th International Conference on Digital Signal Processing (pp. 705-708).

Huang, J.W., Wang, Y., & Shi, Y.Q. (2002). *A blind audio watermarking algorithm with self-synchronization.* Paper presented at the International Symposium on Circuits and Systems 2002.

IFPI. (2006). World sales 2001. Retrieved March 26, 2007, from http://www.ifpi.org/site-content/statistics/worldsales.html

Isabelle, S., & Wornell, G. (1997). Statistical analysis and spectral estimation techniques for one-dimensional chaotic signals. *IEEE Transactions on Signal Processing, 45*(6), 1495-1506.

ISO/IEC 11172-3. (1993). *Coding of moving picture and associated audio for digital storage media at up to about 1.5 Mbits - part 3* (Audio Recording).

ISO/IEC 11172-3. Coding of moving pictures and associated audio for digital storage media at up to about 1.5 Mbit/s-Part 3.

ITU-R. (1997a). *Recommendation BS.1116-1, methods for subjective assessement of small impairments in audio systems including multichannel sound systems.* Paper presented by the International Telecommunications Union Radio Communication Assembly.

ITU-R. (1997b). *Recommendation BS.1284-1, general methods for the subjective assessement of audio quality.* Paper presented by the International Telecommunications Union Radio Communication Assembly.

ITU-R. (1998). *Recommendation BS.1387, method for objective measurements of perceived audio quality (PEAQ).* Paper presented by the International Telecommunications Union Radio Communication Assembly.

ITU-R. (2001). *Recommendation BS.1534, method for the subjective assessment of intermediate quality level of coding systems.* Paper presented by the International Telecommunications Union Radio Communication Assembly.

Jaffard, S., Meyer, Y., & Ryan, R. D. (2001). *Wavelets tools for science and technology.* SIAM.

Jayant, N., Johnston, J., & Safranek, R. (1993). Signal compression based on method of human perception. *Proceedings IEEE, 81,* 1385-1422.

Johnson, K.F., Duric, Z., & Jajodia, S. (2000). *Information hiding: Steganography and watermarking: Attacks and countermeasures.* MA: Kluwer Academic Publishers.

Johnston, J. (1998). Estimation of perceptual entropy using noise masking criteria. In *Proceedings of the IEEE International Conference on Acoustics, Speech, and Signal Processing* (pp. 2524-2527). New York.

Johnston, J. D. (1998). Transfrom coding of audio signals using perceptual noise criteria. *IEEE Journal on Selected Areas in Communications, 6*(2), 314-323.

Jung, S., Seok, J., & Hong, J. (2003). An improved detection technique for spread spectrum audio watermarking with a spectral envelope filter. *ETRI Journal, 25*(1), 52-54.

Kabal, P. (2003, December). *An examination and interpretation of ITU-R BS.1387: Perceptual evaluation of audio quality* (Tech. Rep.). Montreal, Canada: McGill University, Department of Electrical & Computer Engineering.

Kahrs, M., & Brandenburg, K. (Eds.). (1998). *Applications of digital signal processing to audio and acoustics.* Kluwer Academic Publishers.

Kalker, T. (2001, October). Considerations on watermarking security. In J. L. Dugelay & K. Rose (Eds.), *Proceedings of the Fourth Workshop on Multimedia Signal Processing (MMSP)* (pp. 201-206), Cannes, France.

Kalker, T., & Haitsma, J. (2000). Efficient detection of a spatial spread-spectrum watermark in MPEG video streams. In *Proceedings of the IEEE International Conference on Image Processing* (pp. 407-410). Vancouver, British Columbia.

Kalker, T., Depovere, G., Haitsma, J., & Maes, M. (1999, January). *A video watermarking system for broadcast monitoring.* Paper presented at the IS\&T/SPIE's 11th Annual Symposium on Electronic Imaging '99: Security and Watermarking of Multimedia Contents, San Jose, California.

Kalker, T., Linnartz, J.P., & Depovere, G. (1998). On the reliability of detecting electronic watermarks in digital images. In *Proceedings of the European Signal Processing Conference.*

Kaporis, A.C., Kirousis, L.M., & Stamatiou, Y.C. (2000). A note on the non-colorability threshold of a random graph. *Electronic Journal of Combinatorics, 7*(R29).

Karp, R.M. (1972). Reducibility among combinatorial problems. In Miller & Thatcher (Eds.), *Complexity of computer computations* (pp. 85-103). Plenum Press.

Katzenbeisser, S. (2005). Computational security models for digital watermarks. In *Proceedings of the Workshop on Image Analysis for Multimedia Interactive Services (WIAMIS).*

Katzenbeisser, S., & Petitcolas, F.A.P. (2000). *Information hiding techniques for steganography and digital watermarking.* Norwood, MA: Artech House.

Kazaa. (2006). Retrieved March 26, 2007, from http://www.kazaa.com

Keiler, F. (2006, May). Real-time subband-ADPCM low-delay audio coding approach. In *Proceedings of the 120th Audio Engineering Society Convention.* Paris: AES.

Kerckhoffs, A. (1883, January). La cryptographie militaire. *Journal des Sciences Militaires, 9,* 5-38.

Kilian, J., Leighton, F. T., Matheson, L. R., Shamoon, T. G., Tarjan, R. E., & Zane, F. (1998). Resistance of digital watermarks to collusive attacks. In *Proceedings of the 1998 International Symposium on Information Theory* (p. 271). Cambridge, UK.

Kim, H. J. (2003). Audio watermarking techniques. In *Proceedings of the Pacific Rim Workshop on Digital Steganography,* Kyushu Institute of Technology, Kitakyushu, Japan.

Kim, H.J., & Choi, Y.H. (2003). A novel echo-hiding scheme with backward and forward kernels. *IEEE Transactions on Circuits and Systems for Video Technology, 13*(8), 885-889.

Kim, H.J., Choi, Y.H., Seok, J.W., & Hong, J.W. (2004). *Audio watermarking techniques: Intelligent watermarking techniques* (pp. 185-218).

Kirby, D.G. (1995, October). ISO/MPEG subjective tests on multichannel audio systems. In *Proceedings of the 99th Audio Engineering Society Convention.* New York: AES.

Kirkpatrick, S., & Selman, B. (1994). Critical behavior in the satisfiability of random Boolean expressions. *Science, 264,* 1297-1301.

Kiroski, D.K., & Malvar, H.M. *Embedding and detecting spread spectrum watermarks under the estimation attack.* Microsoft Research.

Kirovski, D., & Attias, H. (2002). Audio watermark robustness to de-synchronization via beat detection. *Information Hiding Lecture Notes in Computer Science, 2578,* 160-175.

Kirovski, D., & Malvar, H. (2001). Robust spread-spectrum audio watermarking. In *Proceeding of the IEEE International Conference on Acoustics, Speech, and Signal Processing* (Vol. 3, pp. 1345-1348).

Kirovski, D., & Malvar, H. (2001). Spread-spectrum audio watermarking: Requirements, applications, and limitations. In *Proceedings of the IEEE International Workshop on Multimedia Signal Processing* (pp. 219-224). Cannes, France.

Klapuri, A. P. (2003). Multiple fundamental frequency estimation based on harmonicity and spectral smoothness. *IEEE Trans. Speech and Audio Processing, 11,* 804-816.

Ko B., Nishimura R., & Suzuki Y. (2004b). Robust watermarking based on time-spread echo method with subband decomposition. *IEICE Transactions on Fundamentals, E87-A*(6), 1647-1650.

Ko, B., & Kim, H. (1999). Modified circulant feedback delay networks (MCFDNs) for artificial reverberator using a general recursive filter and CFDNs. *J. Acoustical Society of Korea, 18*(4E), 31-36.

Ko, B., Nishimura, R., & Suzuki, Y. (2002a). *Proposal of an echo-spread watermarking method using PN sequence.* Paper presented at the 2002 Spring Meeting of The Acoustical Society of Japan (pp. 535-536).

Ko, B., Nishimura, R., & Suzuki, Y. (2002b). Time-spread echo method for digital audio watermarking using PN sequences. *Proceedings of ICASSP 2002, 2,* 2001-2004.

Ko, B., Nishimura, R., & Suzuki, Y. (2004a). Log-Scaling watermark detection in digital audio watermarking. *Proceedings of ICASSP 2004, 3,* 81-84.

Kohda, T., Fujisaki, H., & Ideue, S. (2000). On distributions of correlation values of spreading sequences based on Markov information sources. In *Proceeding of the IEEE International Symposium on Circuits and Systems* (Vol. 5, pp. 225-228).

Kolmogorov, A. (1965). Three approaches to the concept of the amount of information. *Probl. of Inform. Transm, 1*(1).

Koukopoulos, D., & Stamatiou, Y. C. (2005). A watermarking scheme for MP3 audio files. *International Journal of Signal Processing, 2*(3), 206-213.

Koukopoulos, D., & Stamatiou, Y.C. (2001). *A compressed domain watermarking algorithm for MPEG Layer 3.* Paper presented at the Multimedia and Security Workshop at ACM Multimedia (pp. 7-10). ACM Press.

Krishnan, S. (2001). *Instantaneous mean frequency estimation using adaptive time-frequency distributions.* Paper presented at the IEEE Canadian Conference on Electrical and Computer Engineering (pp. 141-146).

Kubin, G. (1995). *What is a chaotic signal?* Paper presented at the IEEE Workshop on Nonlinear Signal and Image Processing (pp. 141-144).

Kuo, C.J., & Rigas, H.B. (1991). Quasi m-arrays and gold code arrays. *IEEE Transactions on Information Theory, 37*(2), 385-388.

Kuo, C.J., Deller, J.R, & Jain, A.K. (1996). Pre/post-filter for performance improvement of transform coding. *Signal Processing: Image Communication Journal, 8*(3), 229-239.

Kuo, S. S., Johnston, J. D., Turin, W., & Quackenbush, S. R. (2002). Covert audio watermarking using perceptually tuned signal independent multiband phase modulations. In *Proceeding of the IEEE International Conference on Acoustics, Speech, and Signal Processing* (Vol. 2, pp. 1753-1756).

Kutter, M. (1999). Watermarking resisting to translation, rotation, and scaling. In *Proceedings of the SPIE Mutimedia Systems and Applications* (Vol. 3528, pp. 423-431). International Society for Optical Engineering.

Kutter, M., Voloshynovskiy, S., & Herrigel, A. (2000, January). *The watermark copy attack.* Paper presented at SPIE Security and Watermarking of Multimedia Contents II, San Jose, California.

Kuttruf, H. (2000, October). *Room acoustics* (1st ed.). Spon Press.

Laftsidis, C., Tefas, A., Nikolaidis, N., & Pitas, I. (2003). Robust multibit audio watermarking in the temporal domain. In *Proceedings of the IEEE International Symposium on Circuits and Systems* (pp. 944-947).

Lathi, B.P. (1998). *Modern digital and analog communication system* (3rd ed., pp. 728-737). Oxford University Press.

Le, L., Krishnan, S., & Ghoraani, B. (2006.). *Discrete polynomial transform for digital image watermarking application.* Paper presented at the IEEE International Conference on Multimedia and Expo (pp. 1569-1572).

Lee, K., Kim, D.S., Kim, T., & Moon, K.A. (2003). EM estimation of scale factor for quantization-based audio watermarking. In *Proceedings of the 2nd International Workshop on Digital Watermarking* (pp. 316-327). Seoul, Korea.

Lee, S., & Ho, Y. (2000). Digital audio watermarking in the cepstrum domain. *IEEE Transactions on Consumer Electronics, 46*(3), 744-750.

Lemma, A.N., Aprea, J., Oomen, W., & Kerkhof, L. van de. (2003, April). A temporal domain audio watermarking technique. *IEEETASSP: IEEE Transactions on Signal Processing, 51*(4), 1088–1097.

Leventhal, L. (1986, June). Type 1 and type 2 errors in the statistical analysis of listening tests. *AES Journal, 34*(6), 437-453.

Levin, L. (1973). Universal'nyĭe perebornyĭe zadachi [Universal search problems]. *Problemy Peredachi Informatsii, 9*(3), 265-266.

Levin, L. A. (1986). Average case complete problems. *SIAM Journal on Computing, 15*, 285-286.

Levine, S.N. (1998). *Audio representations for data compression and compressed domain processing.* Unpublished doctoral dissertation, Stanford University.

Li, W., & Xue, X. (2003). An audio watermarking technique that is robust against random cropping. *Computer Music Journal, 27*(4), 58-68.

Li, W., Xue, X.Y., & Li, X.Q. (2003). Localized robust audio watermarking in regions of interest. *ICICS-PCM 2003.*

Li, X., & Yu, H.H. (2000). Transparent and robust audio data hiding in cepstrum domain. In *Proceedings of the IEEE International Conference on Multimedia* (pp. 397-400). New York.

Li, X., Zhang, M., & Zhang, R. (2004). *A new adaptive audio watermarking algorithm*. Paper presented at the Fifth World Congress, Intelligent Control and Automation (Vol. 5, pp. 4357-4361).

Liang, J., Xu, P., & Tran, T.D. (2000). A Robust DCT-based low frequency watermarking scheme. In *Proceedings of the 34th Annual Conference in Information Systems and Science* (Vol. 1, pp. 1-6).

Lichtenauer, J., Setyawan, I., Kalker, T., & Lagendijk, R. (2003). Exhaustive geometrical search and false positive watermark detection probability. In *Proceedings of the SPIE Electronic Imaging on Security and Watermarking of Multimedia Contents* (Vol. 5020, pp. 203-214).

Licks, V., & Jordan, R. (2005). Geometric attacks on image watermarking systems. *IEEE Transactions on Multimedia, 12*(3), 68-78.

Lie, W.N., & Chang, L.C. (2001). *Robust high quality time-domain audio watermarking subject to psychoacoustic masking.* Paper presented at the IEEE International Symposium on Circuits and Systems (Vol. 2, pp. 45-48).

Lin, C. Y., Bloom, J. A., Cox, I. J. Miller, M. L., & Liu, Y. M. (2000). Rotation, scale and translation-resilient public watermarking for images. *Proceedings of SPIE, 3971*, 90-98.

Lin, E.T., Eskicioglu, A.M., Lagendijk, R.L., & Delp, E.J. (2005). Advances in digital video content protection. *Proceedings of the IEEE, 93*(1), 171-183.

Lincoln, B. (1998). An experimental high fidelity perceptual audio coder. *Project in MUS420 Win97.* Retrieved March 18, 2007, from http://www-ccrma.stanford.edu/jos/bosse/

Linnartz, J., & van Dijk, M. (1998, April). Analysis of the sensitivity attack against electronic watermarks in images. In D. Aucsmith (Ed.), *Proceedings of the Second International Workshop on Information Hiding* (Vol. 1525). Portland, OR: Springer-Verlag.

Linnartz, J.P.M.G., Kalker, A.C.C., & Depovere, G.F. (1998). Modeling the false-alarm and missed detection rate for electronic watermarks. In L. D. Aucsmith (Ed.), *Notes in computer science* (Vol. 1525, pp. 329-343). Springer-Verlag.

Liu, Q. (2004). *Digital audio watermarking utilizing discrete wavelet packet transform.* Unpublished master's thesis, Chaoyang University of Technology, Institute of Networking and Communication, Taiwan.

Liu, Y.W. (2005). *Audio watermarking through parametric signal representations.* Unpublished doctoral dissertation, Stanford University.

Liu, Y.W., & Smith, J.O. (2003). Watermarking parametric representations for synthetic audio. In *Proceedings of the IEEE International Conference on Acoustics, Speech, and Signal Processing* (pp. V660-663). Hong Kong, China.

Liu, Y.W., & Smith, J.O. (2004a). Watermarking sinusoidal audio representations by quantization index modulation in multiple frequencies. In *Proceedings of the IEEE International Conference on Acoustics, Speech, and Signal Processing* (pp. 373-376). Montréal, Canada.

Liu, Y.W., & Smith, J.O. (2004b). Multiple watermarking: Is power-sharing better than time-sharing? In *Proceedings of the IEEE International Conference Multimedia, Expo* (pp. 1939-1942). Taipei, Taiwan.

Liu, Y.W., & Smith, J.O. (2004c). Audio watermarking based on sinusoidal analysis and synthesis. In *Proceedings of the International Symposium on Musical Acoustics* (CD-ROM), Nara, Japan.

Lu, C.S., & Liao, H.Y.M. (2001). Multipurpose watermarking for image authentication and protection. *IEEE Trans. Image Processing, 10*, 1579-1592.

Lu, C.S., Liao, H.Y.M., & Chen, L.H. (2000). Multipurpose audio watermarking. In *Proceedings of the IEEE 15th International Conference on Pattern Recognition* (pp. 282-285). Barcelona, Spain.

Maillard, T., & Furon, T. (2004, July). Towards digital rights and exemptions management systems. *Computer Law and Security Report, 20*(4), 281-287.

Malvar, H.S., & Florencio, D.F. (2003). Improved spread spectrum: A new modulation technique for robust watermarking. *IEEE Transactions on Signal Processing, 51*(4), 898-905.

Mansour, M.F., & Tewfik, A.H. (2003). Time-scale invariant audio data embedding. *EURASIP Journal on Applied Signal Processing, 10*, 993-1000.

Markel, J.D., & Gray, A.H. (1976). *Linear prediction of speech.* New York: Springer-Verlag.

McAulay, R.J., & Quatieri, T.F. (1986). Speech analysis/synthesis based on a sinusoidal representation. *IEEE Trans. Acoustics, Speech, Signal Processing, 34*, 744-754.

Meel, I. J. (1999). Spread spectrum (SS) introduction. *Sirius Communications, 2*, 1-33.

Menezes, A., Van Oorschot, P., & Vanstone, S. (1996). *Handbook of applied cryptography.* CRC Press.

Miaou, S.G., Hsu, C.H., Tsai, Y.S., & Chao, H. M. (2000, July). *A secure data hiding technique with heterogeneous data-combining capability for electronic patient records.* Paper presented at the World Congress on Medical Physics and Biomedical Engineering: Electronic Healthcare Records, Chicago.

Miller, M.L., & Bloom, M.A. (1999, September). *Computing the probability of false watermark detection.* Paper presented at the Third Workshop on Information Hiding, Dresden, Germany.

Miller, M.L., Doërr, G.J., & Cox, I.J. (2002). Dirty-paper trellis codes for watermarking. In *Proceedings of the 2002 IEEE International Conference on Image Processing* (pp. 129-132).

Mintzer, F., & Braudaway, G.W. (1999). If one watermark is good, are more better? In *Proceedings of the IEEE International Conference on Acoustics, Speech, and Signal Processing* (pp. 2067-2069), Phoenix, Arizona.

Mitchell, D., Selman, B., & Levesque, H. (1992*). Hard and easy distributions of SAT problems.* Paper presented at the Tenth National Conference on Artificial Intelligence (pp. 459-465).

Mittelholzer, T. (1999, September). An information-theoritic approach to steganography and watermarking. In A. Pfitzmann (Ed.), *Proceedings of the Third International Workshop on Information Hiding* (pp. 1-17). Dresden, Germany: Springer-Verlag.

Molines, E., & Charpentier, F. (1990). Pitch-synchronous waveform processing techniques for text-to-speech synthesis using diphones. *Speech Communication*, 453-467.

Moore, J.C.B. (1997). *An Introduction to the psychology of hearing* (4th ed.). Academic Press.

Motoki, M., & Uehara, R. (1999). Unique solution instance generation for the 3-satisfiability (3SAT) problem (Tech. Rep. No. C-129). Japan, Sciences Tokyo Institute of Technology, Dept. of Math. and Comp.

Motwani, R., & Raghavan, P. (1995). *Randomized algorithms.* Cambridge University Press.

Moulin, P., & Koetter, R. (2005). Data-hiding codes. *Proceedings IEEE, 93*, 2083-2126.

Neubauer, C., & Herre, J. (1998, September). Digital watermarking and its influence on audio quality. In *Proceedings of the 105th Audio Engineering Society Convention.* San Francisco: AES.

Neubauer, C., & Herre, J. (2000). *Advanced audio watermarking and its application.* Paper presented at the 109th AES Convention (AES preprint 5176).

Nishimura, R., Suzuki, M., & Suzuki, Y. (2001). Detection threshold of a periodic phase shift in music sound. In *Proceedings of the 17th International Congress on Acoustics.*

Noll, P. (1993). Wideband speech and audio coding. *IEEE Communications Magazine, 31*(11), 34-44.

Oh, H.O., Kim, H.W., Seok, J.W., Hong, J.W., & Youn, D.H. (2001). Transparent and robust audio watermarking with a new echo embedding technique. In *Proceeding of the IEEE International Conference on Multimedia and Expo* (pp. 433-436).

Oh, H.O., Seok, J.W., Hong, J.W., & Youn, D.H. (2001). New echo embedding technique for robust and imperceptible audio watermarking. In *Proceeding of IEEE International*

Conference Acoustic, Speech, and Signal Processing (Vol. 3, pp. 1341-1344).

Oppenheim, A.V., & Schafer, R.W. (1989) *Discrete-time signal processing.* Englewood Cliffs, NJ: Prentice Hall.

Oppenheim, V.A., & Schaffer, W.R. (1989). *Discrete-time signal processing.* NJ: Prentice Hall.

P′erez-Freire, L., Comesana, P., & P′erez-Gonz′alez, F. (2005, June). Information-theoretic analysis of security in side-informed data hiding. In *Proceedings of the 7th Information Hiding Workshop (IH05).* Barcelona, Spain: Springer-Verlag.

P′erez-Freire, L., P′erez-Gonz′alez, F., Furon, T., & Na, P.C. (in press). Security of lattice-based data hiding against the known message attack. *IEEE Transactions on Information Forensics and Security.*

Painter, T., & Spanias, A. (1997). A review of algorithms for perceptual coding of digital audio signals. In *Proceeding of the International Conference on Digital Signal Processing* (pp. 179-205).

Painter, T., & Spanias, A. (2000). Perceptual coding of digital audio. *Proceedings of the IEEE, 88*(4), 451-513.

Pan, D. (1995). A tutorial on mpeg/audio compression. *IEEE Multimedia, 2*(2), 60-74.

Papadimitriou, C.H. (1994). *Computational complexity.* Addison-Wesley.

Papoulis, A. (1977). *Signal analysis.* New York: McGraw-Hill.

Pereira, M.S., Voloshynovskiy, S., & Pun, T. (2001). *Second generation benchmarking and application oriented evaluation.* Paper presented at the Information Hiding Workshop.

Petitcolas, A.P., et al. (2001). *StirMark benchmark: Audio watermarking attacks.* Paper presented at the International Conference on Information Technology: Coding and Computing (pp. 49-55).

Petitcolas, F. (1998). *MP3Stego.* Computer Laboratory, Cambridge.

Petitcolas, F.A.P. (2000). Watermarking schemes evaluation. *IEEE Signal Processing Magazine, 17.*

Petitcolas, F.A.P., & Anderson, R.J. (1999, June). *Evaluation of copyright marking systems.* Paper presented at the IEEE Multimedia Systems, Florence, Italy.

Petitcolas, F.A.P., Anderson, R.J., & Kuhn, M. G. (1998, April). *Attacks on copyright marking systems.* Paper presented at the Second Workshop on Information Hiding, Portland, Oregon.

Petrovic, R. (2001). Audio signal watermarking based on replica modulation. In *Proceedings of the IEEE TELSIKS* (pp. 227-234). Niš, Yugoslavia.

Pitas, I., & Kaskalis, T. H. (1995). Applying signatures on digital images. In *Proceedings of the IEEE Workshops on Nonlinear Image and Signal Processing* (pp. 460-463).

Podilchuk, I.C., & Delp, J.D. (2001). Digital watermarking: Algorithms and applications. *IEEE Signal Processing Magazine, 18*(4), 33-46.

Pohlmann, K.C. (1991). *Advanced digital audio.* Carmel.

Polikar, R. (2006). *The wavelet tutorial.* Retrieved March 18, 2007, from http://users.rowan.edu/ polikar/Wavelets/wtpart1.html

Poor, H.V. (1994). An *introduction to signal detection and estimation* (2nd ed.). Springer-Verlag.

Proakis, G.J. (2001). *Digital communications* (4th ed.). McGraw-Hill.

Purnhagen, H., & Meine, N. (2000) HILN: The MPEG-4 parametric audio coding tools. In *Proceedings of the IEEE International Symposium on Circuits and Systems* (pp. 201-204), Geneva, Switzerland.

Qian, S., & Chen, D. (1996). *Joint time-frequency analysis: Method and application.* New York: Prentice Hall.

Qiao, L., & Nahrstedt, K. (1998). *Non-invertible watermarking methods for MPEG video and audio.* Paper presented at the Multimedia and Security Workshop at ACM Multimedia (pp. 93-98).

Quackenbush, S. R., Barnwell, T. P., & Clements, M. A. (1988). *Objective measures of speech quality.* NJ: Prentice Hall.

Rangayyan, R., & Krishnan, S. (2001). Feature identification in the time-frequency plane by using the Hough-Radon transform. *IEEE Transactions on Pattern Recognition, 34*, 1147-1158.

Reyes, N. R., Zurera, M. R., Ferreras, F. L., & Amores, P. J. (2003). Adaptive wavelet-packet analysis for audio coding purposes. *Signal Processing, 83*, 919-929.

RIAA. (2006). Market data pages of the web site. Retrieved March 26, 2007, from http://www.riaa.org

Ruandaidh, J. J. K. O., & Pun, T. (1998). Rotation, scale and translation invariant spread spectrum digital image watermarking. *Signal Processing, 66*(3), 303-317.

Ruiz, F.J., & Deller, J.R., Jr. (2000, June). *Digital watermarking of speech signals for the national gallery of the spoken word.* Paper presented at the International Conference on Acoustics, Speech and Signal Processing, Istanbul, Turkey.

Sandford, S. et.al. (1997). *Compression Embedding.* [US Patent 5,778,102].

Sasaki, N., Nishimura, R., & Suzuki, Y. (in press). Audio watermarking based on association analysis. *Proceedings of ICSP 2006.*

Scavone, G.P., & Cook, P. (2005) RtMIDI, RtAudio, and synthesis tookKit (STK) update. In *Proceedings of the International Computer Music Conference,* Barcelona, Spain. Retrieved March 19, 2007, from http://ccrma.stanford.edu/software/stk/papers/stkupdate.pdf

Schimming, T., Gotz, M., & Schwarz, W. (1998). Signal modeling using piecewise linear chaotic generators. In *Proceedings of the European Signal Processing Conference* (pp. 1377-1380).

Schneier, B. (1996). *Applied cryptography.* John Wiley & Sons.

Schroeder, M.R., & Atal, B.S. (1985). Code-excited linear prediction (CELP): High-quality speech at very low bit rates. In *Proceedings of the IEEE International Conference on Acoustics, Speech, and Signal Processing* (pp. 937-940). Tampa, Florida.

Seadle, M.S., Deller, J.R., Jr., & Gurijala, A. (2002, July). *Why watermark? The copyright need for an engineering solution.* Paper presented at the ACM/IEEE Joint Conference on Digital Libraries, Portland, Oregon.

Selman, B. (1995). *Stochastic search and phase transitions.* Paper presented at the International Joint Conference on Artificial Intelligence (Vol. 2, 998-1002).

Sener, S., & Gunsel, B. (2004). Blind audio watermark decoding using independent component analysis. In *Proceedings of the IEEE International Conference on Pattern Recognition* (Vol. 2, pp. 875-878).

Seo, J. S., & Haitsma, J., & Kalker, T. (2002). Linear speed-change resilient audio fingerprinting. In *Proceedings of the First IEEE Benelus Workshop on Model based Processing and Coding of Audio.*

Seok, J., Hong, J., & Kim, J. (2002). A novel audio watermarking algorithm for copyright protection of digital audio. *ETRI Journal, 24*(3), 181-189.

Seok, J.W., & Hong, J.W. (2001). Audio watermarking for copyright protection of digital audio data. *Electronics Letters, 37*(1), 60-61.

Serra, X., & Smith, J. (1990). *Spectral modeling synthesis: A sound analysis/synthesis based on a deterministic plus stochastic decomposition. Computer Music Journal, 14*(4), 12-24.

Shannon C. (1949). Communication in the presence of noise. *Proceedings Institute of Radio Engineers, 37*(1), 10-21.

Shin, S., Kim, O., Kim, J., & Choil, J. (2002). A robust audio watermarking algorithm using pitch scaling. In *Proceedings of the IEEE 10th DSP Workshop* (pp. 701-704). Pine Mountain, Georgia.

Shower, E.G., & Biddulph, R. (1931). Differential pitch sensitivity of the ear. *Journal of the Acoustical Society of America, 3*, 275-287.

Silvestre, G.C.M., Hurley, N.J., Hanau, G.S., & Dowling, W.J. (2001). Informed audio watermarking scheme using digital chaotic signals. In *Proceedings of the IEEE International Conference on Acoustics, Speech, and Signal Processing* (pp. 1361-1364). Salt Lake City, Nevada.

Sinha, D., & Tewfik, A. (1993). Low bit rate transparent audio compression using adapted wavelets. *IEEE Transactions on Signal Processing, 41*(12), 3463-3479.

Sklar, B. (1998). *Digital communications fundamentals and applications.* NJ: Prentice-Hall.

Smith, J., & Serra, X. (1987). PARSHL: An analysis/synthesis program for non-harmonic sounds based on a sinusoidal representation. In *Proceedings of the 1987 International Computer Music Conference,* Urbana-Champaign, Illinois.

Solomonoff, R.J (1964). A formal theory of inductive inference. *Information and Control, 7*(1), 1-22.

Soulodre, G.A., & Lavoie, M.C. (2004, October). Subjective evaluation of MPEG layer II with spectral band replication. In *Proceedings of the 117th Audio Engineering Society Convention.* San Francisco: AES.

Spanias, A. (1994). Speech coding: A tutorial review. *Proceedings of the IEEE, 82*(10), 1541-1582.

Stamatiou, Y.C. (2003). Threshold phenomena: The computer scientist's viewpoint. *European Association of Theoretical Computer Science Bulletin (EATCS), 80,* 199-234.

Steinebach, M. (2004). *Digitale Wasserzeichen für Audiodaten.* Shaker Verlag Aachen.

Steinebach, M., & Zmudzinsiki, S. (2006). Robustheit digitaler Audiowasserzeichen gegen Pitch-Shifting und Time-Stretching, *Sicherheit 2006. Sicherheit - Schutz und Zuverlässigkeit. Beiträge der 3. Jahrestagung des Fachbereichs Sicherheit der Gesellschaft für Informatik e.V. (GI),* ISBN 3885791714, Kölln Verlag.

Steinebach, M., Lang, A., & Dittmann, J. (2002, January). *StirMark benchmark: Audio watermarking attacks based on lossy compression.* Paper presented at the Security and Watermarking of Multimedia Contents IV, Electronic Imaging 2002, Photonics West, San Jose, California.

Steinebach, M., Lang, A., Dittmann, J., & Neubauer, C. (2002, April 8-10). Audio watermarking quality evaluation: Robustness to DA/AD processes. Paper presented at the *International Conference on Information Technology: Coding and Computing, ITCC,* Las Vegas, Nevada (pp. 100-103). Piscataway, NJ: IEEE Computer Society.

Steinebach, M., Petitcolas, F., Raynal, F., Dittmann, J., Fontaine, C., Seibel, S., et al. (2001). Stirmark benchmark: Audio watermarking attacks. In *Proceedings of the International Conference on Information Technology: Coding and Computing* (pp. 49-54). Las Vegas, Nevada.

Steinebach, M., Zmudzinski, S., & Lang, A. (2003). Robustheitsevaluierung von digitalen Audiowasserzeichen im Rundfunkszenario, *CD-Rom zum 2. Thüringer Medienseminar der FKTG zum Thema Rechte Digitaler Medien Intellectual Properties & Content Management,* Erfurt

Steinebach, M., Zmudzinski, S., & Neichtadt, S. (2006). *Robust-audio-hash synchronized audio watermarking.* Paper presented at the 4th International Workshop on Security in Information Systems – (WOSIS 2006) (pp. 58-66). Paphos, Cyprus.

Stern, J., & Craver, S. (2001, October). Lessons learned from the SDMI. In J. L. Dugelay & K. Rose (Eds.), *Proceedings of the Fourth Workshop on Multimedia Signal Processing (MMSP)* (pp. 213-218), Cannes, France.

Stoica, P., & Ng, B.C. (1998). On the Cram`er-Rao bound under parametric constraints. *IEEE Signal Processing Letters, 5*(7), 177–179.

Suzuki, Y., Asano, F., Kim, H., & Sone, T. (1995). An optimum computer-generated pulse signal suitable for the measurement of very long impulse response. *J. Acoustical Society of America, 97*(2), 1119-1123.

Swanson, M. D., Zhu, B., Tewfik, A. H., & Boney, L. (1998). Robust audio watermarking using perceptual masking. *Elsevier Signal Processing, Special Issue on Copyright Protection and Access Control, 66*(3), 337-355.

Swanson, M., Zhu, B., & Tewfik, A. (1999). Current state-of-the-art, challenges and future directions for audio watermarking. In *Proceedings of the IEEE International*

Conference on Multimedia Computing and Systems (pp. 19-24). Florence, Italy.

Takahashi, A., Nishimura, R., & Suzuki, Y. (2005). Multiple watermarks for stereo audio signals using phase-modulation techniques. *IEEE Transactions on Signal Processing, 53*(2), 806-815.

Tefas, A., Giannoula, A., Nikolaidis, N., & Pitas, I. (2005). Enhanced transform-domain correlation-based audio watermarking. In *Proceedings of the IEEE International Conference on Acoustics, Speech, and Signal Processing* (Vol. 2, pp. 1049-1052).

Tefas, A., Nikolaidis, A., Nikolaidis, N., Solachidis, V., Tsekeridou, S., & Pitas, I. (2003). Performance analysis of correlation-based watermarking schemes employing Markov chaotic sequences. *IEEE Transactions on Signal Processing, 51*(7), 1979-1994.

Terhardt, E. (1979). Calculating virtual pitch. *Hearing Research, 1*, 155-182.

Termont, P., De Strycker, L., Vandewege, J., Haitsma, J., Kalker, T., Maes, M., et al. (1999). Performance measurements of a real-time digital watermarking system for broadcast monitoring. In *Proceedings of the IEEE International Conference on Multimedia Computing and Systems* (pp. 220-224). Florence, Italy.

Termont, P., De Stycker, L., Vandewege, J., Op de Beeck, M., Haitsma J., Kalker, T., et al. (2000). How to achieve robustness against scaling in a real-time digital watermarking system for broadcast monitoring. In *Proceedings of the IEEE International Conference on Image Processing* (pp. 407-410). Vancouver, British Columbia.

Thiede, T., Treurniet, W.C., Bitto, R., Schmidmer, C., Sporer, T., Beerends, J.G., et al. (2000). Evaluation of the ITU-R objective audio quality measurement method. *AES Journal, 48*(1/2), 3-29.

Thornburg, H. (2005) *On the detection and modeling of transient audio signals with prior information.* Unpublished doctoral dissertation, Stanford University.

Trappe, W., Wu, M., Wang, Z., & Liu, K. (2003). Anti-collusion fingerprinting for multimedia. *IEEE Transactions on Signal Processing, 51*(4), 1069-1087.

Trappe,W., Wu, M., Wang, Z., & Liu, K. (2003, April). Anti-collusion fingerprinting for multimedia. *IEEE Transactions on Signal Processing, 51*(4), 1069-1087.

Treurniet, W.C., & Soulodre, G.A. (2000). Evaluation of the ITU-R objective audio quality measurement method. *AES Journal, 48*(3), 164-173.

Tsekeridou, S., Nikolaidis, N., Sidiropoulos, N., & Pitas, I. (2000). Copyright protection of still images using self-similar chaotic watermarks. In *Proceeding of the IEEE International Conference on Image Processing* (Vol. 1, pp. 411-414).

Tsekeridou, S., Solachidis, V., Nikolaidis, N., Nikolaidis, A., & Pitas, I. (2001). Statistical analysis of a watermarking system based on Bernoulli chaotic sequences. *Signal Processing: Special Issue on Information Theoretic Issues in Digital Watermarking, 81*(6), 1273-1293.

Tzanetakis, G., Essl, G., & Cook, P. (2001). Audio analysis using the discrete wavelet transform. In *Proceedings of the 2001 International Conference of Acoustics and Music: Theory and Applications,* Skiathos, Greece. Retrieved March 23, 2007, from http://www.cs.princeton.edu/~gessl/papers/amta2001.pdf

Valenti, J. (2002, February). *Piracy threatens to destroy movie industry and U.S. economy.* Testimony before the US Senate foreign relations committee.

Van der Veen, M., Bruekers, F., Haitsma, J., Kalker, T., Lemma, A. N., & Oomen, W. (2001). *Robust, multi-functional and high-quality audio watermarking technology.* Paper presented at the AES 110th Convention, Amsterdam, Netherlands.

Van der Veen, M., Bruekers, F., van Leest, A., & Cavin, S. (2003). High capacity reversible watermarking for audio. In *Proceedings of the SPIE Security and Watermarking of Multimedia Contents* (pp. 1-11). San Jose, California.

Vaseghi, S. V. (2000). *Advanced digital signal processing and noise reduction.* John Wiley & Sons, Ltd.

Veldhuis, R. N. J., Breeuwer, M., & van der Wall, R. G. (1998). Subband coding of digital audio signals. *Philips Res. Rep., 44*(2-3), 329-343.

Venkatachalam, V., Cazzanti, L., Dhillon, N., & Wells, M. (2004). Automatic identification of sound recordings. *IEEE Signal Processing Magazine, 2*(2), 92-99.

Vercoe, B.L., Gardner, W.G., & Scheirer, E.D. (1998). Structured audio: Creation, transmission, and rendering of parametric sound representations. *Proceedings IEEE, 86*, 922-940.

Vladimir, B., & Rao, K.R. (2001). An efficient implementation of the forward and inverse MDCT in MPEG audio coding. *IEEE Signal Processing Letters, 8*(2).

Voloshynovskiy, S., Pereira, S., Pun, T., Su, J.K., & Eggers, J. J. (2001). Attacks and Benchmarking. *IEEE Communication Magazine, 39*(8).

Wang, A. (1995). Instantaneous and frequency-warped techniues for source separation and signal parameterization. In *Proceedings of the IEEE Workshop on Applications of Signal Processing to Audio and Acoustics,* New Paltz, New York.

Wang, S., Sekey, A., & Gersho, A. (1992). An objective measure for predicting subjective quality of speech coders. *IEEE Journal on Selected Areas in Communications, 10*(5), 819-829.

Wang, Y. (2004). Estimation-based patchwork image watermarking technique. *Journal of Digital Information Management, 1*, 154-161.

Wessel, D., & Wright, M., (Eds.). (2004). In *Proceedings of the Open Sound Control Conference*. Berkeley, CA: Center For New Music and Audio Technology (CNMAT). Retrieved March 23, 2007, from http://www.opensoundcontrol.org/proceedings

Wickens, D.T. (2002). *Elementary signal detection theory.* Oxford University Press.

Wier, C.C., Jesteadt, W., & Green, D.M. (1977). Frequency discrimination as a function of frequency and sensation level. *Journal of the Acoustical Society of America, 61*, 178-184.

Wong, P. W., & Memon, N. (2001, October). Secret and public key image watermarking schemes for images authentication and ownership verification. *IEEE Transactions on Image Processing, 10*(10), 1593-1601.

Wu, C. P., Su, P. C., & Kuo, C. C. J. (1999). Robust audio watermarking for copyright protection. *Proceeding of SPIE, 3807*, 387-397.

Wu, C. P., Su, P. C., & Kuo, C. C. J. (2000). Robust and efficient digital audio watermarking using audio content analysis. *Proceedings of SPIE, 3971*, 382-392.

Wu, M., & Liu, B. (2004). Data hiding in binary image for authentication and annotation. *IEEE Transactions on Multimedia, 6*(4), 528-538.

Wu, M., Craver, S., Felten, E.W., & Liu, B. (2001). Analysis of attacks on SDMI audio watermarks. In *Proceedings of the IEEE International Conference on Acoustics, Speech, and Signal Processing* (pp. 1369-1372). Salt Lake City, Nevada.

Wu, M., Trappe, W., Wang, Z., & Liu, K. (2004). Collusion-resistant fingerprinting for multimedia. *IEEE Signal Processing Magazine, 21*(2), 15-27.

Wu, S., Huang, J., Huang, D., & Shi, Y. Q. (2005). Efficiently self-synchronized audio watermarking for assured audio data transmission. *IEEE Transactions on Broadcasting, 51*(1), 69-76.

Wu, W.C., & Chen, O.T.C (2006). Analysis-by-synthesis echo hiding scheme using mirrored kernels. In *Proceeding of IEEE International Conference on Acoustic, Speech & Signal Processing,* II 325-II 328.

Wu, W.C., Chen, O.T.C, & Wang, Y.H. (2003). An echo watermarking method using an analysis-by-synthesis approach. In *Proceeding of the 5th IASTED International Conference on Signal and Image Processing* (pp. 365-369).

Xiang, S. J., Huang, J. W., & Yang, R. (2006). Time-scale invariant audio watermarking based on the statistical features in time domain. In *Proceedings of the 8th Information Hiding Workshop.*

Xu, C., Wu, J., Sun, Q., & Xin, K. (1999). Applications of digital watermarking technology in audio signals. *J. Audio Engineering Society, 47*(10), 805-812.

Yeo, I. K., & Kim, H. J. (2003a). Modified patchwork algorithm: The novel audio watermarking scheme. *IEEE Transactions on Speech and Audio Processing, 11*, 381-386.

Yeo, I. K., & Kim, H. J. (2003b). Generalized patchwork algorithm for image watermarking. *ACM Multimedia Systems, 9*, 261-265.

Yuan, S., & Huss, S. (2004, September). *Audio watermarking algorithm for real-time speech integrity and authentication.* Paper presented at the ACM Multimedia and Security Workshop, Magdeburg, Germany.

Zölzer, U. (1997, August). *Digital audio signal processing*, John Wiley & Sons.

Zurera, M. R., Ferreras, F. L., Amores, M. P. J., Bascon, S. M. & Reyes, N. R. (2001). A new algorithm for translating psychoacoustic information to the wavelet domain. *Signal Processing, 81*, 519-531.

Zwicker, E. (1956). Die elementaren Grundlagen zur Bestimmung der Informationskapazität des Gehörs [The elementary bases for the determination of the information capacity of the hearing]. *Acustica, 6*, 365-381.

Zwicker, E., & Fastl, H. (1990) *Psychoacoustics facts and models.* Berlin: Springer-Verlag.

Zwicker, E., & Zwicker, U.T. (1991) Audio engineering and psychoacoustics: Matching signals to the final receiver, the human auditory system. *J. Audio Eng. Soc., 39*, 115-126.

About the Contributors

Nedeljko Cvejic received the Dipl.-Ing. in electrical engineering from the University of Belgrade, Serbia (2000), and the DrTech degree from the University of Oulu, Finland (2004). From 2001 to 2004, he was a research scientist at the Department of Electrical and Information Engineering, University of Oulu, Finland. He is currently a research associate with the Department of Electrical and Electronic Engineering of the University of Bristol, United Kingdom. His research interests include digital watermarking, image and video fusion, and sensor networks.

Tapio Seppänen received an MSc degree in electrical engineering in 1985, and a PhD in computer engineering in 1990, from the University of Oulu, Finland. Currently he is serving as a full professor at the same university. He teaches and conducts research on multimedia signal processing and biomedical signal processing. Special interests include digital watermarking, pattern recognition applications, and content-based multimedia retrieval. He has contributed to some 250 scientific journal and conference papers.

* * * * *

Michael Arnold studied physics at the University of Würzburg where he received his diploma degree in 1994. He received a PhD in informatics from the University of Darmstadt in 2004. From 1996 until 2005, he worked in the Fraunhofer Institute for Computer Graphics as a research scientist in the Department for Security Technology. In 2005, he joined Thomson Corporate Research Hannover to work as senior research and development engineer in the Audio Lab. He is working in the fields of digital content protection and is co-author of the book entitled *Techniques and Applications of Digital Watermarking and Content Protection.*

Peter Baum studied mechanical engineering at the University of Hannover where he received his diploma degree in 1994. From 1995 to 2000, he worked in a consulting company for mechanical engineering, and at the University of Hannover in the field of structural analysis and signal processing. He received a DrIng (PhD) in engineering from the University of Hannover in 2000. He joined Thomson in 2001. Dr. Baum is currently the head of the audio watermarking group in Thomson Corporate Research.

Francois Cayre received the computer science and MS from the Université de Technologie de Compiègne, Compiègne, France (2000). In 2003, he received a PhD from both the Université catholique de Louvain, Louvain-la-Neuve, Belgium and the École Nationale Supérieure des Télécommunications, Paris. He was an INRIA post-doc fellow at the IRISA public research center, Rennes, France. He is now an associate professor with the Laboratoire des Images et des Signaux, Institut National Polytechnique de Grenoble, Grenoble, France. His main research interests include watermarking, 3D watermarking and steganography, and multimedia security.

Oscal Chen received a BS in electrical engineering from National Taiwan University (1987), and an MS and a PhD in electrical engineering from University of Southern California, Los Angeles, USA (1990 and 1994, respectively). Dr. Chen worked Computer Processor Architecture Department of Computer Communication & Research Labs, Industrial Technology Research Institute, Taiwan, for serving a system design engineer, project leader, and section chief from 1994 to 1995. Dr. Chen was an associate professor and a professor in Department of Electrical Engineering, National Chung Cheng University, Taiwan from September 1995 to August 2003, and after August 2003, respectively.

Serhat Erkucuk received a BSc and MASc in electrical engineering from Middle East Technical University, Ankara, Turkey, and from Ryerson University, Toronto, ON, Canada (2001 and 2003, respectively). Since September 2003, he has been pursuing the PhD in the School of Engineering Science, Simon Fraser University, Burnaby, BC, Canada. His main research interests include multimedia signal processing, time-frequency analysis, spread spectrum communications, and ultra wideband communications. Erküçük was awarded the Governor General's Gold Medal as recognition of his outstanding scholastic achievements during his MASc degree program.

Caroline Fontaine received a PhD in computer science from the University of Paris 6, Paris, in 1998, for a work dealing with cryptography, error-correcting codes, and watermarking. She has been with the Computer Science Lab at the University of Lille 1, Lille, France, as an associate professor from 1999 to 2002. Since 2005, she has been a CNRS researcher at the IRISA public research center in Rennes, France, working in the TEMICS project. Her research interests include cryptography and cryptanalysis (mainly of symmetric encryption schemes), digital watermarking, and security of mobile ad hoc networks.

Say Wei Foo received a BEng in electrical engineering from the University of Newcastle, Australia in 1972, an MSc in industrial and systems engineering from the University of Singapore in 1979, and a PhD in electrical engineering from Imperial College, University of London in 1983. He also holds a postgraduate diploma in business administration and a certified diploma in accountancy and finance. From 1973 to 1992, he worked in the Electronics Division of the Defense Science Organization, Singapore, where he conducted research and carried out development work on speech related equipment. From 1992 to 2001, he was the associate professor with the Department of Electrical and Computer Engineering, National University of Singapore. In 2002, he joined the School of Electrical and Electronic Engineering, Nanyang Technological University. His research interests include audio, speech, and image signal processing.

Teddy Furon received an MS in digital communications in 1998 and the PhD in signal and image processing in 2002, from the École Nationale Supérieure des Télécommunications, Paris. From 1998 to 2001, he was a research engineer with the Security Lab of THOMSON Multimedia, Rennes, France, working on digital watermarking in the framework of copy protection. He continued working on digital watermarking as a postdoctoral fellow at TELE Lab., Université catholique de Louvain, Louvain-la-Neuve, Belgium. He is now a researcher with the INRIA institute, working within the TEMICS project of the IRISA public research center, Rennes, France. Dr. Furon was a co-recipient of the IWDW'04 Best Paper Award.

Behnaz Ghoraani received a BSc and MSc in electrical engineering from Sharif University of Technology, and Polytechnique University, Tehran, Iran (1998 and 2000, respectively). Currently, she is pursuing a PhD in electrical engineering, Ryerson University, ON, Canada. She also held a research assistantship position at Ryerson University from 2005 to August 2006, on fingerprinting and watermarking. From 2000 to 2004, she was with Gamma Irradiation Center in Iran as an electronics design engineer. Her research interests include feature extraction, and classification for biometrics, multimedia information forensics, and time-frequency analysis.

Alexia Giannoula received the Diploma degree (Honors) from the Computer Science Department, Aristotle University of Thessaloniki, Thessaloniki, Greece, in 2001, and the Master of Applied Science (MASc) from the Electrical and Computer Engineering Department (ECE), University of Toronto, Toronto, ON, Canada, in 2003. Her Diploma and master work involved research in the areas of image/video processing, wavelets, watermarking, and compression. She is currently at the Institute of Biomaterials and Biomedical Engineering, University of Toronto (ECE), pursuing the PhD in the areas of biomedical ultrasound, medical imaging, and acoustic theory. From September 2001 to August 2002, she served as a research assistant at the Artificial Intelligence and Information Analysis Laboratory, Aristotle University of Thessaloniki, Greece. She also holds several teaching assistantship positions at the ECE Department of the University of Toronto and she is the recipient of the Ontario Graduate Scholarship (OGS) of Canada, the American Hellenic Educational Progressive Association (AHEPA), and the State Scholarship Foundation of Greece.

Aparna Gurijala received an MS degree in electrical engineering from Michigan State University, East Lansing, in 2001. She received her PhD in electrical engineering from Michigan State University in 2006. She served as a research assistant on the National Gallery of Spoken Word (NGSW) project from 2000 to 2004, and has been involved in the development of speech watermarking algorithms for NGSW. Her research interests lie in the areas of multimedia security and signal processing, with a focus on speech, audio, and video watermarking, signal processing based approaches for encryption, and adaptive signal processing. Gurijala is a member of IEEE Signal Processing Society.

Xing He is a PhD candidate with the Digital Audio and Speech Processing Laboratory in the Department of Electrical and Computer Engineering, University of Miami. He is graduating in December 2006. He received his BS and MS in electrical engineering from Northern Jiaotong University, China, (1997 and 2000, respectively). His research interest and publications are in the areas of digital audio watermarking, digital audio coding, psychoacoustics, automatic speech recognition, and digital signal processing.

Hyoung-Joong Kim received a BS, MS, and PhD from Seoul National University, Seoul, Korea (1978, 1986, and 1989, respectively). He joined the faculty of the Department of Control and Instrumentation Engineering, Kangwon National University, Chunchon, Korea, in 1989, where he is currently a professor. He was a visiting scholar at the University of Southern California from 1992-1993. From 1998-2000, he was the prime investigator of the iPCTV Project developing interactive digital television and implemented MHP, ATVEF, and DASE specifications and performed conformance tests. He has been the prime investigator of the iMS (Interactive Media Solution) developing the end-to-end solution for data broadcasting system since 2000. He was the founder of the International Workshop on Digital Watermarking (IWDW) and served as a co-chair of the Technical Program Committee of the workshop. He edited *Digital Watermarking, Lecture Notes in Computer Science*, (Vol. 2613), with Dr. Fabien A.P. Petitcolas. He served as a guest editor of the *IEEE Transactions on Circuits and Systems for Video Technology* in 2003. Since 2003, he has been the director of the Media Service Research Center (MSRC-ITRC) sponsored by the Ministry of Information and Communication. His research interests include parallel computing, multimedia computing, and multimedia security. He is a member of the IEEE.

Byeong-Seob Ko was born in Jeju, Korea, in 1971. He received a BS (Honors) and an MSc in electronics engineering from Chongju University, Chongju, Korea (1998 and 2000, respectively), and a PhD in information science from Tohoku University, Sendai, Japan, in 2004. He is currently working at SAMSUNG Co., Ltd. His present research interests include digital signal processing of acoustic signals, digital watermarking, and psychoacoustics. Dr. Ko was a scholarship student of the Ministry of Education, Culture, Sports, Science, and Technology of Japan and has also received the Kamei Scholarship.

Dimitrios Koukopoulos was born in Karditsa in 1975. He received his engineering diploma from the Department of Computer Engineering & Informatics of the University of Patras in 1998. He received an MSc in communications and signal processing from the Imperial College of the University of London in 1999. He completed successfully his PhD in the Department of Computer Engineering & Informatics of the University of Patras in 2003, with supervisor Professor Paul Spirakis in adversarial queueing theory. He worked as a junior researcher in audio watermarking and quality of service issues of large-scale networks at the Research and Academic Computer Technology Institute in Patras, Greece, from 1999 to 2003. Currently, he is a visiting professor at the Department of Cultural Heritage Management and New Technologies of the University of Ioannina. His research interests cover watermarking, telecommunications, signal processing and network quality of service issues.

Sridhar Krishnan received a BE in electronics and communication engineering from Anna University, Madras, India, in 1993, and an MSc and PhD in electrical and computer engineering from the University of Calgary, Calgary, Alberta, Canada (1996 and 1999, respectively). He joined the Department of Electrical and Computer Engineering, Ryerson University, Toronto, Ontario, Canada in July 1999, and currently he is an associate professor and chairman of the department. Sri Krishnan's research interests include adaptive signal processing, biomedical signal/image analysis, and multimedia processing and communications.

Yi-Wen Liu received a BS from National Taiwan University in 1996, and an MS and a PhD in electrical engineering from Stanford University (2000 and 2006, respectively). He was a summer intern with VerbalTek, Inc., San Jose, CA, in 2000, and with AT&T Research Labs, Middletown, NJ, in

2001. From 2002-05, his dissertation on audio watermarking was supervised by Dr. Julius Smith and Dr. Marina Bosi at the Center for Computer Research in Music and Acoustics (CCRMA). Currently a postdoc fellow and software engineer, he works on middle ear impulse reflectometry at Boys Town National Research Hospital, Omaha, USA.

Subhamoy Maitra received a BS of electronics and telecommunication engineering in 1992, from Jadavpur University, Calcutta, India, an MS of technology in computer science in 1996 from Indian Statistical Institute, Calcutta, and a PhD from the Indian Statistical Institute, Calcutta, in 2001. Currently, he is a faculty member with the Indian Statistical Institute. His research interest is in cryptology and digital watermarking.

Nikos Nikolaidis received the Diploma of Electrical Engineering in 1991, and a PhD in electrical engineering in 1997, both from the Aristotle University of Thessaloniki, Greece. From 1998 to 2002, he was postdoctoral researcher and teaching assistant at the Department of Informatics, Aristotle University of Thessaloniki. He is currently a lecturer in the same department. Dr. Nikolaidis is the co-author of the book *3-D Image Processing Algorithms* (Wiley, 2000). He has co-authored five book chapters, 20 journal papers, and 77 conference papers. His research interests include computer graphics, image and video processing and analysis, copyright protection of multimedia, and 3-D image processing. Dr. Nikolaidis is currently serving as associate editor for the *EURASIP Journal on Image and Video Processing* and the *International Journal of Innovative Computing, Information and Control.*

Ryouichi Nishimura received a BS in information engineering, and an MSc and a PhD in information science, all from Tohoku University, Sendai, Japan (1993, 1995, and 1998, respectively). From 1998 to 2000, he was a visiting researcher at ATR Media Integration and Communications Research Laboratories, Kyoto, Japan, where he worked on auditory virtual reality. He is currently an associate professor at the Research Institute of Electrical Communication, Tohoku University, Sendai, Japan. His present research interests are sound recognition, audio data hiding, and acoustic signal processing. Dr. Nishimura received the Best Paper Award from the Virtual Reality Society of Japan in 2001.

Michael Scordilis is an associate professor of electrical and computer engineering and directs the Digital Audio and Speech Processing Laboratory at the University of Miami. He received his BS of engineering in communication engineering from the Royal Melbourne Institute of Technology, Australia, 1983, and an MS and PhD of philosophy in electrical engineering from Clemson University, USA (1986 and 1990, respectively). His research activities and publications are in the areas of acoustics, speech, and audio signal processing. He is a member of the Audio Engineering Society and the International Speech Communication Association, and senior member of the Institute of Electrical and Electronics Engineers.

Yannis Stamatiou was born in Volos in 1968. He holds a degree of computer engineering and informatics from the University of Patras and a PhD from the same department. He is currently an assistant professor at the University of Ioannina, Mathematics Department, Greece, and a scientific consultant on security and cryptography issues of the Research and Academic Computer Technology Institute (RACTI). His scientific interests lie in the fields of security and cryptography as well as the study of threshold phenomena arising in computationally intractable problems. He is a member of ACM and IEEE.

Martin Steinebach is a research assistant at Fraunhofer IPSI (Integrated Publication and Information Systems Institute). His main research topic is digital audio watermarking. He developed algorithms for mp2, MIDI and PCM data watermarking, content fragile watermarking, and invertible audio watermarking. He also introduced concepts for applying audio watermarks in e-commerce environments and for transaction watermarking. In 2003, he received a PhD at the Technical University of Darmstadt for this work on digital audio watermarking. Since 2002, he has been the head of the Department MERIT (Media Securtiy in IT) and of the C4M Competence Centre for Media Security.

Yôiti Suzuki graduated from Tohoku University, Sendai, Japan, in 1976, where he also received a PhD in electrical and communication engineering in 1981. He is currently a professor at the Research Institute of Electrical Communication, Tohoku University. His research interests include psychoacoustics, high-definition auditory display, and digital signal processing of acoustic signals. Dr. Suzuki has been serving as the president of the Acoustical Society of Japan since 2006 until 2008. He was a recipient of the Takenaka and the RCA David Samoff Scholarships, and has received the Awaya Kiyoshi Award and the Sato Prize from the Acoustical Society of Japan.

Walter Voeßing is a senior development engineer in the audio processing lab of THOMSON in Hannover/FRG. He received the degree of Dipl.-Ing. in electrical communication engineering in 1977. He has worked in the field of car radio development, error correction systems, and computer based simulation of digital audio signal processing. Since 1983, he has worked on audio data reduction, and since 2003 on digital audio watermarking in the Corporate Innovation & Research Center of THOMSON in Hannover/FRG.

Wen-Chih Wu was born in Taiwan in 1964. He received an MS in electrical engineering from National Tsing Hua University in 1992. Since August 1992, he has been a lecturer in the Department of Electrical Engineering, WuFeng Institute of technology, Chiayi, Taiwan. He is currently working toward the PhD in electrical engineering at National Chung Cheng University. His research interests include digital signal processing and information security.

Shijun Xiang received a PhD of computer science at Sun Yat-Sen University of P.R. China on June, 2006. The adviser was Professor Jiwu Huang. He is currently a full time research lecturer in the group of Professor Hyoung-Joong Kim at CIST, Graduate School of Information Management and Security, Korea University. His research interests mainly include robust watermarking and secure watermarking. He is interested in challenging issues in robust audio watermarking, such as desynchronization attacks and DA/AD processing, as well as image watermarking against globally and locally geometric attacks.

In-Kwon Yeo received a PhD in statistics from University of Wisconsin-Madison in 1997. He joined the Department of Control and Instrumentation Engineering, Kangwon National University, as a visiting professor in 2000. He is currently an assistant professor at the Division of Mathematics and Statistical Informatics, Chonbuk National University, Jeonju, Korea. His current research interests include transformations, pattern classifications of time series data, and bioinformatics.

Index

K

L

M

N

O

P

Q

R

S

T

U

V

W

Z